Linear vibrations

MECHANICS: DYNAMICAL SYSTEMS
Editors: L Meirovitch and GÆ Oravas

E.H. Dowell, Aeroelasticity of Plates and Shells. 1974.
ISBN 90-286-0404-9.
D.G.B. Edelen, Lagrangian Mechanics of Nonconservative Nonholonomic Systems. 1977. ISBN 90-286-0077-9.
J.L. Junkins, An Introduction to Optical Estimation of Dynamic Systems. 1978. ISBN 90-286-0067-1.
E.H. Dowell et al., A Modern Course in Aeroelasticity. 1978.
ISBN 90-286-0057-4.
L. Meirovitch, Computational Methods in Structural Dynamics. 19
ISBN 90-286-0580-0.
B. Skalmierski and A. Tylikowski, Stochastic Processes in Dynami
1982. ISBN 90-247-2686-7.
P.C. Müller and W.O. Schiehlen, Linear Vibrations. 1985.
ISBN 90-247-2983-1.

Linear vibrations

A theoretical treatment of multi-degree-of-freedom vibrating systems

P.C. Müller

Bergische Universität Gesamthochschule Wuppertal
Sicherheitstechnische Regelungs- und Messtechnik
Gausstrasse 20, 5600 Wuppertal 1, FRG

W.O. Schiehlen

Institut B für Mechanik
Universität Stuttgart
Pfaffenwaldring 9, 7000 Stuttgart 80, FRG

Translated by S. Swierczkowski

1985 **Springer-Science+Business Media, B.V**

Distributors

for the United States and Canada: Kluwer Academic Publishers, 190 Old Derby Street, Hingham, MA 02043, USA
for the UK and Ireland: Kluwer Academic Publishers, MTP Press Limited, Falcon House, Queen Square, Lancaster LA1 1RN, UK
for all other countries: Kluwer Academic Publishers Group, Distribution Center, P.O. Box 322, 3300 AH Dordrecht, The Netherlands

Library of Congress Cataloging in Publication Data

Müller, P. C. (Peter Christian), 1940-
Linear vibrations.

(Mechanics, dynamical systems ; v. 7)
Includes bibliographical references and index.
1. Vibration--Data processing. I. Schiehlen, W.O. (Werner O.) II. Title. III. Series: Monographs and textbooks on mechanics of solids and fluids. Mechanics, dynamical systems ; v. 7.
TA355.M83 1984 620.3'028'54 84-8189
ISBN 978-94-010-8735-3 ISBN 978-94-009-5047-4 (eBook)
DOI 10.1007/978-94-009-5047-4

ISBN **978-94-010-8735-3** (this volume)

Contents

Preface

In the last decade the development in vibration analysis was characterized by increasing demands on precision and by the growing use of electronic computers. At present, improvements in precision are obtained by a more accurate modelling of technical systems. Thus, for instance, a system with one degree of freedom is often not accepted, as it used to be, as a model for vibration analysis in mechanical engineering. As a rule, vehicles and machines have to be modelled as systems with many degrees of freedom such as multibody systems, finite element systems or continua. The mathematical description of multi-degree-of-freedom systems leads to matrix representations of the corresponding equations. These are then conveniently analyzed by means of electronic computers, that is, by the analog computer and especially by the digital machine. Hence there exists a mutually stimulating interaction between the growing requirements and the increasing computational facilities.

The present book deals with linear vibration analysis of technical systems with many degrees of freedom in a form allowing the use of computers for finding solutions.

Part I begins with the classification of vibrating systems. The main characteristics here are the kind of differential equation, the time dependence of the coefficients and the attributes of the exciting process. Next it is shown by giving examples involving mechanical vibrating systems how to set up equations of motion and how to transform these into state equations. The latter form the basis for the analysis of vibrating systems and these can be conveniently solved by a computer.

In Part II time-invariant vibrating systems are discussed. The general solution of the state equations leads to the fundamental matrix and the eigenvalue problem. The qualitative behaviour of vibrating systems is described by their stability and boundedness. Free vibrations are characterized by their natural modes. Vibrations excited by periodic functions may display resonance, pseudoresonance or absorption. The last two phenomena can occur only in multi-degree-of-freedom systems. Lastly, random vibrations are investigated by means of covariance analysis and spectral-density analysis.

Part III is devoted to time-variant vibrating systems. For periodic time-variant systems finding the general solution and analysing the stability is reduced to integration over a bounded interval and to a modified eigenvalue problem. Parameter-excited vibrations are periodic only under special conditions, just as in the case of forced vibrations, as the period of the excitation usually differs from the period of the response. Parameter-excited random vibrations admit investigation by means of covariance analysis. Also some remarks concerning general time-variant vibrating systems are included.

In the fourth and last part the mathematical prerequisites are presented which go beyond the knowledge of matrix calculus; the reader is assumed to be familiar with the latter. First of all, the important notions of controllability and observability are explained. They are essential in stability analysis. Then, the linear matrix equations are discussed which play an important role in the preceding chapters. Finally, the numerical aspects of solving linear equations, of the eigenvalue problem and of integration are presented.

The present volume has the character of a textbook. The reader is presented with the basic theory of vibrations; moreover, in many cases the wide scope of the presentation saves him the need to consult more specialized literature. His knowledge will be deepened if he works through the interrelated exercises which accompany each section. The multiple-choice questions will allow him to control his progress. Since always one of the given answers is correct, the solutions to these questions are not supplied.

P.C. Müller W.O. Schiehlen

Part I

Mathematical description of vibrating systems

The theoretical analysis of vibrations is based upon an exact mathematical description of the considered technical systems. This mathematical description leads to one or more differential equations. The structure of these equations already discloses some characteristic features of the vibrating system. Thus Section 1 is devoted to the classification of vibrational phenomena and to the outlining of the contents of this book. In Sections 2 and 3 procedures of setting up equations of motion and state equations are illustrated through the application to examples of mechanical vibrating systems.

1. Classification of vibrations

Vibrations are characterized by time-varying, more or less periodical processes. The sinusoidal vibration of an electric oscillator, the periodic swinging of a simple pendulum, or the random motion of a building subjected to gusts of wind are typical examples of vibrating systems. Processes of limited duration are also included in the notion of vibrations. Due to unavoidable energy dissipation, free vibrations will decay in practice and a strongly damped vibrating system will display short-term vibrations under excitation by impulse forces. These few examples show the need for putting some order in the variety of vibration phenomena, i.e., a need for a classification within the large class of vibrating systems. Such a classification will also be useful for a clear definition of the contents of this book.

One finds in literature a number of characteristics employed to classify vibrations. The most important characteristics are as follows:
Number of degrees of freedom,
Engineering applications,
Type of differential equations,
Coefficients of the differential equation,
Excitation mechanisms,
Kinematics of vibrations,
Energy balance.

The number of degrees of freedom is equal to the minimum number of independent coordinates required for the complete description of the time behavior of the vibrating system. Such a system is usually given as an idealized model (mathematical model) which comprises all the essential properties of the technical problem. Depending on whether there is only one degree of freedom or more, one speaks of a *single-degree-of-freedom* vibrating system or of a *multi-degree-of-freedom* vibrating system. These systems will be referred to as a *single-dof-system* and a *multi-dof-system*. If the number of degrees of freedom is infinite, the system is called *continuous*, or a *continuum*. For a continuum, the independent coordinates are functions of both time and space. The simple pendulum is a typical example of a single-dof-system, Fig. 1.1. Its behavior can be completely described by the position angle $\alpha(t)$ as a function of time. The low-pass filter in Fig. 1.2 is an example of a multi-dof-system. Its behavior in time is described by the currents $i_1(t)$, $i_2(t)$, $i_3(t)$. The homogeneous elastic bar in Fig. 1.3 is a continuous vibrating system and the displacement $w(x, t)$ is a function of space and time, so that the elastic bar has infinitely many degrees of freedom.

The order n of a vibrating system is closely related to the number f of its degrees of freedom. The order is the smallest number of state variables needed to completely determine the time behavior of the vibrating system. The order is also equal to the number of initial conditions. From this, it follows

$$1 \leqslant f \leqslant n. \tag{1.1}$$

The simple pendulum, a single-dof-system, serves as an example of a vibrating system of the order n = 2, as both the initial position $\alpha(t_0) = \alpha_0$ and the initial velocity $\dot{\alpha}(t_0) = \dot{\alpha}_0$ are needed for the complete determina-

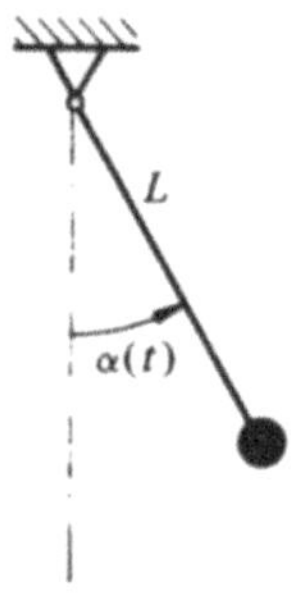

Fig. 1.1. Simple pendulum as an example of a single-dof-vibrating system.

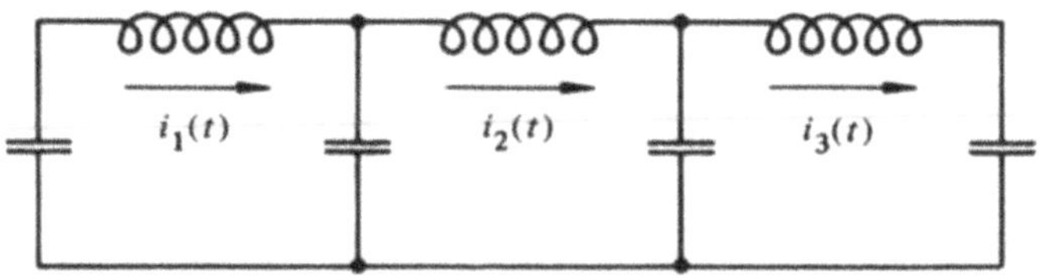

Fig. 1.2. Low-pass as an example of a multi-dof-vibrating system.

Fig. 1.3. Homogeneous elastic bar as an example of a continuous vibrating system.

tion of its behavior in time. For ordinary mechanical systems *, it yields $n = 2f$.

In engineering applications one distinguishes between *electrical* and *mechanical* vibrating systems. High-pass and low-pass filters, oscillating circuits, gyrators, and active filters are important electrical vibrating systems. Among the mechanical vibrating systems are elastic chains, elastic structures, all types of vehicles, gyroscopes, actively controlled systems. The division into electrical and mechanical systems is of secondary importance since many electrical systems can be represented by mechanical ones and vice versa. Fig. 1.4 shows that this analogy can be set up in two different ways represented by typical components. But classifications within one area of applications have a methodical meaning. Thus, in engineering mechanics one speaks of structural vibrations when elastic structures are investigated and only the mass and spring forces are taken into account. Damping forces, for instance, enter into consideration when vibrations of vehicles are investigated. Vibrations of gyroscopic systems are influenced by gyroscopic forces, and in control systems, all of the above mentioned forces may appear.

The differential equation of a vibrating system may be linear or nonlinear, homogeneous or nonhomogeneous. Accordingly, one speaks of *linear* or *nonlinear* vibrations. Most vibrations with large amplitudes are nonlinear, and for those with small amplitudes it is possible to linearize the equations, thus reducing the discussion to the case of linear vibra-

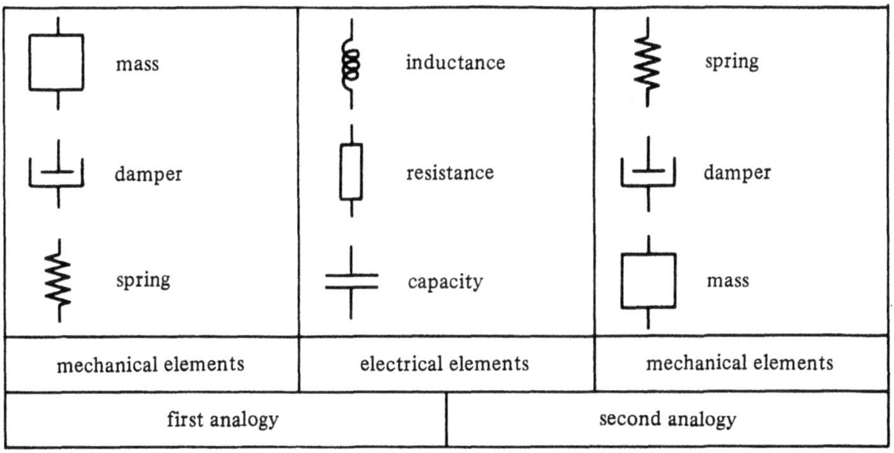

Fig. 1.4. Electrical-mechanical analogies.

* An ordinary mechanical system is given if only holonomic constraints occur and the forces depend on position and velocity only, see Section 2.3.

tions. If the simple pendulum of Fig. 1.1 undergoes large amplitudes, then it is a nonlinear system described by the differential equation

$$\ddot{\alpha}(t)+\frac{g}{L}\sin\alpha(t)=0, \tag{1.2}$$

whereas for small amplitudes, it may be viewed as a linear system described by

$$\ddot{\alpha}(t)+\frac{g}{L}\alpha(t)=0, \qquad \alpha(t)\ll 1. \tag{1.3}$$

Here g denotes the gravitational constant and L the length of the pendulum. For vibrating systems with a homogeneous differential equation a typical name is not used. They are classified according to the properties of the excitation process. When the differential equation is nonhomogeneous, the vibrations are said to be forced. Thus, for the simple pendulum acted upon by a disturbance function $h(t)$, one has

$$\ddot{\alpha}(t)+\frac{g}{L}\sin\alpha(t)=h(t), \tag{1.4}$$

and the vibration is forced and nonlinear.

The differential equation of a vibrating system can have *time-variant* or *time-invariant* coefficients. Accordingly, the system is called time-variant or time-invariant. For mechanical systems, one encounters time-invariant coefficients when the constraints between the Cartesian coordinates are scleronomic and the force coefficients are time-invariant. Conversely, time-variant coefficients are caused by rheonomic constraints or time-variant force coefficients. Free and self-excited vibrations lead to homogeneous differential equations with time-invariant coefficients, whereas parametrically excited vibrations are associated with homogeneous differential equations with time-variant coefficients. The simple pendulum with a moving point of suspension, Fig. 1.5, is a time-variant vibrating system with the differential equation

$$\ddot{\alpha}(t)+\left(\frac{g}{L}-\frac{\Omega^2}{L}\cos\Omega t\right)\alpha(t)=0, \qquad \alpha(t)\ll 1, \tag{1.5}$$

Hence, for small amplitudes, one has linear parametrically excited vibrations.

The following classification of vibrations according to the generating mechanism is due to Magnus. Accordingly, one distinguishes between *free, self-excited, parametrically excited* and *forced* vibrations. A vibrating system which receives no energy is called free. Free systems can then be further divided into damped or undamped systems. Free vibrations, frequently called natural vibrations, always lead to homogeneous differential equations with time-invariant coefficients. These equations are often linear or they permit linearization. Thus a free, undamped, linear vibrating system with one independent coordinate $y(t)$ has the differential equation

$$\ddot{y}(t)+\omega^2 y(t)=0, \tag{1.6}$$

where ω denotes the natural frequency of the system. A system which is supplied by energy in step with the oscillations is called a self-excited vibrating system. Self-excited vibrations are also determined by homogeneous differential equations with time-invariant coefficients but these equations are always nonlinear. Though a dedamped vibration may begin with small amplitudes, these will grow with time thus causing the system to become nonlinear. The following differential equation is typical of a self-excited nonlinear vibrating system:

$$\ddot{y}(t)+f(y(t),\ \dot{y}(t))=0. \tag{1.7}$$

If energy is supplied to the system so that its coefficients undergo variations in time, these being mostly of periodic nature, the system is called parametrically excited. Such a system may exhibit damped as well as undamped or dedamped vibrations. It is described by a linear homogeneous differential equation with time-dependent coefficients, e.g. by

$$\ddot{y}(t)+p(t)\dot{y}(t)+q(t)y(t)=0. \tag{1.8}$$

Moreover, there are nonlinear parametrically excited vibrations also known. A vibrating system is called externally excited if energy is supplied from an external source. In this case a nonhomogeneous perturbing term appears in the differential equation. For a linear, single-

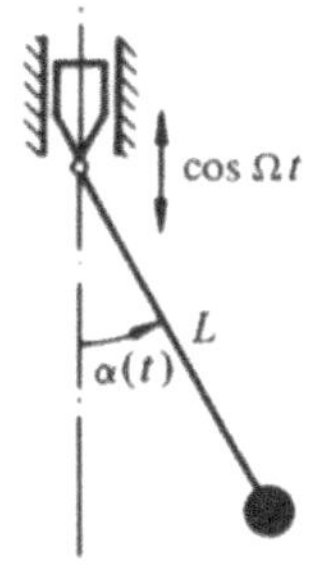

Fig. 1.5. Simple pendulum with moving point of suspension.

dof-vibrating system, one obtains the following differential equation with time-invariant coefficients

$$\ddot{y}(t)+\omega^2 y(t)=h(t). \tag{1.9}$$

Impulse function, step function, sinusoidal or harmonic excitation function are important perturbing functions $h(t)$. If a vibrating system is externally excited by a stochastic process, then its vibrations result in a stochastic process and one speaks of random vibrations. Correspondingly, a nonlinear vibrating system can be subjected to external perturbances as well.

Free and self-excited vibrations are typical of autonomous vibrating systems, as their frequency depends only on the properties of the system and not on any external time dependent influences. On the other hand, parametrically excited and forced vibrations appear only in heteronomic systems which are characterized by the presence of external excitation in the coefficients or from a perturbing term in the differential equation.

The kinematics of vibrations are concerned with investigating the time dependence of various state variables of the system. If this time dependence can be described by a sine or cosine function, the vibration is called *sinusoidal*. As can be seen in Fig. 1.6, a sinusoidal vibration can always be represented by an equation of the type

$$x(t)=a\cos(\omega t+\psi_0)=\operatorname{Re}\{\boldsymbol{a}\,\mathrm{e}^{\mathrm{i}\omega t}\}. \tag{1.10}$$

Here a is the amplitude, ω the natural frequency, $T=2\pi/\omega$ the period and ψ_0 the null phase angle of the vibration. The complex quantity

$$\underline{a}=a\,\mathrm{e}^{\mathrm{i}\psi_0} \tag{1.11}$$

is called the complex amplitude or vector of the sinusoidal vibration, Fig. 1.7. The linear undamped natural vibration of a single-dof-system is an example of a sinusoidal vibration. Vibrations are called nearly sinusoidal if they can be viewed as sinusoidal vibrations with an amplitude and null

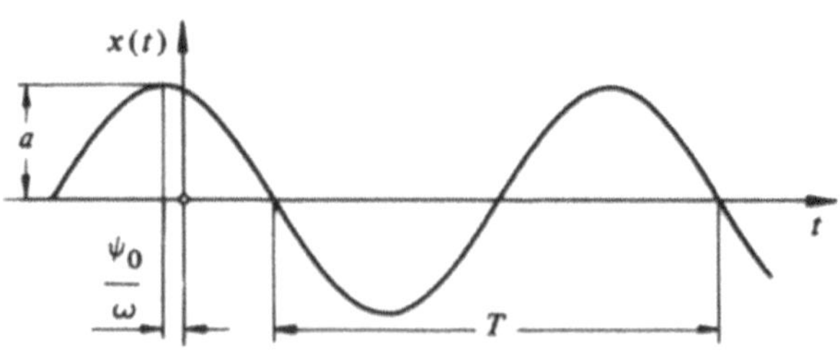

Fig. 1.6. Sinusoidal vibration.

phase angle that changes slowly with time, Fig. 1.8. Lightly damped, linear natural vibrations are nearly sinusoidal. Any process exhibiting a time dependence of periodical nature will be called a *periodic* vibrating system, Fig. 1.9. The period T of such a system determines the corresponding fundamental frequency $\omega = 2\pi/T$. Each periodic vibration can be represented by a Fourier series consisting of cosine vibrations:

$$x(t) = \tfrac{1}{2}a_0 + \sum_{k=1}^{\infty} a_k \cos(k\omega t + \psi_{0k}). \tag{1.12}$$

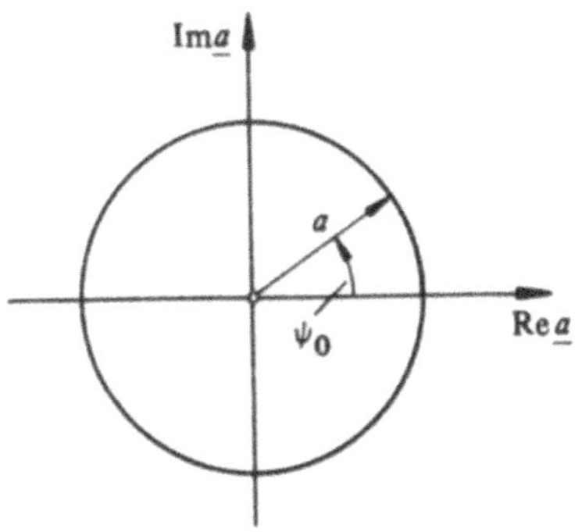

Fig. 1.7. Vector of a sinusoidal vibration in the complex plane.

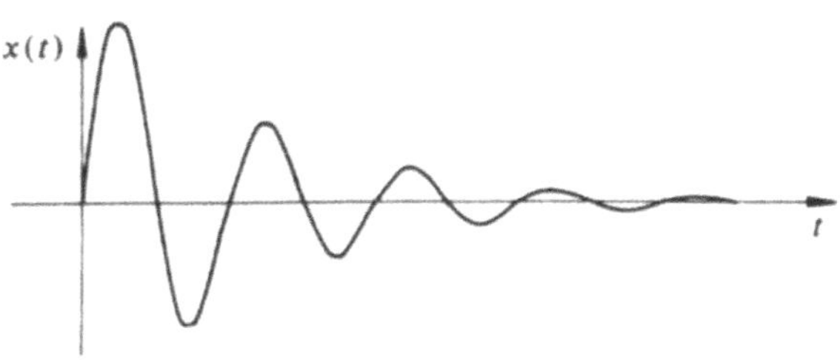

Fig. 1.8. Nearly sinusoidal vibration.

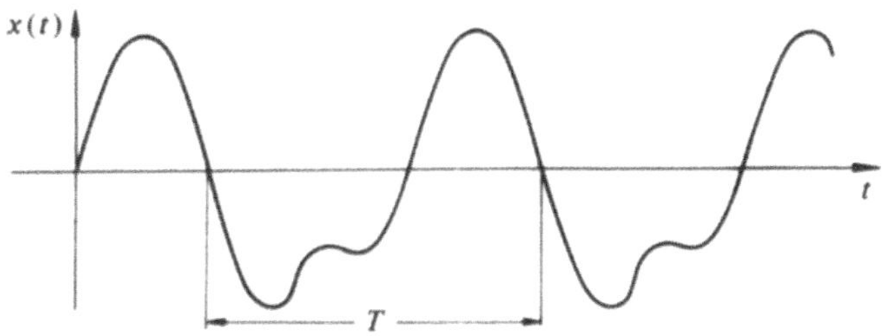

Fig. 1.9. Periodic vibration.

Conversely, to decompose a periodic vibration into cosine vibrations with frequencies $k\omega$, one may use the following formula for the amplitudes and phase angles:

$$a_k\, e^{i\psi_{0k}} = \frac{2}{T}\int_0^T x(t)\, e^{-ik\omega t} dt, \qquad k = 0(1)\infty. \tag{1.13}$$

For example, self-excited vibrations are often periodic.

A continuing, non periodic process will be called a *nonperiodic* vibration. If it is of small duration in comparison to a characteristic duration, it is called an *impulsive vibration*, Fig. 1.10. Impulsive vibrations belong to the class of decaying, nonperiodic vibrations. A continuing, nonperiodic, random vibration is called a *noise process*, Fig. 1.11. A steady-state noise process can be characterized by its power spectral density

$$S(\omega) = \lim_{T\to\infty} \frac{1}{T}\left|\int_{-T}^{+T} x(t)\, e^{-i\omega t} dt\right|^2. \tag{1.14}$$

A final possibility of classifying mechanical vibrations is provided by the forces influencing the vibration. If only conservative forces are acting on the vibrating system, one calls the system *conservative*. If *nonconservative* forces are acting, one speaks of a *nonconservative* vibrating system. Thus, free undamped vibrations are conservative, while free damped vibrations are nonconservative.

Fig. 1.10. Impulsive vibration.

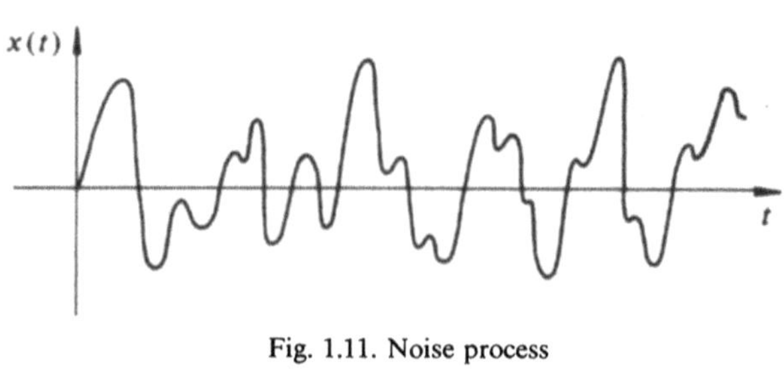

Fig. 1.11. Noise process

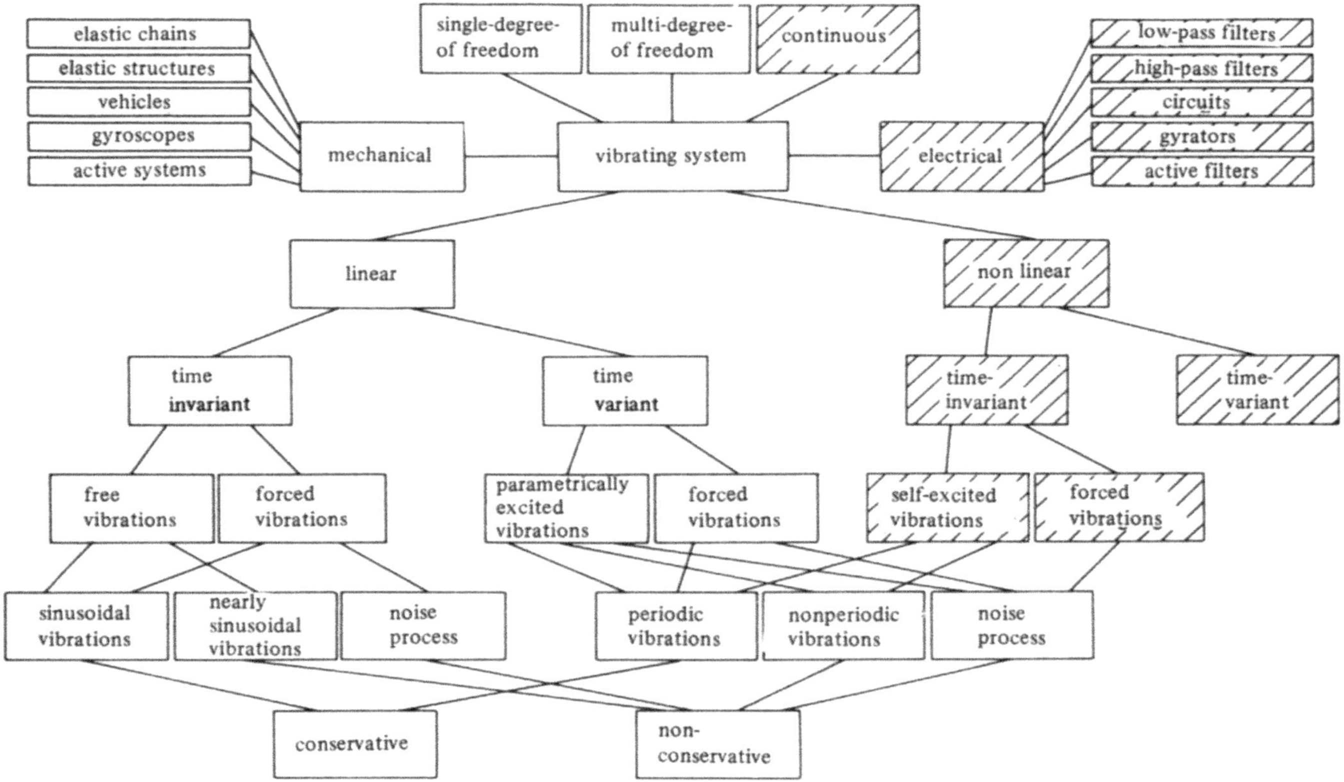

Fig. 1.12. Classification of vibrations.

The above classification of vibrating systems is represented in Fig. 1.12. The contents of this book will concern only the non-shaded boxes. Thus, only single-dof and multi-dof linear vibrating systems with time-variant and time-invariant coefficients performing free, forced, or parametrically excited vibrations will be treated. Among the self-excited vibrations, only the linear approximations resulting in dedamped vibrations are included. Most of the examples are taken from mechanics, simply because the authors have a long standing familiarity with this field. The mathematical presentation uses exclusively the *matrix calculus* which, apart from being compact, allows for direct accessibility to digital computers. The equations of motion for mechanical vibrating systems will also be given in matrix notation. Thus a very uniform presentation is obtained.

One encounters vibrating systems not only in engineering but also in biology, chemistry, cybernetics, economics and sociology. These vibrations can also be described mathematically, provided that they appear sufficiently distinct. Often the problems of constructing a suitable theoretical model and finding the state equations cannot be satisfactorily solved since the natural laws governing such processes are not sufficiently known. Hence, it is necessary to validate the theoretical considerations of such vibrations by experiments.

Problem 1.1: Vibrating chain

Consider the vibrating chain with masses m_1, m_2 and spring constants k_1, k_2 as shown in Fig. 1.13. Its motion is described by the following system of differential equations:

$$\left.\begin{aligned} m_1\ddot{y}_1(t)+(k_1+k_2)y_1(t)-k_2y_2(t)=0,\\ m_2\ddot{y}_2(t)+k_2(y_2(t)-y_1(t))=0.\end{aligned}\right\} \tag{1.15}$$

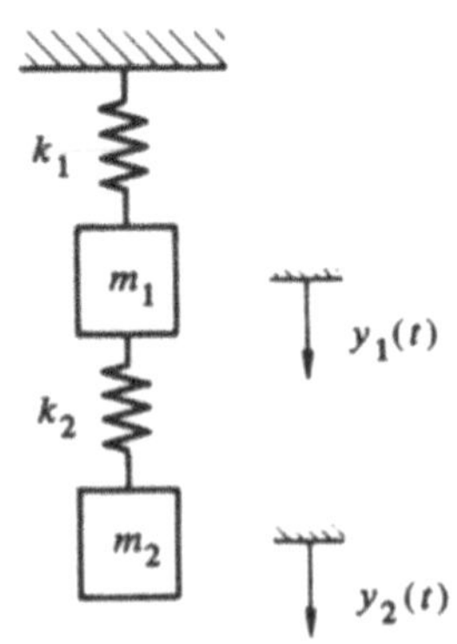

Fig. 1.13. Vibrating chain.

Which properties does the system (1.15) possess and what kind of vibrations can it perform?

□ single-dof,	□ multi-dof,
□ linear,	□ nonlinear,
□ time-invariant,	□ time-variant,
□ free,	□ forced.

Problem 1.2: Heavy gyroscope

Consider a heavy gyroscope rotating with a constant angular velocity Ω, Fig. 1.14. It performs vibrations measured by the small rotations α and β about the x-axis and y-axis. These are described by the differential equations

$$\left.\begin{aligned} I_x\ddot{\alpha}(t) - k\alpha(t) + I_z\Omega\dot{\beta}(t) &= I_{yz}\Omega^2 \cos \Omega t, \\ I_x\ddot{\beta}(t) - k\beta(t) - I_z\Omega\dot{\alpha}(t) &= I_{yz}\Omega^2 \sin \Omega t. \end{aligned}\right\} \tag{1.16}$$

Here I_x and I_z denote the moments of inertia, k describes the influence of gravity and I_{yz} is the dynamic unbalance.

Which properties does this vibrating system possess and what kind of vibrations can it perform?

□ single-dof,	□ multi-dof,
□ linear,	□ nonlinear,
□ time-variant,	□ time-invariant,
□ free,	□ forced.

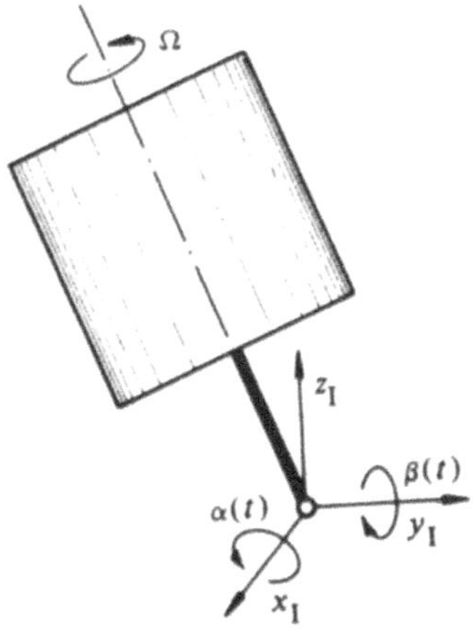

Fig. 1.14. Heavy gyroscope.

2. Mechanical vibrating systems

The mathematical description of vibrating systems is achieved by means of the *equations of motion* or *equations of vibration*. These are determined from the fundamental physical laws governing the system under consideration. Thus it is not possible to give within the scope of this book a full presentation of the equations of motion of all vibrating systems. Instead, the methods and procedures will be illustrated on examples of mechanical multi-dof-systems. This section begins with the discussion of the choice of suitable models which contain all the essential technical properties of the vibrating system. Then follows a kinematical description of the possible motions of the system, and finally the fundamental laws of dynamics are applied, especially Lagrange's equations of motion and the Newton-Euler equations.

2.1. Modeling of technical systems

A technical system rarely admits a total description by the equations of motion. Usually, certain idealizations are necessary to limit the amount of mathematical effort needed to solve the technical problem. The need to make sensible simplifications forces the engineer ever again to ask: which are the properties of vibrations having essential influence on the proper operation and the quality of the given system? Let us illustrate this with an example from automotive engineering.

A motor car, Fig. 2.1, is a very complex system with many possible modes of vibration. The vibrations are generated by the engine, the road and other external forces. The choice of the theoretical model is now determined by the character of the technical problem. If the low frequency vertical vibrations responsible for riding comfort and safety are to be investigated, one can use as a model the *three-body system* shown in Fig. 2.2. In this model the car body and the axles, together with the wheels, are regarded as rigid bodies connected to each other by springs and shock absorbers. The elasticity of the tires is represented by undamped springs. The corresponding spring constant is much larger than the constant for the springs supporting the car body and thus, we distinguish in automotive engineering between sprung and unsprung masses. For the analysis of the high frequency bending vibrations of the car body the most suitable model is a *lumped mass system,* Fig. 2.3. According to the finite element method, the car body is now represented by a system of interconnected bars and triangular or rectangular plates. Each of these elements is an elastic body whose mass, elasticity, and damping are concentrated in the nodes. One thus obtains a mechanical system with many particles. One gets its equations of motion by combining the usual differential equations which pertain to the various particles. Using matrix calculus, a digital computer can cope with this type of problem, even when many degrees of freedom are involved. To investigate the torsional

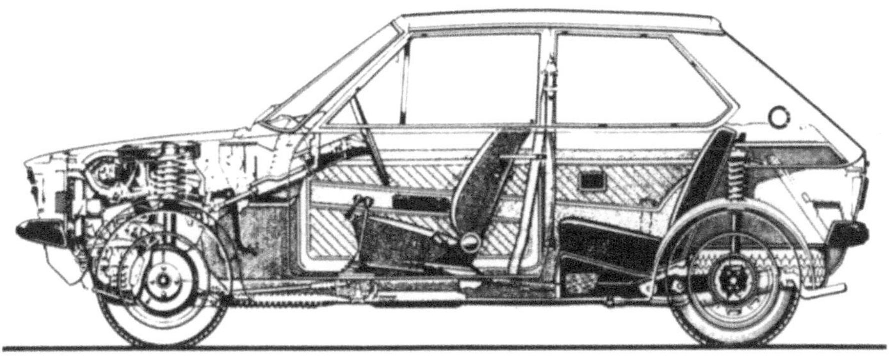

Fig. 2.1. Cross-section of a motor car.

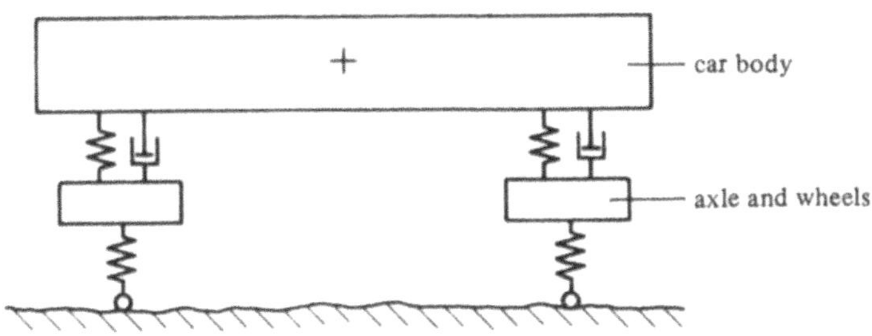

Fig. 2.2. Three-body system for the analysis of vertical vibrations of a motor car.

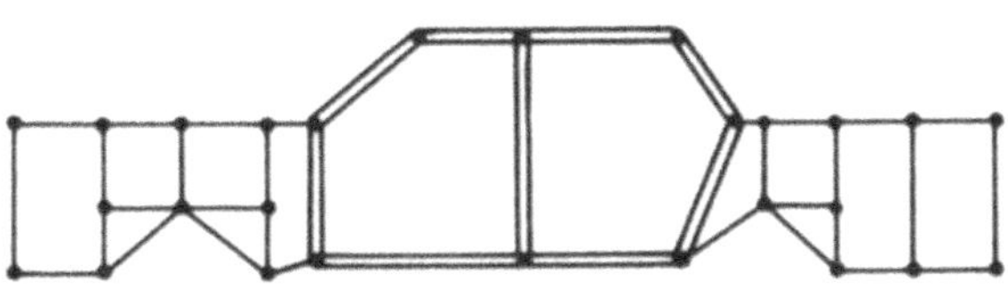

Fig. 2.3. Multibody system for the analysis of bending vibrations of a motor car.

vibrations of the steering wheel, Fig. 2.4, one may consider the steering spindle as a continuum endowed with mass and elasticity, and the wheel as a rigid body. Such a *continuous model* leads to partial differential equations which fall beyond the scope of this book. Let us mention, however, that the general solution for continuous vibrating systems can be reduced by means of the *separation method* to solving systems of ordinary differential equations. This establishes a close connection between continuous and multi-dof-systems.

An important model for a multi-dof mechanical system is a *multibody system* composed of *rigid bodies, springs, dampers*, and *servomotors*. The servomotors generate in the multibody system time-dependent steering or perturbing forces, state dependent control forces, or other forces. The notion of servomotor will thus be considered here in its widest generality. For instance, an unbalance excitation is also regarded to be generated by a servomotor.

2.2. Kinematics of multibody systems

The motion of a multibody system, Fig. 2.5, is composed by the individual motions of each constituent body, where the latter motions are essentially influenced by the *constraints* imposed upon each individual body by all the others. These constraints are due to rigid bearings, joints, or supports connected with the various bodies.

The *position* of a rigid body K_1 in space can be fully described by a Cartesian coordinate system rigidly connected with the body. Let this be the system x_1, y_1, z_1 with its origin attached in the center of mass C_1. This system can be described relative to an inertial system x_I, y_I, z_I by means of a 3×1 *translation vector* $\boldsymbol{r}_1$, and a 3×3 rotation tensor $\boldsymbol{S}_1$.

To begin with, let us consider a free body. The translation vector given by its three Cartesian coordinates with respect to the inertial system reads as

$$\boldsymbol{r}_1 = [r_x \ r_y \ r_z]^T, \tag{2.1}$$

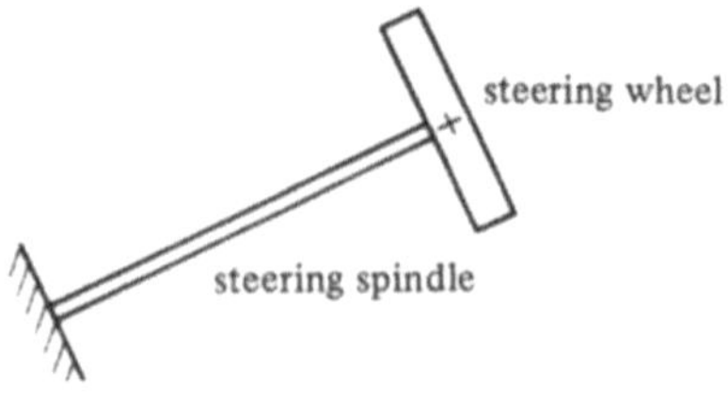

Fig. 2.4. Continuum for the analysis of rotary vibrations of a steering wheel.

and the rotation tensor is described by three generalized coordinates, e.g. by the Kardan angles α, β, γ, as

$$S_1 = \begin{bmatrix} \cos\beta\cos\gamma & -\cos\beta\sin\gamma & \sin\beta \\ \cos\alpha\sin\gamma + \sin\alpha\sin\beta\cos\gamma & \cos\alpha\cos\gamma - \sin\alpha\sin\beta\sin\gamma & -\sin\alpha\cos\beta \\ \sin\alpha\sin\gamma - \cos\alpha\sin\beta\cos\gamma & \sin\alpha\cos\gamma + \cos\alpha\sin\beta\sin\gamma & \cos\alpha\cos\beta \end{bmatrix}. \quad (2.2)$$

The tensor (2.2) is obtained as a result of three consecutive rotations about the x-axis, the y-axis, and the z-axis. Thus (2.2) may be represented as a product of three matrices

$$S_1 = \begin{bmatrix} 1 & 0 & 0 \\ 0 & \cos\alpha & -\sin\alpha \\ 0 & \sin\alpha & \cos\alpha \end{bmatrix} \begin{bmatrix} \cos\beta & 0 & \sin\beta \\ 0 & 1 & 0 \\ -\sin\beta & 0 & \cos\beta \end{bmatrix} \begin{bmatrix} \cos\gamma & -\sin\gamma & 0 \\ \sin\gamma & \cos\gamma & 0 \\ 0 & 0 & 1 \end{bmatrix}. \quad (2.3)$$

The rotation tensors are orthogonal matrices satisfying $S_1^{-1} = S_1^T$ and $\det S_1 = 1$. It is thus evident that the position of a free body is determined by six generalized coordinates corresponding to the six degrees of freedom of a rigid body. These generalized coordinates may be arranged to form a 6×1 position vector

$$z = \left[r_x \; r_y \; r_z \; \alpha \; \beta \; \gamma\right]^T. \quad (2.4)$$

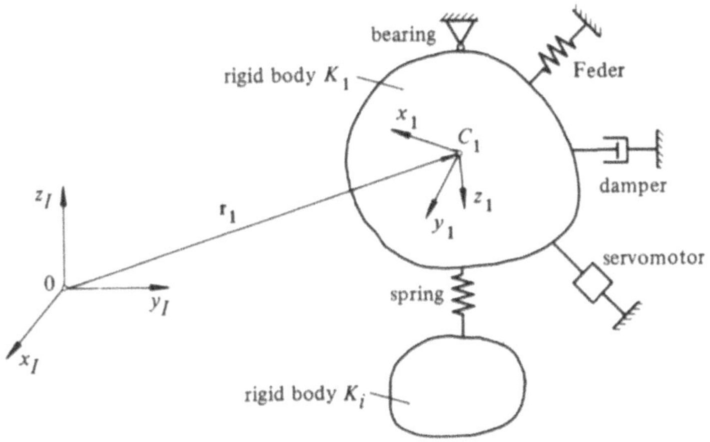

Fig. 2.5. Multibody system.

The translation vector and the rotation tensor can be considered as vector functions of the position vector z:

$$\boldsymbol{r}_1 = \boldsymbol{r}_1(\boldsymbol{z}), \qquad \boldsymbol{S}_1 = \boldsymbol{S}_1(\boldsymbol{z}). \tag{2.5}$$

If the motion of such a body is restricted to rotation about a fixed axis, Fig. 2.6, the body is subject to five constraints and there remains one degree of freedom. The constraints are, according to Fig. 2.6,

$$\left.\begin{aligned}
r_x &= (\cos\beta_0 \cos\gamma)a + (\sin\beta_0)b,\\
r_y &= (\cos\alpha_0 \sin\gamma + \sin\alpha_0 \sin\beta_0 \cos\gamma)a - (\sin\alpha_0 \cos\beta_0)b,\\
r_z &= (\sin\alpha_0 \sin\gamma - \cos\alpha_0 \sin\beta_0 \cos\gamma)a + (\cos\alpha_0 \cos\beta_0)b,\\
\alpha &= \alpha_0,\\
\beta &= \beta_0.
\end{aligned}\right\} \tag{2.6}$$

Thus there remains one generalized coordinate

$$z = \gamma \tag{2.7}$$

and the translation vector and rotation tensor depend only on the angle γ,

$$\boldsymbol{r}_1 = \boldsymbol{r}_1(\gamma), \qquad \boldsymbol{S}_1 = \boldsymbol{S}_1(\gamma). \tag{2.8}$$

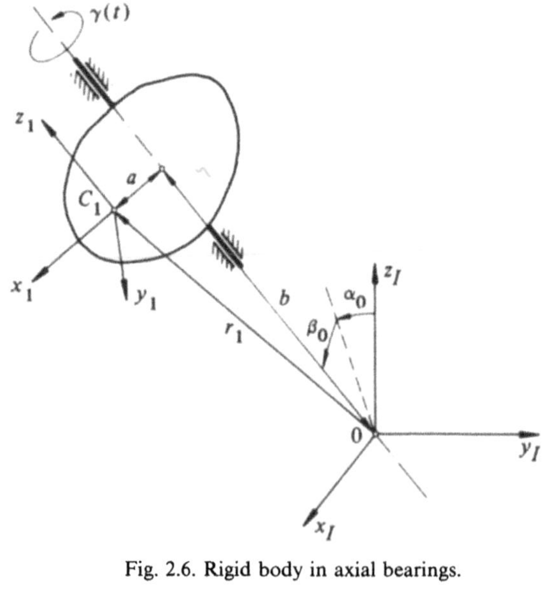

Fig. 2.6. Rigid body in axial bearings.

The constraints (2.6) are called *holonomic* since they depend only on the coordinates. If the velocities also appear in a nonintegrable form in the constraints, we call the latter *nonholonomic.*

Having discussed the motion of one rigid body, let us now consider a multibody system. For a system of p rigid bodies subject to q holonomic time-variant constraints, the translation vector and the rotation tensor of each body K_i, $i = 1(1)p$, may depend on all f *generalized coordinates* of the system. Here again the number of generalized coordinates equals the number of degrees of freedom,

$$f = 6p - q. \tag{2.9}$$

Collecting the generalized coordinates to form one $f \times 1$ *position vector*, one obtains for the translation vector and rotation tensor of a body K_i

$$\boldsymbol{r}_i = \boldsymbol{r}_i(z, t), \qquad \boldsymbol{S}_i = \boldsymbol{S}_i(z, t), \qquad i = 1(1)p. \tag{2.10}$$

Thus the position of a multibody system is totally described by a minimum set of independent coordinates. The translation vector and rotation tensor of a rigid body determine the velocity and the angular velocity of the body respectively. Differentiating (2.10) with respect to time, one obtains the 3×1 *velocity vector* $\boldsymbol{v}_i$, and the 3×1 *angular velocity vector* $\boldsymbol{\omega}_i$, of the body K_i relative to the inertial system

$$\left.\begin{aligned} \boldsymbol{v}_i &= \boldsymbol{J}_{Ti}(z, t)\dot{z} + \bar{\boldsymbol{v}}_i(z, t), & \bar{\boldsymbol{v}}_i &= \frac{\partial \boldsymbol{r}_i}{\partial t}, \\ \boldsymbol{\omega}_i &= \boldsymbol{J}_{Ri}(z, t)\dot{z} + \bar{\boldsymbol{\omega}}_i(z, t), & \bar{\boldsymbol{\omega}}_i &= \frac{\partial \boldsymbol{s}_i}{\partial t}. \end{aligned}\right\} \tag{2.11}$$

In calculating the velocity $\boldsymbol{v}_i$, the rules for the total differential have been used, and the $3 \times f$ Jacobian translation matrix

$$\boldsymbol{J}_{Ti} = \frac{\partial \boldsymbol{r}_i}{\partial z} = \begin{bmatrix} \frac{\partial r_{xi}}{\partial z_1} & \frac{\partial r_{xi}}{\partial z_2} & \cdots & \frac{\partial r_{xi}}{\partial z_f} \\ \frac{\partial r_{yi}}{\partial z_1} & \frac{\partial r_{yi}}{\partial z_2} & \cdots & \frac{\partial r_{yi}}{\partial z_f} \\ \frac{\partial r_{zi}}{\partial z_1} & \frac{\partial r_{zi}}{\partial z_2} & \cdots & \frac{\partial r_{zi}}{\partial z_f} \end{bmatrix} \tag{2.12}$$

together with the local velocity $\bar{\boldsymbol{v}}_i$ were introduced. In the calculation of the angular velocity $\boldsymbol{\omega}_i$, the $3 \times f$ Jacobian rotation matrix

$$\boldsymbol{J}_{Ri} = \frac{\partial \boldsymbol{s}_i}{\partial \boldsymbol{z}} = \begin{bmatrix} \frac{\partial s_{xi}}{\partial z_1} & \frac{\partial s_{xi}}{\partial z_2} & \cdots & \frac{\partial s_{xi}}{\partial z_f} \\ \frac{\partial s_{yi}}{\partial z_1} & \frac{\partial s_{yi}}{\partial z_2} & \cdots & \frac{\partial s_{yi}}{\partial z_f} \\ \frac{\partial s_{zi}}{\partial z_1} & \frac{\partial s_{zi}}{\partial z_2} & \cdots & \frac{\partial s_{zi}}{\partial z_f} \end{bmatrix} \tag{2.13}$$

and the local angular velocity $\bar{\boldsymbol{\omega}}_i$ were introduced. For the angular velocities the 3×1 vector of infinitesimal rotation

$$\partial \boldsymbol{s}_i = \left[\partial s_{xi} \; \partial s_{yi} \; \partial s_{zi}\right]^T \tag{2.14}$$

was used.

First, the local angular velocity $\bar{\boldsymbol{\omega}}_i$ is evaluated. Defining by means of the coordinates in (2.14) the skewsymmetric 3×3 tensor

$$\partial \tilde{\boldsymbol{s}} = \begin{bmatrix} 0 & -\partial s_z & \partial s_y \\ \partial s_z & 0 & -\partial s_x \\ -\partial s_y & \partial s_x & 0 \end{bmatrix}, \tag{2.15}$$

characterized by the symbol ~ due to Roberson *, one can write the tensor of infinitesimal rotation for $\boldsymbol{S}(t=0) = \boldsymbol{E}$ in the form

$$\boldsymbol{S}(\partial t) = \boldsymbol{E} + \partial \tilde{\boldsymbol{s}}(t). \tag{2.16}$$

This can be obtained immediately from (2.2) if one introduces the infinitesimal angles $\partial\alpha(t)$, $\partial\beta(t)$, $\partial\gamma(t)$. Here $\boldsymbol{E}$ denotes the 3×3 unit tensor. According to the product rule for rotation tensors, one thus obtains for an infinitesimal rotation at an arbitrary instant

$$\boldsymbol{S}(t + \partial t) = (\boldsymbol{E} + \partial \tilde{\boldsymbol{s}}(t))\boldsymbol{S}(t). \tag{2.17}$$

Hence, the differential of the rotation tensor is given by

$$\partial \boldsymbol{S}(t) = \boldsymbol{S}(t + \partial t) - \boldsymbol{S}(t) = \partial \tilde{\boldsymbol{s}}(t) \cdot \boldsymbol{S}(t) \tag{2.18}$$

and, using the identity $\boldsymbol{S}^{-1} = \boldsymbol{S}^T$, one gets the following equation for the infinitesimal rotation

$$\partial \tilde{\boldsymbol{s}}(t) = \partial \boldsymbol{S}(t) \boldsymbol{S}^T(t). \tag{2.19}$$

* Roberson applies the symbol ~ to vectors. Hence a small letter is used in (2.15) to denote a tensor.

In other words, the tensor of local angular velocity can be obtained by formally differentiating the rotation tensor with respect to time and multiplying by its transpose

$$\frac{\partial \tilde{s}}{\partial t} = \frac{\partial \boldsymbol{S}}{\partial t} \boldsymbol{S}^T = \begin{bmatrix} 0 & -\frac{\partial s_z}{\partial t} & \frac{\partial s_y}{\partial t} \\ \frac{\partial s_z}{\partial t} & 0 & -\frac{\partial s_x}{\partial t} \\ -\frac{\partial s_y}{\partial t} & \frac{\partial s_x}{\partial t} & 0 \end{bmatrix}. \tag{2.20}$$

Rewriting this and using (2.14) and (2.15), one finds the vector of local angular velocity to be

$$\bar{\boldsymbol{\omega}} = \begin{bmatrix} \frac{\partial s_x}{\partial t} & \frac{\partial s_y}{\partial t} & \frac{\partial s_z}{\partial t} \end{bmatrix}^T. \tag{2.21}$$

The Jacobian rotation matrix in (2.11), and respectively in (2.13), is determined analogously

$$\frac{\partial \tilde{s}_i}{\partial z_j} = \frac{\partial \boldsymbol{S}_i}{\partial z_j} \boldsymbol{S}_i^T. \tag{2.22}$$

Here z_j denotes the j-th generalized coordinate, $j = 1(1)f$.

To conclude, it should be mentioned that the matrix multiplication of the skewsymmetric 3×3 tensor $\tilde{\boldsymbol{s}}$ by the 3×1 vector $\boldsymbol{u}$ corresponds exactly to the *vector product* of $\boldsymbol{s}$ by $\boldsymbol{u}$:

$$\tilde{\boldsymbol{s}}\boldsymbol{u} \equiv \boldsymbol{s} \times \boldsymbol{u}. \tag{2.23}$$

In the future, this matrix notation will always be used for the vector product since the latter is not introduced explicitly in the matrix calculus.

Example 2.1. The angular velocity of a body

Consider a body K_1, axially suspended in its centre of mass C_1. Its angular velocity will be determined from the rotation tensor. According to (2.3), when a rotation is applied about the fixed x-axis, Fig. 2.7, the rotation tensor is given by

$$\boldsymbol{S}_1 = \begin{bmatrix} 1 & 0 & 0 \\ 0 & \cos\alpha & -\sin\alpha \\ 0 & \sin\alpha & \cos\alpha \end{bmatrix}. \tag{2.24}$$

Using this and (2.20) one finds the tensor of angular velocity to be

$$\frac{\mathrm{d}\tilde{s}_1}{\mathrm{d}t} = \frac{\mathrm{d}\boldsymbol{S}_1}{\mathrm{d}t}\boldsymbol{S}_1^T = \begin{bmatrix} 0 & 0 & 0 \\ 0 & -\dot{\alpha}\sin\alpha & -\dot{\alpha}\cos\alpha \\ 0 & \dot{\alpha}\cos\alpha & -\dot{\alpha}\sin\alpha \end{bmatrix} \begin{bmatrix} 1 & 0 & 0 \\ 0 & \cos\alpha & \sin\alpha \\ 0 & -\sin\alpha & \cos\alpha \end{bmatrix}$$

$$= \begin{bmatrix} 0 & 0 & 0 \\ 0 & 0 & -\dot{\alpha} \\ 0 & \dot{\alpha} & 0 \end{bmatrix}. \tag{2.25}$$

Rewriting this, one obtains the expected result

$$\boldsymbol{\omega}_1 = [\dot{\alpha} \quad 0 \quad 0]^T. \tag{2.26}$$

Example 2.2. The kinematics of a double pendulum

Consider a planar double pendulum with equal lengths L and equal particles m. The system has two degrees of freedom and one may use angles α_1 and α_2 as general coordinates, Fig. 2.8. The position vector is

$$z(t) = [\alpha_1 \quad \alpha_2]^T. \tag{2.27}$$

The translation vectors for the points K_1 and K_2 are

$$\boldsymbol{r}_1(z) = \begin{bmatrix} 0 \\ L\sin\alpha_1 \\ -L\cos\alpha_1 \end{bmatrix}, \quad \boldsymbol{r}_2(z) = \begin{bmatrix} 0 \\ L(\sin\alpha_1 + \sin\alpha_2) \\ -L(\cos\alpha_1 + \cos\alpha_2) \end{bmatrix}. \tag{2.28}$$

Since the masses are particles, there are no rotation tensors. Performing partial differentiation with respect to the generalized coordinates, one

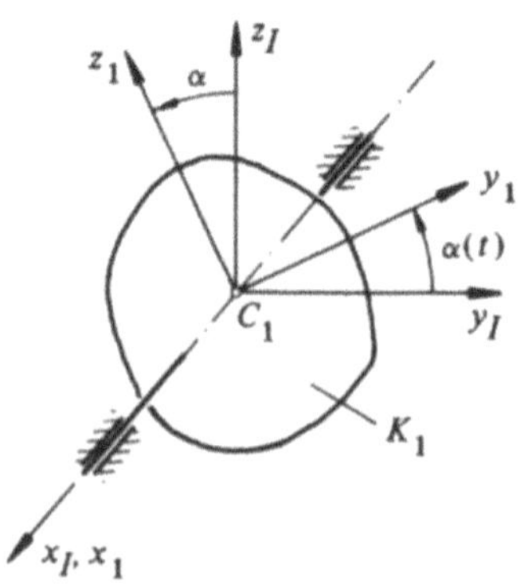

Fig. 2.7. Body rotating about a fixed axis.

obtains the 3×2 Jacobian matrices (2.12) for the two particles

$$J_{T1}(z) = \begin{bmatrix} 0 & 0 \\ \cos\alpha_1 & 0 \\ \sin\alpha_1 & 0 \end{bmatrix} L, \qquad J_{T2}(z) = \begin{bmatrix} 0 & 0 \\ \cos\alpha_1 & \cos\alpha_2 \\ \sin\alpha_1 & \sin\alpha_2 \end{bmatrix} L. \tag{2.29}$$

Further, one has $\bar{v}_1 = \bar{v}_2 = \mathbf{0}$, since the translation vectors (2.28) do not depend explicitly on time. From (2.11) one thus gets the velocity vectors

$$v_1(z, \dot{z}) = L \begin{bmatrix} 0 \\ \cos\alpha_1 \\ \sin\alpha_1 \end{bmatrix} \dot{\alpha}_1, \quad v_2(z, \dot{z}) = L \begin{bmatrix} 0 & 0 \\ \cos\alpha_1 & \cos\alpha_2 \\ \sin\alpha_1 & \sin\alpha_2 \end{bmatrix} \begin{bmatrix} \dot{\alpha}_1 \\ \dot{\alpha}_2 \end{bmatrix}. \tag{2.30}$$

This result can be confirmed by differentiating (2.28).

2.3. Lagrange's equations of motion

Lagrange's equations of motion for a system of p rigid bodies under holonomic constraints are given by

$$\frac{\mathrm{d}}{\mathrm{d}t}\frac{\partial T}{\partial \dot{z}} - \frac{\partial T}{\partial z} = \boldsymbol{q}^T. \tag{2.31}$$

Here differentiation with respect to a vector is defined according to (2.12). The scalar *kinetic energy*

$$T = \tfrac{1}{2} \sum_{i=1}^{p} \left(\boldsymbol{v}_i^T m_i \boldsymbol{v}_i + \boldsymbol{\omega}_i^T \boldsymbol{I}_i \boldsymbol{\omega}_i \right) \tag{2.32}$$

of the vibrating system is given by the kinematic quantities $\boldsymbol{v}_i$ and $\boldsymbol{\omega}_i$, the

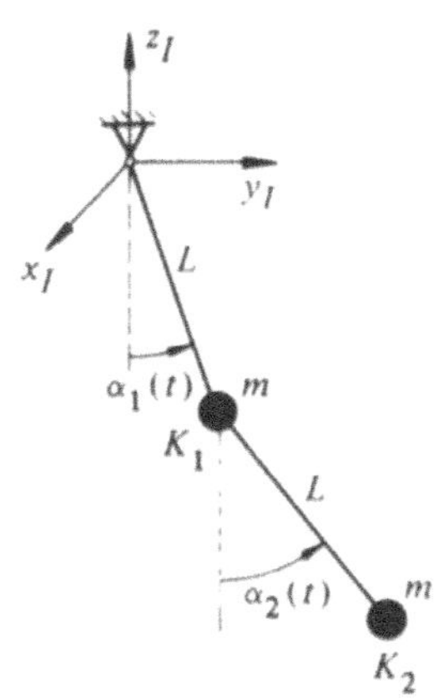

Fig. 2.8. Double pendulum.

masses m_i, and the tensors of inertia $\boldsymbol{I}_i$ of the bodies K_i, $i = 1(1)p$. The symmetric 3×3 *tensor of inertia*

$$\boldsymbol{I}_i = \int_{m_i} \left(\boldsymbol{s}_i^T \boldsymbol{s}_i \boldsymbol{E} - \boldsymbol{s}_i \boldsymbol{s}_i^T \right) \mathrm{d}m_i \tag{2.33}$$

has to be taken with respect to the center of mass C_i, relative to the inertial system, Fig. 2.9. Frequently however, the tensor of inertia is given in a body-fixed coordinate system. Denoting this tensor by $\boldsymbol{I}_i'$, one can obtain the tensor of inertia $\boldsymbol{I}_i$ relative to an inertial system by applying a *coordinate transformation* by means of the rotation tensor $\boldsymbol{S}_i$:

$$\boldsymbol{I}_i = \boldsymbol{S}_i \boldsymbol{I}_i' \boldsymbol{S}_i^T. \tag{2.34}$$

The tensor of inertia $\boldsymbol{I}_i'$ is always time-invariant while the tensor $\boldsymbol{I}_i$ may depend on the generalized coordinates $\boldsymbol{z}$ and the time t.

Using the Jacobian matrices one obtains from the external forces $\boldsymbol{f}_i$ and external torques $\boldsymbol{l}_i$ acting on the various bodies, according to the *free body principle,* the $f \times 1$ vector of generalized forces

$$\boldsymbol{q} = \sum_{i=1}^{p} \left(\boldsymbol{J}_{Ti}^T \boldsymbol{f}_i + \boldsymbol{J}_{Ri}^T \boldsymbol{l}_i \right). \tag{2.35}$$

Here $\boldsymbol{f}_i$ and $\boldsymbol{l}_i$ are always 3×1 vectors relative to the inertial system. The external forces and torques can be split into their applied and constraint components

$$\boldsymbol{f}_i = \boldsymbol{f}_{Ei} + \boldsymbol{f}_{Zi}, \qquad \boldsymbol{l}_i = \boldsymbol{l}_{Ei} = \boldsymbol{l}_{Zi}, \tag{2.36}$$

where the constraint forces are precisely those which arise as reactions to the constraints. According to the virtual work principle, the constraint

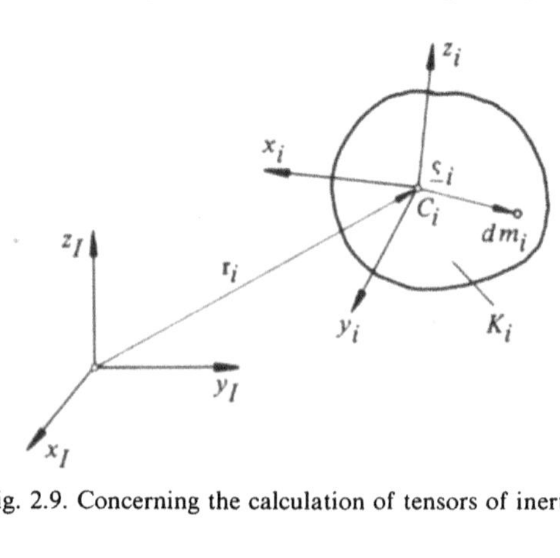

Fig. 2.9. Concerning the calculation of tensors of inertia.

forces do not contribute to the generalized forces

$$\sum_{i=1}^{p} \left(\boldsymbol{J}_{Ti}^T \boldsymbol{f}_{Zi} + \boldsymbol{J}_{Ri}^T \boldsymbol{l}_{Zi} \right) = \mathbf{0}. \tag{2.37}$$

Thus, the generalized forces are entirely determined by the applied forces, and (2.35) may be replaced by

$$\boldsymbol{q} = \boldsymbol{q}(z, \dot{z}, t) = \sum_{i=1}^{p} \left(\boldsymbol{J}_{Ti}^T \boldsymbol{f}_{Ei} + \boldsymbol{J}_{Ri}^T \boldsymbol{l}_{Ei} \right). \tag{2.38}$$

The generalized forces (2.39) are called position and velocity dependent if they depend only on the generalized coordinates z, the generalized velocities $\dot{z}$, and the time (see Schiehlen and Kreuzer). This holds particularly for conservative forces in which all generalized forces can be obtained from the potential energy U:

$$\boldsymbol{q}^T = \boldsymbol{q}^T(z) = -\frac{\partial U}{\partial z}. \tag{2.39}$$

Hence, the generalized forces depend only on the generalized coordinates z.

Lagrange's equations (2.31) appear at first sight to be rather simple, yet further processing of these equations may prove to be cumbersome. This can be seen if one introduces the velocities and angular velocities (2.11) in the expression (2.32) for the kinetic energy

$$T = \tfrac{1}{2} \sum_{i=1}^{p} \left(\dot{z}^T \boldsymbol{J}_{Ti}^T m_i \boldsymbol{J}_{Ti} \dot{z} + 2\, \dot{z}^T \boldsymbol{J}_{Ti}^T m_i \bar{\boldsymbol{v}}_i + \bar{\boldsymbol{v}}_i^T m_i \bar{\boldsymbol{v}}_i \right.$$
$$\left. + z^T \boldsymbol{J}_{Ri}^T \boldsymbol{I}_i \boldsymbol{J}_{Ri} \dot{z} + 2\, \dot{z}^T \boldsymbol{J}_{Ri}^T \boldsymbol{I}_i \bar{\boldsymbol{\omega}}_i + \bar{\boldsymbol{\omega}}_i^T \boldsymbol{I}_i \bar{\boldsymbol{\omega}}_i \right). \tag{2.40}$$

This formula for the kinetic energy must undergo partial differentiation with respect to z and $\dot{z}$, and subsequently total differentiation with respect to time. This is arduous and may easily lead to mistakes. If this differentiation is correctly performed, one obtains as result the equations of motion of an *ordinary mechanical system*

$$\boldsymbol{M}(z, t)\ddot{z}(t) + \boldsymbol{g}(z, \dot{z}, t) = \boldsymbol{q}(z, \dot{z}, t), \tag{2.41}$$

where

$$\boldsymbol{M}(z, t) = \sum_{i=1}^{p} \left(\boldsymbol{J}_{Ti}^T m_i \boldsymbol{J}_{Ti} + \boldsymbol{J}_{Ri}^T \boldsymbol{I}_i \boldsymbol{J}_{Ri} \right) \tag{2.42}$$

is the symmetric $f \times f$ inertia matrix, and $\boldsymbol{g}$ is an $f \times 1$ vector comprising the generalized Coriolis and centrifugal forces. A mechanical system is called ordinary if the constraints are holonomic and the forces only depend on position, velocity and/or time.

Example 2.3. Lagrange's equations for a double pendulum

The kinematic relations for a double pendulum, Fig. 2.8, are known from Example 2.2. According to (2.40), the kinetic energy is given by

$$T = \tfrac{1}{2}\dot{z}^T\left(J_{T1}^T m_1 J_{T1} + J_{T2}^T m_2 J_{T2}\right)\dot{z} \tag{2.43}$$

or, more explicitly by

$$T = \frac{mL^2}{2}\begin{bmatrix}\dot{\alpha}_1 & \dot{\alpha}_2\end{bmatrix}\begin{bmatrix} 2 & \cos(\alpha_1 - \alpha_2) \\ \cos(\alpha_1 - \alpha_2) & 1 \end{bmatrix}\begin{bmatrix}\dot{\alpha}_1 \\ \dot{\alpha}_2\end{bmatrix}$$

$$= \frac{mL^2}{2}\left(2\dot{\alpha}_1^2 + 2\cos(\alpha_1 - \alpha_2)\dot{\alpha}_1\dot{\alpha}_2 + \dot{\alpha}_2^2\right). \tag{2.44}$$

The formula for the potential energy is

$$U = mgh_1 + mgh_2 = mgL(3 - 2\cos\alpha_1 - \cos\alpha_2). \tag{2.45}$$

Next, one has to carry out the partial and total differentiations. In detail, one obtains

$$\frac{\partial T}{\partial \alpha_1} = -mL^2 \sin(\alpha_1 - \alpha_2)\dot{\alpha}_1\dot{\alpha}_2, \tag{2.46}$$

$$\frac{\partial T}{\partial \alpha_2} = mL^2 \sin(\alpha_1 - \alpha_2)\dot{\alpha}_1\dot{\alpha}_2, \tag{2.47}$$

$$\frac{\partial T}{\partial \dot{\alpha}_1} = 2mL^2\dot{\alpha}_1 + mL^2\cos(\alpha_1 - \alpha_2)\dot{\alpha}_2, \tag{2.48}$$

$$\frac{\partial T}{\partial \dot{\alpha}_2} = mL^2\cos(\alpha_1 - \alpha_2)\dot{\alpha}_1 + mL^2\dot{\alpha}_2, \tag{2.49}$$

$$\frac{\mathrm{d}}{\mathrm{d}t}\frac{\partial T}{\partial \dot{\alpha}_1} = 2mL^2\ddot{\alpha}_1 + mL^2\cos(\alpha_1 - \alpha_2)\ddot{\alpha}_2$$

$$-mL^2\sin(\alpha_1 - \alpha_2)(\dot{\alpha}_1 - \dot{\alpha}_2)\dot{\alpha}_2, \tag{2.50}$$

$$\frac{\mathrm{d}}{\mathrm{d}t}\frac{\partial T}{\partial \dot{\alpha}_2} = mL^2\cos(\alpha_1 - \alpha_2)\ddot{\alpha}_1 - mL^2\sin(\alpha_1 - \alpha_2)(\dot{\alpha}_1 - \dot{\alpha}_2)\dot{\alpha}_1$$

$$+mL^2\ddot{\alpha}_2, \tag{2.51}$$

$$\frac{\partial U}{\partial \alpha_1} = 2mgL\sin\alpha_1, \tag{2.52}$$

$$\frac{\partial U}{\partial \alpha_2} = mgL \sin \alpha_2. \tag{2.53}$$

Thus, Lagrange's equations (2.31) together with (2.39) yield the following equations of motion

$$2mL^2\ddot{\alpha}_1 + mL^2 \cos(\alpha_1 - \alpha_2)\ddot{\alpha}_2 + mL^2 \sin(\alpha_1 - \alpha_2)\dot{\alpha}_2^2$$
$$- mL^2 \sin(\alpha_1 - \alpha_2)\dot{\alpha}_1\dot{\alpha}_2 + mL^2 \sin(\alpha_1 - \alpha_2)\dot{\alpha}_1\dot{\alpha}_2 = -2mgL \sin \alpha_1, \tag{2.54}$$

$$mL^2 \cos(\alpha_1 - \alpha_2)\ddot{\alpha}_1 + mL^2\ddot{\alpha}_2 - mL^2 \sin(\alpha_1 - \alpha_2)\dot{\alpha}_1^2$$
$$+ mL^2 \sin(\alpha_1 - \alpha_2)\dot{\alpha}_1\dot{\alpha}_2 - mL^2 \sin(\alpha_1 - \alpha_2)\dot{\alpha}_1\dot{\alpha}_2 = -mgL \sin \alpha_2. \tag{2.55}$$

These equations of motion can be written in matrix form as

$$\underbrace{mL^2\begin{bmatrix} 2 & \cos(\alpha_1 - \alpha_2) \\ \cos(\alpha_1 - \alpha_2) & 1 \end{bmatrix}}_{M(z)} \underbrace{\begin{bmatrix} \ddot{\alpha}_1 \\ \ddot{\alpha}_2 \end{bmatrix}}_{\ddot{z}(t)} + \underbrace{mL^2\begin{bmatrix} \sin(\alpha_1 - \alpha_2)\dot{\alpha}_2^2 \\ -\sin(\alpha_1 - \alpha_2)\dot{\alpha}_1^2 \end{bmatrix}}_{g(z,\, \dot{z})}$$
$$= \underbrace{-mgL\begin{bmatrix} 2 \sin \alpha_1 \\ \sin \alpha_2 \end{bmatrix}}_{q(z)}. \tag{2.56}$$

Thus the equations of motion have been found in the form of (2.41) with a symmetric inertia matrix $M(z)$. ◆

2.4. Newton-Euler equations of motion

Lagrange's equations (2.31) can be set up directly for the entire multibody system. In contrast, the Newton-Euler equations have to be set up according to the free body principle, i.e. separately for each constituent body of the system. Thus for a system of p rigid bodies one has to solve p simple subproblems, which corresponds to engineering practice.

Newton's equation for each body K_i with mass m_i reads as

$$m_i\dot{v}_i = f_i, \qquad i = 1(1)p, \tag{2.57}$$

and Euler's equation is given by

$$I_i\dot{\omega}_i + \tilde{\omega}_i I_i \omega_i = l_i, \qquad i = 1(1)p \tag{2.58}$$

Newton's equation is also called the *momentum principle* and Euler's equation is also known as the *angular momentum principle.* In Newton's equation (2.57) f_i denotes the sum of the external forces in the inertial system. In Euler's equation (2.58) there appear the tensor of inertia I_i and the sum of all the external torques l_i relative to the center of mass C_i in the inertial system. Further, $\tilde{\omega}_i$ is the skewsymmetric 3×3 tensor corresponding to the angular velocity ω_i which yields the vector product. Introducing the velocities and angular velocities (2.11) in (2.57) and (2.58), one finds

$$m_i J_{Ti}\ddot{z} + m_i \dot{J}_{Ti}\dot{z} + m_i \dot{\bar{v}}_i = f_i, \qquad i = 1(1)p \tag{2.59}$$

and

$$I_i J_{Ri}\ddot{z} + I_i \dot{J}_{Ri}\dot{z} + I_i \dot{\bar{\omega}}_i + (\widetilde{J_{Ri}\dot{z}} + \tilde{\bar{\omega}}_i) I_i (J_{Ri}\dot{z} + \bar{\omega}_i) = l_i, \qquad i = 1(1)p. \tag{2.60}$$

This gives $6p$ scalar equations for f generalized coordinates z, $j = 1(1)f$. These equations can be simultaneously written in matrix notation

$$\bar{M}(z, t)\ddot{z}(t) + \bar{g}(z, \dot{z}, t) = \bar{q}(z, \dot{z}, t), \tag{2.61}$$

where

$$\bar{M}(z, t) = \begin{bmatrix} m_1 J_{T1} \\ m_2 J_{T2} \\ \vdots \\ I_{(p-1)} J_{R(p-1)} \\ I_p J_{Rp} \end{bmatrix} \tag{2.62}$$

is a $6p \times f$ inertia matrix, $\bar{g}$ is a $6p \times 1$ vector, and

$$\bar{q}(z, \dot{z}, t) = \bar{q}_E + \bar{q}_Z = \begin{bmatrix} f_1 \\ f_2 \\ \vdots \\ l_{(p-1)} \\ l_p \end{bmatrix} \tag{2.63}$$

is a $6p \times 1$ vector of forces and torques. In an ordinary multibody system the forces and torques depend on z, $\dot{z}$, and t, and can be divided into applied and constraint forces, respectively torques. Using the principle of virtual work (2.37) one can eliminate the constraint forces and at the same time reduce the order of the system (2.61). For this purpose the $6p \times f$ Jacobian matrix

$$\bar{\boldsymbol{J}}(z, t) = \left[\boldsymbol{J}_{T1}^T \; \boldsymbol{J}_{T2}^T \ldots \boldsymbol{J}_{R(p-1)}^T \boldsymbol{J}_{Rp}^T \right]^T \tag{2.64}$$

of the total system is introduced. Then (2.37) can be rewritten as a matrix product

$$\bar{\boldsymbol{J}}^T \bar{\boldsymbol{q}}_Z = \mathbf{0}. \tag{2.65}$$

By a left multiplication of (2.61) by $\bar{\boldsymbol{J}}^T$ one immediately gets the equation of motion *

$$\boldsymbol{M}(z, t)\ddot{z}(t) + \boldsymbol{g}(z, \dot{z}, t) = \boldsymbol{q}(z, \dot{z}, t), \tag{2.66}$$

which is identical to the equation (2.41), which was found by the method of Lagrange's equations. In particular, the $f \times f$ inertia matrix $\boldsymbol{M}(z, t)$ satisfies, by (2.62) and (2.64),

$$\boldsymbol{M}(z, t) = \bar{\boldsymbol{J}}^T(z, t)\overline{\boldsymbol{M}}(z, t) \tag{2.67}$$

and for the $f \times 1$ vector of generalized forces one has

$$\boldsymbol{q}(z, \dot{z}, t) = \bar{\boldsymbol{J}}^T(z, t)\bar{\boldsymbol{q}}_E(z, \dot{z}, t). \tag{2.68}$$

If only the equations of motion in generalized coordinates are desired one may replace in Newton's equation (2.57) the sum of all external forces $\boldsymbol{f}_i$ by the sum of all applied forces $\boldsymbol{f}_{Ei}$, and in Euler's equation (2.58) the sum of all external torques $\boldsymbol{l}_i$ may be replaced by the sum of all applied torques $\boldsymbol{l}_{Ei}$.

Example 2.4. Newton-Euler equations for the double pendulum

The kinematic relations for the double pendulum, Fig. 2.8, are known from Example 2.2. According to (2.59), Newton's equations for the particles K_1 and K_2 of the double pendulum are

$$m_i \boldsymbol{J}_{Ti}\ddot{z} + m_i \dot{\boldsymbol{J}}_{Ti}\dot{z} = \boldsymbol{f}_i, \qquad i = 1, 2, \tag{2.69}$$

* If the constraint forces are eliminated with the help of the Jacobian matrices, one attains at the same time a symmetrization of the inertia matrix. This is by no means guaranteed if the elimination occurs by other methods.

where the applied gravity forces mg and the constraint forces Z_i act as shown in Fig. 2.10. Euler's equations are absent, since the double pendulum consists only of particles.

Newton's equations can now be written in matrix notation according to (2.61) as follows:

$$\underbrace{mL\begin{bmatrix} 0 & 0 \\ \cos\alpha_1 & 0 \\ \sin\alpha_1 & 0 \\ \hdashline 0 & 0 \\ \cos\alpha_1 & \cos\alpha_2 \\ \sin\alpha_1 & \sin\alpha_2 \end{bmatrix}}_{\bar{\boldsymbol{M}}(z)} \underbrace{\underbrace{\begin{bmatrix} \ddot{\alpha}_1 \\ \ddot{\alpha}_2 \end{bmatrix}}_{\ddot{z}(t)}}_{\ddot{z}(t)} + \underbrace{mL\begin{bmatrix} 0 \\ -\sin\alpha_1\dot{\alpha}_1^2 \\ \cos\alpha_1\dot{\alpha}_1^2 \\ \hdashline 0 \\ -\sin\alpha_1\dot{\alpha}_1^2-\sin\alpha_2\dot{\alpha}_2^2 \\ \cos\alpha_1\dot{\alpha}_1^2+\cos\alpha_2\dot{\alpha}_2^2 \end{bmatrix}}_{\bar{\boldsymbol{g}}(z,\,\dot{z})}$$

$$= \underbrace{\begin{bmatrix} 0 \\ 0 \\ -mg \\ \hdashline 0 \\ 0 \\ -mg \end{bmatrix}}_{=\bar{\boldsymbol{q}}_E} + \underbrace{\begin{bmatrix} 0 \\ -Z_1\sin\alpha_1+Z_2\sin\alpha_2 \\ Z_1\cos\alpha_1-Z_2\cos\alpha_2 \\ \hdashline 0 \\ -Z_2\sin\alpha_2 \\ Z_2\cos\alpha_2 \end{bmatrix}}_{\bar{\boldsymbol{q}}_Z(z)}. \tag{2.70}$$

To eliminate the constraint forces and to reduce the order of the system,

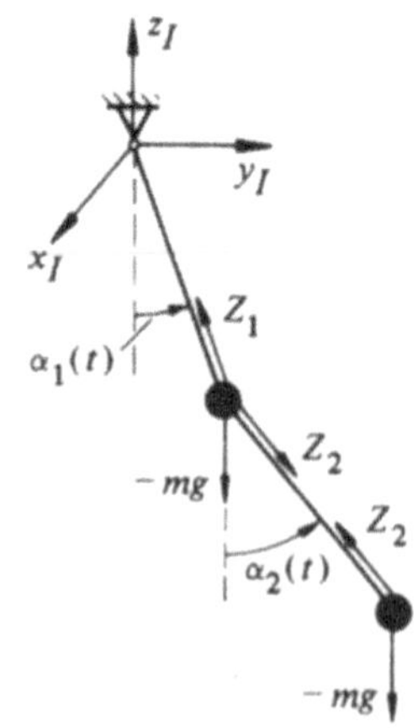

Fig. 2.10. Forces acting on the double pendulum.

one multiplies (2.70) from the left by the transpose of the Jacobian matrix of the system

$$\bar{J}^T(z) = L\begin{bmatrix} 0 & \cos\alpha_1 & \sin\alpha_1 & 0 & \cos\alpha_1 & \sin\alpha_1 \\ 0 & 0 & 0 & 0 & \cos\alpha_2 & \sin\alpha_2 \end{bmatrix} \tag{2.71}$$

and obtains the equations of motion

$$\underbrace{mL^2\begin{bmatrix} 2 & \cos(\alpha_1-\alpha_2) \\ \cos(\alpha_1-\alpha_2) & 1 \end{bmatrix}}_{M(z)}\underbrace{\begin{bmatrix}\ddot{\alpha}_1 \\ \ddot{\alpha}_2\end{bmatrix}}_{\ddot{z}(t)} + \underbrace{mL^2\begin{bmatrix}\sin(\alpha_1-\alpha_2)\dot{\alpha}_2^2 \\ -\sin(\alpha_1-\alpha_2)\dot{\alpha}_1^2\end{bmatrix}}_{g(z,\dot{z})}$$

$$= \underbrace{-mgL\begin{bmatrix} 2\sin\alpha_1 \\ \sin\alpha_2 \end{bmatrix}}_{=\; q(z)}, \tag{2.72}$$

which coincides exactly with the result (2.56). One should observe that the constraint forces have disappeared, and thus it would have been admissible to cancel the constraint forces $\bar{q}_Z$ already in (2.70).

Comparing Lagrange's and Newton-Euler's equations, one is inclined to favor the latter. The kinematics are not different, but finding the kinetic energy and carrying out the subsequent differentiations for Lagrange's equations are replaced for Newton-Euler's equations by simple matrix operations, admitting evident physical interpretation. Since both methods lead to identical results, it is thus to some extent a question of habit or proficiency which of the methods should be applied.

Problem 2.1. Centrifuge models

In designing a centrifuge as in Fig. 2.11 one has to analyse the following vibration phenomena:

A) spatial vibrations of the centrifuge rotor to prevent damages to the casing by the rotor
B) rotational vibrations of the axis to prevent the breaking of the axis through fatigue.

1. Which of the following construction elements are responsible for the appearance of the following vibrations?

	spatial vibrations	rotational vibrations
rigid centrifuge rotor	☐	☐
elastic axis	☐	☐
rigid axis	☐	☐
rigid driving rotor	☐	☐
motor suspension	☐	☐

2. Which models are suitable for the vibration analysis?

	spatial vibrations	rotational vibrations
multibody systems	☐	☐
finite element systems	☐	☐
continuous systems	☐	☐

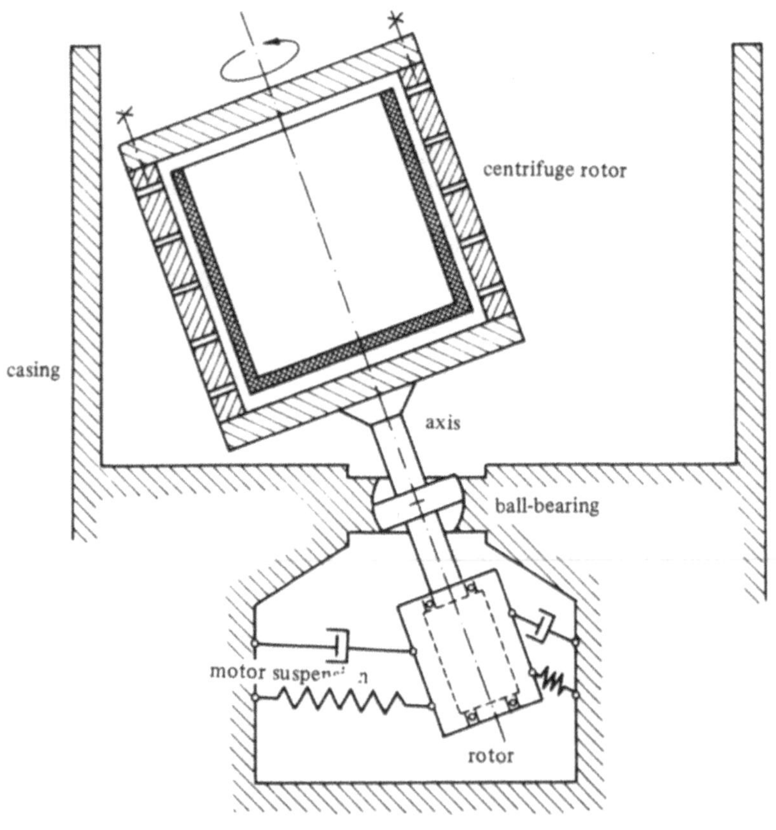

Fig. 2.11. Cross-section of a centrifuge.

Problem 2.2. Equations of motion of the vertical vibrations of a motor car

1. Which of the following methods is most suitable for designing a model for analysing the vertical body and wheel vibrations of a motor car?
   □ method of multibody systems,
   □ method of finite element systems,
   □ method of continuous systems.
2. Find the equations of motion for the model in Fig. 2.12 for the vertical vibrations of a motor car by means of Lagrange's equations. Let m_1 be the mass of the car body, m_2 the joint mass of the axles and wheels connected to the body mass by means of the body spring with spring constant k_1, and a shock absorber with damping constant d_1. The elasticity of the tires is characterized by the spring constant k_2 and the damping of the tires by the damping constant d_2. The generalized coordinates $y_1(t)$, $y_2(t)$, and the excitation coordinate $y_e(t)$ are taken relative to an equilibrium position under the assumption of a horizontal road

$$\left.\begin{aligned} r_{z1}(t) &= r_{z1S} + y_1(t), & r_{z1S} &= \text{const},\\ r_{z2}(t) &= r_{z2S} + y_2(t), & r_{z2S} &= \text{const},\\ r_{ze}(t) &= r_{zeS} + y_e(t), & r_{zeS} &= \text{const}. \end{aligned}\right\} \tag{2.73}$$

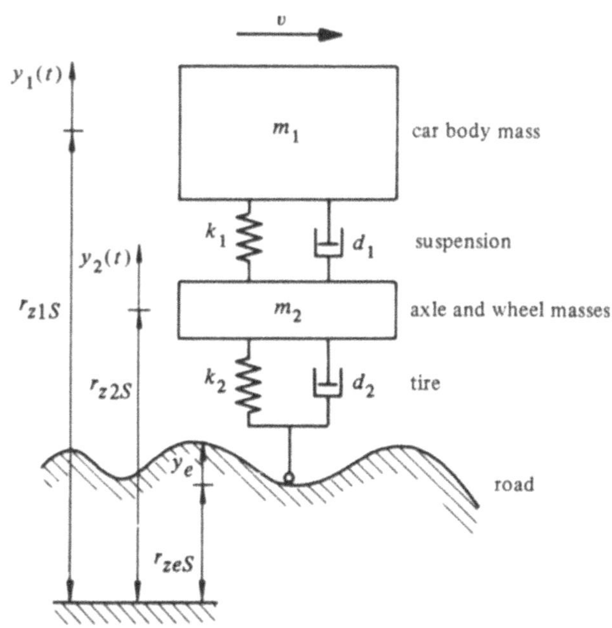

Fig. 2.12. Model for the vertical vibrations of a motor car.

The excitation coordinate $y_e(t)$ is a function of time provided that the motor car travels with nonzero velocity, $v \neq 0$.

a) Which is the correct formula for kinetic energy?

□ $T = \frac{1}{2}(m_1 \dot{y}_1 + m_2 \dot{y}_2)^2$,

□ $T = \frac{1}{2} m_1 (\dot{y}_1 + \dot{y}_2)^2 + \frac{1}{2} m_2 \dot{y}_2^2$,

□ $T = \frac{1}{2} m_1 \dot{y}_1^2 + \frac{1}{2} m_2 \dot{y}_2^2$.

b) Which is the potential energy?

□ $U = \frac{1}{2} k_1 (y_1 + y_2)^2 + \frac{1}{2} k_2 (y_2 + y_e)^2$,

□ $U = \frac{1}{2} k_1 (y_1 - y_2)^2 + \frac{1}{2} k_2 (y_2 - y_e)^2$,

□ $U = \frac{1}{2} k_1 y_1^2 + \frac{1}{2} k_2 y_2^2 + \frac{1}{2} k_2 y_e^2$.

c) Which is the vector of the generalized damping forces?

□ $\boldsymbol{q} = [-d_1(\dot{y}_1 - \dot{y}_2) \quad d_1(\dot{y}_1 - \dot{y}_2) - d_2(\dot{y}_2 - \dot{y}_e)]^T$,

□ $\boldsymbol{q} = [-d_1(\dot{y}_1 + \dot{y}_2) \quad d_1(\dot{y}_1 + \dot{y}_2) + d_2(\dot{y}_2 + \dot{y}_e)]^T$,

□ $\boldsymbol{q} = [\ d_1 \dot{y}_1 \quad (d_1 + d_2)(\dot{y}_2 + \dot{y}_e)]^T$.

3. What is the form of the equations of motion?

□ $$\begin{bmatrix} m_1 & 0 \\ 0 & m_2 \end{bmatrix} \begin{bmatrix} \ddot{y}_1 \\ \ddot{y}_2 \end{bmatrix} + \begin{bmatrix} d_1 & 0 \\ 0 & d_2 \end{bmatrix} \begin{bmatrix} \dot{y}_1 \\ \dot{y}_2 \end{bmatrix} + \begin{bmatrix} k_1 & -k_1 \\ -k_1 & k_2 \end{bmatrix} \begin{bmatrix} y_1 \\ y_2 \end{bmatrix} = \begin{bmatrix} 0 \\ k_2 y_e \end{bmatrix},$$

□ $$\begin{bmatrix} m_1 & 0 \\ 0 & m_2 \end{bmatrix} \begin{bmatrix} \ddot{y}_1 \\ \ddot{y}_2 \end{bmatrix} + \begin{bmatrix} d_1 & -d_1 \\ -d_1 & d_1 + d_2 \end{bmatrix} \begin{bmatrix} \dot{y}_1 \\ \dot{y}_2 \end{bmatrix} + \begin{bmatrix} k_1 & -k_1 \\ -k_1 & k_1 + k_2 \end{bmatrix} \begin{bmatrix} y_1 \\ y_2 \end{bmatrix}$$
$$= \begin{bmatrix} 0 \\ k_2 y_e + d_2 \dot{y}_e \end{bmatrix},$$

□ $$\begin{bmatrix} m_1 & m_1 \\ m_1 & m_1 + m_2 \end{bmatrix} \begin{bmatrix} \ddot{y}_1 \\ \ddot{y}_2 \end{bmatrix} + \begin{bmatrix} d_1 & -d_1 \\ -d_1 & d_1 + d_2 \end{bmatrix} \begin{bmatrix} \dot{y}_1 \\ \dot{y}_2 \end{bmatrix} + \begin{bmatrix} k_1 & -k_1 \\ -k_1 & k_2 \end{bmatrix} \begin{bmatrix} y_1 \\ y_2 \end{bmatrix}$$
$$= \begin{bmatrix} 0 \\ d_2 \dot{y}_e \end{bmatrix}.$$ ◆

3. State equations of linear vibrating systems

When arbitrarily large displacements are considered, the equations of vibrating systems are, as a rule, nonlinear. However in mechanical engineering large displacements in vibrations are seldom desirable, indeed, just the opposite holds true. The engineer makes efforts to arrange the vibrating system so that its motion is kept within the vicinity of a prescribed reference motion. When this is the case, the equations of motion relative to this reference motion can often be linearized. The existing *linearization* possibilities will be illustrated in examples of mechanical systems. The linearization can be carried out either subsequently to generating the equations of motion or can be introduced into the kinematical relations prior to generating the equations. In both cases one obtains the same linear equations of motion, i.e. a second order differential equation for the position vector. The equations of motion can be transformed into the *state equation*, i.e. into a first order differential equation for the state vector of the system. Ordinary mechanical vibrating systems and the analogous electrical systems may be equally represented by their equations of motion or their state equations. This is however not the case for general vibrating systems whose state equations are preferred since the equations of motion do not offer any special advantage for the analysis of vibrations.

In this chapter the treatment of linear equations of motion and state equations of ordinary mechanical systems will be considered first with some remarks concerning general mechanical and electromagnetic systems. Lastly, a discussion of the transformation laws for the state equations which allow transitions from one state vector to another will be presented.

3.1. Ordinary mechanical systems

According to (2.41) or (2.66), the equation of motion of a nonlinear mechanical system has the form

$$\boldsymbol{M}(z, t)\ddot{z}(t) + \boldsymbol{g}(z, \dot{z}, t) = \boldsymbol{q}(z, \dot{z}, t), \tag{3.1}$$

where $z(t)$ is the $f \times 1$ position vector. The engineering problem provides the reference motion

$$z_S = z_S(t). \tag{3.2}$$

A typical reference motion is, for instance, the equilibrium position

$$z_S(t) \equiv \mathbf{0}, \tag{3.3}$$

but in general, this does not hold true. A reference motion can just as easily be provided by a steady rotation of a rigid body about an inertial axis or by the trajectory of a vehicle. The vibrations in the neighborbood of the reference motion are then given by

$$z(t) = z_S(t) + y(t) \tag{3.4}$$

where the $f \times 1$ position vector $y(t)$ and the $f \times 1$ velocity vector $\dot{y}(t)$ should remain small relative to comparison vectors $\boldsymbol{a}$ and $\boldsymbol{b}$, $y(t) \ll \boldsymbol{a}$ and $\dot{y}(t) \ll \boldsymbol{b}$. The coordinates of these comparison vectors are usually obtained from the maximum values of certain nonlinear functions and a characteristic frequency. Substituting (3.4) into (3.1) one now obtains, with the use of the Taylor series expansion and assuming differentiability of the vector functions,

$$\boldsymbol{M}(z, t) = \boldsymbol{M}_0(t) + \boldsymbol{M}_1(y, t) + \ldots, \tag{3.5}$$

$$\boldsymbol{g}(z, \dot{z}, t) = \boldsymbol{g}_0(t) + \boldsymbol{G}_{1y}(t)\, y(t) + \boldsymbol{G}_{1\dot{y}}(t)\, \dot{y}(t) + \ldots, \tag{3.6}$$

$$\boldsymbol{q}(z, \dot{z}, t) = \boldsymbol{q}_0(t) + \boldsymbol{Q}_{1y}(t)\, y(t) + \boldsymbol{Q}_{1\dot{y}}(t)\, \dot{y}(t) + \ldots, \tag{3.7}$$

where only the independent position and velocity terms $\boldsymbol{M}_0(t)$, $\boldsymbol{g}_0(t)$, and $\boldsymbol{q}_0(t)$ and the linear $y(t)$ and $\dot{y}(t)$ terms have been written. If one now substitutes (3.5 to 3.7) into (3.1) and neglects the terms of second order and higher in $y(t)$ and $\dot{y}(t)$, one obtains the linear *equations of motion*

$$\boldsymbol{M}(t)\, \ddot{y}(t) + \boldsymbol{P}(t)\, \dot{y}(t) + \boldsymbol{Q}(t)\, y(t) = \boldsymbol{h}(t). \tag{3.8}$$

Here, for simplicity, the following notation has been introduced:

$$\left.\begin{aligned}
&\boldsymbol{M}(t) = \boldsymbol{M}_0(t) = \boldsymbol{M}(z_S(t), t),\\
&\boldsymbol{P}(t) = \boldsymbol{G}_{1\dot{y}}(t) - \boldsymbol{Q}_{1\dot{y}}(t),\\
&\boldsymbol{Q}(t) = \boldsymbol{M}_{1S}(t) + \boldsymbol{G}_{1y}(t) - \boldsymbol{Q}_{1y}(t),\\
&\boldsymbol{M}_{1S}(t)\, y(t) = \boldsymbol{M}_1(y, t)\ddot{z}_S(t),\\
&\boldsymbol{h}(t) = \boldsymbol{q}_0(t) - \boldsymbol{g}_0(t) - \boldsymbol{M}_0(t)\ddot{z}_S(t).
\end{aligned}\right\} \tag{3.9}$$

The equation of motion (3.8) is a second order vector differential equation for the $f \times 1$ position vector $y(t)$ characterizing a multi-dof time-variant vibrating system. A *time-invariant* vibrating system

$$\boldsymbol{M}\ddot{\boldsymbol{y}}(t) + (\boldsymbol{D} + \boldsymbol{G})\dot{\boldsymbol{y}}(t) + (\boldsymbol{K} + \boldsymbol{N})\boldsymbol{y}(t) = \boldsymbol{h}(t). \tag{3.10}$$

is given whenever the symmetric $f \times f$ inertia matrix $\boldsymbol{M}$, the $f \times f$ matrix $\boldsymbol{P}$ of the velocity dependent forces, and the $f \times f$ matrix $\boldsymbol{Q}$ of the position dependent forces are time-invariant. In (3.10) the matrix $\boldsymbol{P}$ and $\boldsymbol{Q}$ have been decomposed into the symmetric $f \times f$ matrix $\boldsymbol{D}$, the skewsymmetric $f \times f$ matrix $\boldsymbol{G}$, the symmetric $f \times f$ matrix $\boldsymbol{K}$, and the skewsymmetric $f \times f$ matrix $\boldsymbol{N}$. On the right hand side one again has the time-dependent $f \times 1$ excitation vector $\boldsymbol{h}(t)$.

For free vibrations, $\boldsymbol{h}(t) \equiv \boldsymbol{0}$, the matrices in (3.10) permit physical interpretation. A left multiplication of (3.10) by $\dot{\boldsymbol{y}}^T$ yields the scalar expression

$$\underbrace{\dot{\boldsymbol{y}}^T\boldsymbol{M}\ddot{\boldsymbol{y}}}_{\frac{\mathrm{d}}{\mathrm{d}t}T} + \underbrace{\dot{\boldsymbol{y}}^T\boldsymbol{D}\dot{\boldsymbol{y}}}_{2R} + \underbrace{\dot{\boldsymbol{y}}^T\boldsymbol{G}\dot{\boldsymbol{y}}}_{0} + \underbrace{\dot{\boldsymbol{y}}^T\boldsymbol{K}\boldsymbol{y}}_{\frac{\mathrm{d}}{\mathrm{d}t}U} + \underbrace{\dot{\boldsymbol{y}}^T\boldsymbol{N}\boldsymbol{y} = 0.}_{2S = 0} \tag{3.11}$$

The inertia matrix $\boldsymbol{M}$ determines the kinetic energy and thus the *inertial forces*, the damping matrix $\boldsymbol{D}$ represents via the Rayleigh function R the *damping forces*, and the matrix $\boldsymbol{G}$ describes the *gyroscopic forces* which do not alter the energy balance. The matrix $\boldsymbol{K}$ determines the potential energy, and thus the conservative *position forces*, while the matrix $\boldsymbol{N}$ describes the *nonconservative position forces*. If $\boldsymbol{D} = \boldsymbol{0}$ and $\boldsymbol{N} = \boldsymbol{0}$, then the vibrating system is conservative, i.e. the total energy $T + U$ is constant.

Example 3.1. Linearization of the equations of motion of the double pendulum

Consider the equation of motion (2.72) of the double pendulum derived in Example 2.4. In this example the equations will be linearized relative the equilibrium position. Therefore $z_S(t) \equiv \boldsymbol{0}$, $\boldsymbol{y}(t) \ll \boldsymbol{a}$, $\boldsymbol{y}(t) \ll \boldsymbol{b}$ with $\boldsymbol{a} = [1 \;\; 1]^T$ and $\boldsymbol{b} = \boldsymbol{a}\sqrt{g/L}$, i.e. the generalized coordinates are $\alpha_1 \ll 1$, $\alpha_2 \ll 1$ and the generalized velocities are $\dot{\alpha}_1 \ll \sqrt{g/L}$, $\dot{\alpha}_2 \ll \sqrt{g/L}$. The Taylor series expansions up to the first term are

$$\boldsymbol{M}(z) = mL^2\begin{bmatrix} 2 & \cos(\alpha_1 - \alpha_2) \\ \cos(\alpha_1 - \alpha_2) & 1 \end{bmatrix} = mL^2\begin{bmatrix} 2 & 1 \\ 1 & 1 \end{bmatrix} + \boldsymbol{0} + \ldots, \tag{3.12}$$

$$g(z, \dot{z}) = mL^2 \begin{bmatrix} \sin(\alpha_1 - \alpha_2)\dot{\alpha}_2^2 \\ -\sin(\alpha_1 - \alpha_2)\dot{\alpha}_1^2 \end{bmatrix} = \mathbf{0} + \mathbf{0} + \mathbf{0} + \dots, \tag{3.13}$$

$$q(z, \dot{z}) = -mgL \begin{bmatrix} 2 \sin \alpha_1 \\ \sin \alpha_2 \end{bmatrix} = \mathbf{0} - mgL \begin{bmatrix} 2 & 0 \\ 0 & 1 \end{bmatrix} \begin{bmatrix} \alpha_1 \\ \alpha_2 \end{bmatrix} + \mathbf{0} + \dots . \tag{3.14}$$

Substituting (3.12) to (3.14) into (2.72) and neglecting terms of the second order and higher, one obtains the linearized equations of motion

$$\underbrace{\begin{bmatrix} 2 & 1 \\ 1 & 1 \end{bmatrix}}_{M} \underbrace{\begin{bmatrix} \ddot{\alpha}_1 \\ \ddot{\alpha}_2 \end{bmatrix}}_{\ddot{y}(t)} + \underbrace{\frac{g}{L}\begin{bmatrix} 2 & 0 \\ 0 & 1 \end{bmatrix}}_{K} \underbrace{\begin{bmatrix} \alpha_1 \\ \alpha_2 \end{bmatrix}}_{y(t)} = \mathbf{0}. \tag{3.15}$$

Hence small motions of the double pendulum about its equilibrium position constitute a time-invariant vibrating system. Also, since the damping matrix $\boldsymbol{D}$ and the matrix $\boldsymbol{N}$ of the nonconservative position forces do not appear in (3.15), the double pendulum is a conservative vibrating system. ◆

There is not much practical value in linearizing the equations of motion (3.1), for one does first generate nonlinear equations of motion only to linearize them. One should rather attempt to attain the simplification due to linearization as early as possible, i.e. already in the kinematics of the problem. For a nonlinear mechanical system the translation vector and the rotation tensor are, according to (2.10),

$$\left.\begin{aligned} r_i &= r_i(z, t), \\ S_i &= S_i(z, t), \end{aligned}\right\} \quad i = 1(1)p, \tag{3.16}$$

where z again denotes the $f \times 1$ position vector. Applying Taylor expansions and using (3.4) one may write

$$\left.\begin{aligned} r_i(z, t) &= r_{i0}(t) + r_{i1}(y, t) + r_{i2}(y^2, t) + \dots, \\ S_i(z, t) &= S_{i0}(t) + S_{i1}(y, t) + S_{i2}(y^2, t) + \dots, \end{aligned}\right\} \tag{3.17}$$

where $r_{i0}(t)$ and $S_{i0}(t)$ describe the reference motion. Further, one gets from (3.17) the Jacobian matrices for the translation and rotation

$$\left.\begin{aligned} J_{Ti}(z, t) &= J_{Ti0}(t) + J_{Ti1}(y, t) + \dots, \\ J_{Ri}(z, t) &= J_{Ri0}(t) + J_{Ri1}(y, t) + \dots \end{aligned}\right\} \tag{3.18}$$

and the local velocities

$$\left.\begin{aligned}\bar{\boldsymbol{v}}_i(z,t) &= \bar{\boldsymbol{v}}_{i0}(t) + \bar{\boldsymbol{V}}_{i1}(t)\,y + \ldots, \\ \bar{\boldsymbol{\omega}}_i(z,t) &= \bar{\boldsymbol{\omega}}_{i0}(t) + \bar{\boldsymbol{\Omega}}_{i1}(t)\,y + \ldots.\end{aligned}\right\} \tag{3.19}$$

If (3.18) and (3.19) are substituted into (2.11) and terms of second order and higher are neglected, one obtains

$$\left.\begin{aligned}\boldsymbol{v}_i &= \boldsymbol{J}_{Ti0}(t)\,\dot{y}(t) + \boldsymbol{K}_{Ti0}(t)\,y(t) + \boldsymbol{v}_{iS}(t), \\ \boldsymbol{\omega}_i &= \boldsymbol{J}_{Ri0}(t)\,\dot{y}(t) + \boldsymbol{K}_{Ri0}(t)\,y(t) + \boldsymbol{\omega}_{is}(t),\end{aligned}\right\} \tag{3.20}$$

where

$$\left.\begin{aligned}\boldsymbol{K}_{Ti0}(t)\,y(t) &= \boldsymbol{J}_{Ti1}(y,t)\dot{z}_S(t) + \bar{\boldsymbol{V}}_{i1}(t)\,y(t), \\ \boldsymbol{v}_{iS}(t) &= \boldsymbol{J}_{Ti0}(t)\dot{z}_S(t) + \bar{\boldsymbol{v}}_{i0}(t)\end{aligned}\right\} \tag{3.21}$$

and also the corresponding equations for $\boldsymbol{K}_{Ri0}(t)$ and $\boldsymbol{\omega}_{iS}(t)$ hold. This result states that in the general case the zero term and the first term of the Jacobian matrices (3.18) are needed to determine the velocities $\boldsymbol{v}_i$ and the angular velocities $\boldsymbol{\omega}_i$. For a constant reference motion $z_S = \text{const}$, e.g. for the position of equilibrium, it would be sufficient to use only the zero term of the Jacobian matrices (3.18). But, as will be seen later, the first term of the Jacobian matrices (3.18) is relevant in the reduction of the order of the system. The first or linear term of the Jacobian matrices is obtained from the second term of the translation vector and rotation tensor expansion (3.17). Hence, for the purpose of linearizing the equations of motion one may terminate the series expansions (3.17), as a rule, only after the second term.

Substituting (3.20) into the Newton-Euler equations (2.57) and (2.58), and expanding the vector (2.63) of the applied forces and moments under the assumption of differentiability into a Taylor series up to the first term, one gets the linearized equation (2.61) with the $6p \times f$ matrices $\bar{\boldsymbol{M}}_0(t)$, $\bar{\boldsymbol{P}}(t)$, and $\bar{\boldsymbol{Q}}(t)$, as well as the $6p \times 1$ excitation function $\bar{\boldsymbol{h}}(t)$

$$\bar{\boldsymbol{M}}_0(t)\,\ddot{y}(t) + \bar{\boldsymbol{P}}(t)\,\dot{y}(t) + \bar{\boldsymbol{Q}}(t)\,y(t) = \bar{\boldsymbol{h}}(t). \tag{3.22}$$

To reduce the order of the system, one still must multiply (3.22) by the transposed Jacobian matrix (2.64)

$$\bar{\boldsymbol{J}}^T(z,t) = \bar{\boldsymbol{J}}_0^T(t) + \bar{\boldsymbol{J}}_1^T(y,t) + \ldots. \tag{3.23}$$

If one again neglects the quadratic and higher order terms, one gets the linearized equations of motion

$$\boldsymbol{M}(t)\ddot{\boldsymbol{y}}(t)+\boldsymbol{P}(t)\dot{\boldsymbol{y}}(t)+\boldsymbol{Q}(t)\boldsymbol{y}(t)=\boldsymbol{h}(t), \tag{3.24}$$

where

$$\left.\begin{aligned}&\boldsymbol{M}(t)=\bar{\boldsymbol{J}}_0^T(t)\bar{\boldsymbol{M}}_0(t),\ \boldsymbol{P}(t)=\bar{\boldsymbol{J}}_0^T(t)\bar{\boldsymbol{P}}(t),\\&\boldsymbol{Q}(t)\boldsymbol{y}(t)=\bar{\boldsymbol{J}}_0^T(t)\bar{\boldsymbol{Q}}(t)\boldsymbol{y}(t)-\bar{\boldsymbol{J}}_1^T(\boldsymbol{y},t)\bar{\boldsymbol{h}}(t),\\&\boldsymbol{h}(t)=\bar{\boldsymbol{J}}_0^T(t)\bar{\boldsymbol{h}}(t).\end{aligned}\right\} \tag{3.25}$$

Thereby it has been shown that one may not neglect the linear terms of the Jacobian matrices $\boldsymbol{J}_{Ti1}(\boldsymbol{y}, t)$ and $\boldsymbol{J}_{Ri1}(\boldsymbol{y}, t)$, and correspondingly the quadratic terms $\boldsymbol{r}_{i2}(\boldsymbol{y}^2, t)$ and $\boldsymbol{S}_{i2}(\boldsymbol{y}^2, t)$ of the translation vector and the rotation tensor. As for the rest, the remarks following (3.9) apply accordingly.

Example 3.2. Linearization of the kinematics of the double pendulum

In this example the translation vectors of the particles K_1 and K_2 of the double pendulum, discussed in Example 2.2, will be linearized with respect to the equilibrium position $\boldsymbol{z}_S(t)\equiv\boldsymbol{0}$. The Taylor series expansion of (2.28) up to the second term yields

$$\boldsymbol{r}_1(\boldsymbol{z})=L\begin{bmatrix}0\\ \sin\alpha_1\\ -\cos\alpha_1\end{bmatrix}=\begin{bmatrix}0\\0\\-L\end{bmatrix}+\begin{bmatrix}0\\L\alpha_1\\0\end{bmatrix}+\begin{bmatrix}0\\0\\ \frac{1}{2}L\alpha_1^2\end{bmatrix}+\dots \tag{3.26}$$

and

$$\boldsymbol{r}_2(\boldsymbol{z})=L\begin{bmatrix}0\\ \sin\alpha_1+\sin\alpha_2\\ -\cos\alpha_1-\cos\alpha_2\end{bmatrix}=\begin{bmatrix}0\\0\\-2L\end{bmatrix}+\begin{bmatrix}0\\L(\alpha_1+\alpha_2)\\0\end{bmatrix}$$

$$+\begin{bmatrix}0\\0\\ \frac{1}{2}L(\alpha_1^2+\alpha_2^2)\end{bmatrix}+\dots. \tag{3.27}$$

The Jacobian matrices are obtained from the linearized translation vectors (3.26) and (3.27),

$$\boldsymbol{J}_{T1}(\boldsymbol{y})=\begin{bmatrix}0&0\\L&0\\0&0\end{bmatrix}+\begin{bmatrix}0&0\\0&0\\L\alpha_1&0\end{bmatrix} \tag{3.28}$$

and

$$J_{T2}(y) = \begin{bmatrix} 0 & 0 \\ L & L \\ 0 & 0 \end{bmatrix} + \begin{bmatrix} 0 & 0 \\ 0 & 0 \\ L\alpha_1 & L\alpha_2 \end{bmatrix}, \tag{3.29}$$

or from the Taylor series expansion up to the first term of (2.29). The linearized velocities are obtained from (3.20) by taking $\boldsymbol{K}_{Ti0}(t) = \mathbf{0}$, $\boldsymbol{v}_{iS}(t) = \mathbf{0}$, $i = 1, 2$:

$$\boldsymbol{v}_1(\dot{y}) = \begin{bmatrix} 0 \\ L\dot{\alpha}_1 \\ 0 \end{bmatrix}, \quad \boldsymbol{v}_2(\dot{y}) = \begin{bmatrix} 0 \\ L(\dot{\alpha}_1 + \dot{\alpha}_2) \\ 0 \end{bmatrix}. \tag{3.30}$$

This shows that in this case the velocities are not influenced by the linear terms of the Jacobian matrices. Thus Newton's equations (2.57) imply

$$mL \begin{bmatrix} 0 & 0 \\ 1 & 0 \\ 0 & 0 \\ \hline 0 & 0 \\ 1 & 1 \\ 0 & 0 \end{bmatrix} \begin{bmatrix} \ddot{\alpha}_1 \\ \ddot{\alpha}_2 \end{bmatrix} = \begin{bmatrix} 0 \\ 0 \\ -mg \\ \hline 0 \\ 0 \\ -mg \end{bmatrix} + \bar{\boldsymbol{q}}_Z. \tag{3.31}$$

Note that the matrices $\overline{\boldsymbol{P}}(t)$ and $\overline{\boldsymbol{Q}}(t)$ do not appear at all in (3.31). To reduce the order of the system the transposed linearized Jacobian matrix (3.23) is used

$$\bar{\boldsymbol{J}}^T = L \left[\begin{array}{ccc|ccc} 0 & 1 & \alpha_1 & 0 & 1 & \alpha_1 \\ 0 & 0 & 0 & 0 & 1 & \alpha_2 \end{array}\right]. \tag{3.32}$$

Multiplying (3.31) from the left by (3.32) one obtains the linearized equations of motion

$$\underbrace{\begin{bmatrix} 2 & 1 \\ 1 & 1 \end{bmatrix}}_{\boldsymbol{M}} \underbrace{\begin{bmatrix} \ddot{\alpha}_1 \\ \ddot{\alpha}_2 \end{bmatrix}}_{\ddot{y}(t)} + \underbrace{\frac{g}{L}\begin{bmatrix} 2 & 0 \\ 0 & 1 \end{bmatrix}}_{\boldsymbol{K}} \underbrace{\begin{bmatrix} \alpha_1 \\ \alpha_2 \end{bmatrix}}_{y(t)} = \mathbf{0}, \tag{3.33}$$

which again coincides with (3.15). It is precisely the linear term of the Jacobian matrix which gives rise to the matrix $\boldsymbol{K}$ of the conservative position forces. This shows that it is absolutely necessary to go as far as the quadratic term in the series expansion of the translation vector. ◆

The equations of motion (3.8), and respectively (3.24), for ordinary mechanical systems must still be supplemented with the initial conditions

$$\left.\begin{aligned} &\boldsymbol{M}(t)\ddot{\boldsymbol{y}}(t)+\boldsymbol{P}(t)\dot{\boldsymbol{y}}(t)+\boldsymbol{Q}(t)\boldsymbol{y}(t)=\boldsymbol{h}(t),\\ &\boldsymbol{y}(t_0)=\boldsymbol{y}_0,\ \dot{\boldsymbol{y}}(t_0)=\dot{\boldsymbol{y}}_0. \end{aligned}\right\} \tag{3.34}$$

Whereas the time behaviour of the system is uniquely described by the $f\times 1$ position vector $\boldsymbol{y}(t)$, the stating of the initial conditions requires in addition the $f\times 1$ velocity vector $\dot{\boldsymbol{y}}(t_0)$. Forming one vector from both the position and velocity vectors one obtains the *state vector*

$$\boldsymbol{x}(t)=\begin{bmatrix}\boldsymbol{y}(t)\\ \hline \dot{\boldsymbol{y}}(t)\end{bmatrix}. \tag{3.35}$$

The order of the state vector $\boldsymbol{x}(t)$ is thus two times the order of the position vector $\boldsymbol{y}(t)$:

$$n=2f. \tag{3.36}$$

Its components are also called state variables.

One can use the state vector (3.35) to write the equations of motion (3.34), supplemented by the trivial equation $\dot{\boldsymbol{y}}(t)=\dot{\boldsymbol{y}}(t)$, in the form of the *state equation*

$$\dot{\boldsymbol{x}}(t)=\boldsymbol{A}(t)\boldsymbol{x}(t)+\boldsymbol{b}(t),\qquad \boldsymbol{x}(t_0)=\boldsymbol{x}_0. \tag{3.37}$$

Thus an $n\times 1$ vector differential equation is obtained. For ordinary mechanical systems and the state vector given by (3.35), the $n\times n$ system matrix $\boldsymbol{A}(t)$ has the special structure

$$\boldsymbol{A}(t)=\left[\begin{array}{c|c}\boldsymbol{0} & \boldsymbol{E}\\ \hline -\boldsymbol{M}^{-1}(t)\boldsymbol{Q}(t) & -\boldsymbol{M}^{-1}(t)\boldsymbol{P}(t)\end{array}\right], \tag{3.38}$$

i.e. it is built up of the $f\times f$ matrices $\boldsymbol{M}(t)$, $\boldsymbol{Q}(t)$, $\boldsymbol{P}(t)$, the $f\times f$ zero matrix $\boldsymbol{0}$, and the $f\times f$ identity matrix $\boldsymbol{E}$. The $n\times 1$ excitation vector $\boldsymbol{b}(t)$ has a specified structure

$$\boldsymbol{b}(t)=\begin{bmatrix}\boldsymbol{0}\\ \hline \boldsymbol{M}^{-1}(t)\boldsymbol{h}(t)\end{bmatrix}, \tag{3.39}$$

where in addition to the $f\times 1$ excitation function $\boldsymbol{h}(t)$ there also appears the $f\times f$ matrix $\boldsymbol{M}(t)$ and the $f\times 1$ zero vector $\boldsymbol{0}$. From (2.42), the inertia matrix is always regular in the case of ordinary mechanical systems, i.e. for such systems the inverse matrix $\boldsymbol{M}^{-1}(t)$ always exists.

The state equation (3.37) has the advantage that it can be investigated by applying various results of the general *theory of linear systems*. Many of these results were obtained in recent years from the theory of control systems and by mathematical research. The state equation is also readily solved and compatible with computer programs. Its definite disadvantage is the doubling of the system's order, according to (3.36), whereas the equations of motion avoid this disadvantage. In particular, due to its special structure (3.10), the time-invariant vibrating system possesses a whole series of properties which do not appear in the general theory of systems. Thus it is important to have at ones disposal both the equations of motion and the state equation.

Example 3.3. State equation of a double pendulum

The linear equations of motion of a double pendulum were derived in Example 3.1 and Example 3.2. Here they will be transformed into the state equation.

The 4×1 state vector is

$$\boldsymbol{x}(t) = \left[\, \boldsymbol{y}^T \,\vdots\, \dot{\boldsymbol{y}}^T \right]^T = \left[\alpha_1 \;\; \alpha_2 \,\vdots\, \dot{\alpha}_1 \;\; \dot{\alpha}_2\right]^T. \tag{3.40}$$

It satisfies the state equation

$$\underbrace{\begin{bmatrix} \dot{\alpha}_1 \\ \dot{\alpha}_2 \\ \hline \ddot{\alpha}_1 \\ \ddot{\alpha}_2 \end{bmatrix}}_{\dot{\boldsymbol{x}}(t) =} = \underbrace{\left[\begin{array}{cc|cc} 0 & 0 & 1 & 0 \\ 0 & 0 & 0 & 1 \\ \hline -2\dfrac{g}{L} & \dfrac{g}{L} & 0 & 0 \\ 2\dfrac{g}{L} & -2\dfrac{g}{L} & 0 & 0 \end{array}\right]}_{\boldsymbol{A}} \underbrace{\begin{bmatrix} \alpha_1 \\ \alpha_2 \\ \hline \dot{\alpha}_1 \\ \dot{\alpha}_2 \end{bmatrix}}_{\boldsymbol{x}(t)}, \tag{3.41}$$

obtained through the use of the inverse inertia matrix

$$\boldsymbol{M}^{-1} = \begin{bmatrix} 1 & -1 \\ -1 & 2 \end{bmatrix}. \tag{3.42}$$

The symmetry properties of the matrices $\boldsymbol{M}$ and $\boldsymbol{K}$ are lost in $\boldsymbol{A}$. On the other hand, the right lower submatrix of $\boldsymbol{A}$ is, for undamped nongyroscopic systems, always an $f \times f$ zero matrix. ◆

3.2. General linear systems

A characteristic property of ordinary mechanical vibrating systems and the corresponding electrical systems is that they can be represented in a closed form by their equations of motion. This property, due to the special structure of these systems, does not carry over to general systems. General linear vibrating systems can be described in closed form only by their state equation

$$\dot{x}(t) = A(t)x(t) + b(t), \qquad x(t_0) = x_0 \tag{3.43}$$

where, however, the specific properties (3.35), (3.38), and (3.39) of the $n \times 1$ state vector $x(t)$, the $n \times n$ system matrix $A(t)$, and the $n \times 1$ excitation vector $b(t)$ do not hold. The superiority of the state equation, as compared with the equations of motion, lies in the fact that it can be applied for the mathematical description of all linear systems. However, in engineering practice the appearance of general linear vibrating systems is spread over a wide area. Hence the setting up of state equations for general linear systems will not be presented in this book systematically. Instead, the procedures will be illustrated in two examples.

Example 3.4. Vibrating system with elastic damper

The general mechanical system shown in Fig. 3.1 is a vibrating system with an elastic damper. The combined action of the spring and the damper is intended to limit the forces that may act on the mass m as a result of disturbing impulses. Nevertheless the system will be damped. The vibrating system has one degree of freedom marked by the coordinate $y_1(t)$, but this coordinate is not sufficient to describe the system uniquely. According to Newton's equation (2.57), with the additional state variable $y_2(t)$ of Fig. 3.1, it is obtained

$$m\ddot{y}_1(t) = -b(y_1(t) - y_2(t)) - ky_1(t). \tag{3.44}$$

However this equation must still be supplemented with the damper equation

$$d\,\dot{y}_2(t) = b(y_1(t) - y_2(t)) \tag{3.45}$$

in which the inertia of the damper has been neglected. Thus the state vector is

$$x(t) = [y_1 \quad \dot{y}_1 \quad y_2]^T \tag{3.46}$$

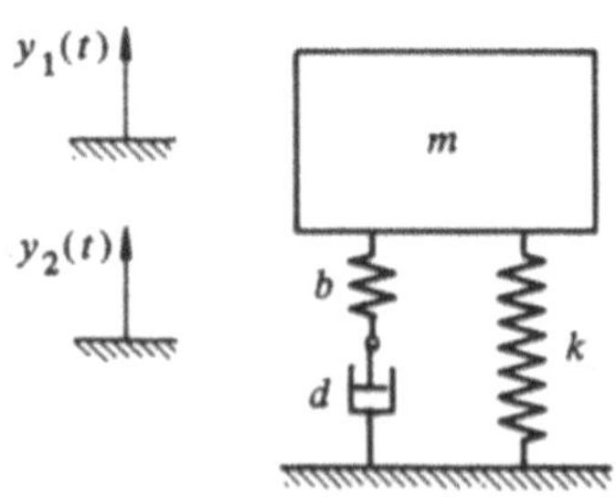

Fig. 3.1. Vibrating system with elastic damper

and the state equation has the form

$$\underbrace{\begin{bmatrix} \dot{y}_1 \\ \ddot{y}_1 \\ \dot{y}_2 \end{bmatrix}}_{\dot{x}(t)} = \underbrace{\begin{bmatrix} 0 & 1 & 0 \\ -\frac{b+k}{m} & 0 & \frac{b}{m} \\ \frac{b}{d} & 0 & -\frac{b}{d} \end{bmatrix}}_{A} \underbrace{\begin{bmatrix} y_1 \\ \dot{y}_1 \\ y_2 \end{bmatrix}}_{x(t)}, \tag{3.47}$$

which contains the trivial equation $\dot{y}_1(t)=\dot{y}_1(t)$ added to (3.44). It can be easily seen that the system matrix A in (3.47) does not possess the characteristic form (3.38) of ordinary systems. ◆

Example 3.5. Magnetically levitated vehicle

A simple model for an electromagnetically levitated vehicle is composed of a single mass and an electromagnet. The supporting electromagnetic force acting on the mass of the vehicle depends on the distance between the electromagnet and the rail, and on the excitation current. Hence Newton's equation (2.57) for the vertical motion, Fig. 3.2, is

$$m\ddot{y}(t) = cy(t) + di(t) - m\ddot{y}_e(t), \tag{3.48}$$

under the assumption that only small deviations from equilibrium occur. The current $i(t)$ in the electromagnet is determined by means of a voltage control by the distance variation $y(t)$ and the velocity $\dot{y}(t)$,

$$T\dot{i}(t) + i(t) + k\dot{y}(t) = ay(t) + b\dot{y}(t), \tag{3.49}$$

where T, k are electromagnetic constants and a, b are control gains. Thus one again gets a general linear vibrating system. The state vector is

$$x(t) = [y \quad \dot{y} \quad i]^T \tag{3.50}$$

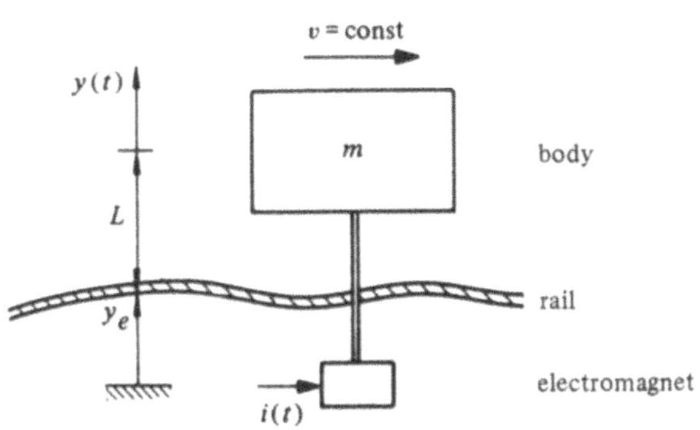

Fig. 3.2. Model of a magnetically levitated vehicle

and for the state equation one gets from (3.48) and (3.49), with the use of the trivial equation $\dot{y}(t)=\dot{y}(t)$,

$$\underbrace{\begin{bmatrix} \dot{y} \\ \ddot{y} \\ \dot{i} \end{bmatrix}}_{\dot{x}(t)} = \underbrace{\begin{bmatrix} 0 & 1 & 0 \\ \frac{c}{m} & 0 & \frac{d}{m} \\ \frac{a}{T} & \frac{b-k}{T} & -\frac{1}{T} \end{bmatrix}}_{A} \underbrace{\begin{bmatrix} y \\ \dot{y} \\ i \end{bmatrix}}_{x(t)} + \underbrace{\begin{bmatrix} 0 \\ -\ddot{y}_e \\ 0 \end{bmatrix}}_{b(t)} \tag{3.51}$$

The excitation vector $\boldsymbol{b}(t)$ in (3.51) is determined by the acceleration $\ddot{y}_e(t)$ caused by roughness of the rail. The system (3.51) does not satisfy the special conditions (3.38) and (3.39) which hold for ordinary mechanical systems.

Thus it has been shown in two examples that no universally valid statements can be made about the structure of the system matrix and the excitation vector of a general linear system.

3.3. Transformation of linear state equations

The state of a vibrating system is uniquely described by a state vector $\boldsymbol{x}(t)$. The choice, however, of the state variables, i.e. of the components of the state vector remains *arbitrary*. The state variables $x_i(t)$, $i=1(1)n$, can always arise by means of linearly independent combinations from other state variables $x_j(t)$, $j=1(1)n$,

$$x_i = \sum_{j=1}^{n} t_{ij}\hat{x}_j, \qquad i=1(1)n. \tag{3.52}$$

In matrix notation, (3.52) yields a transformation of state vectors

$$\boldsymbol{x}(t) = \boldsymbol{T}\hat{\boldsymbol{x}}(t), \tag{3.53}$$

where $\boldsymbol{T}$ is a regular time-invariant $n \times n$ transformation matrix. Substituting (3.53) into the state equation (3.43), one gets

$$\boldsymbol{T}\dot{\hat{\boldsymbol{x}}}(t) = \boldsymbol{A}(t)\boldsymbol{T}\hat{\boldsymbol{x}}(t) + \boldsymbol{b}(t) \tag{3.54}$$

or

$$\dot{\hat{\boldsymbol{x}}}(t) = \hat{\boldsymbol{A}}(t)\hat{\boldsymbol{x}}(t) + \hat{\boldsymbol{b}}(t) \tag{3.55}$$

with the transformation laws

$$\hat{\boldsymbol{A}}(t)=\boldsymbol{T}^{-1}\boldsymbol{A}(t)\boldsymbol{T} \tag{3.56}$$

$$\hat{\boldsymbol{b}}(t)=\boldsymbol{T}^{-1}\boldsymbol{b}(t). \tag{3.57}$$

for the system matrix and the excitation vector. The transformation law (3.56) is called a *similarity transformation*.

For ordinary mechanical systems the position vector $\boldsymbol{y}(t)$ suffices to describe the time behavior of the system. The choice of the generalized coordinates remains arbitrary which can be expressed by a nonsingular transformation

$$\boldsymbol{y}(t)=\boldsymbol{U}\hat{\boldsymbol{y}}(t). \tag{3.58}$$

Here $\boldsymbol{U}$ is a regular, time-invariant, $f\times f$ transformation matrix. Hence the translation vector and the rotation tensor transform according to

$$\left.\begin{aligned}\boldsymbol{r}_i(\boldsymbol{y},t)&=\boldsymbol{r}_i(\boldsymbol{U}\hat{\boldsymbol{y}},t)=\hat{\boldsymbol{r}}_i(\hat{\boldsymbol{y}},t),\\ \boldsymbol{S}_i(\boldsymbol{y},t)&=\boldsymbol{S}_i(\boldsymbol{U}\hat{\boldsymbol{y}},t)=\hat{\boldsymbol{S}}_i(\hat{\boldsymbol{y}},t)\end{aligned}\right\} \tag{3.59}$$

and the corresponding Jacobian matrices satisfy

$$\hat{\boldsymbol{J}}_{Ti}(\hat{\boldsymbol{y}},t)=\boldsymbol{J}_{Ti}(\boldsymbol{y},t)\boldsymbol{U},\quad \hat{\boldsymbol{J}}_{Ri}(\hat{\boldsymbol{y}},t)=\boldsymbol{J}_{Ri}(\boldsymbol{y},t)\boldsymbol{U}. \tag{3.60}$$

By substituting (3.60) into (2.42) one now obtains the transformation law for the inertia matrix with $\boldsymbol{y}\ll\boldsymbol{a}$,

$$\hat{\boldsymbol{M}}(t)=\boldsymbol{U}^T\boldsymbol{M}(t)\boldsymbol{U}. \tag{3.61}$$

This transformation law, also called a *congruence transformation*, holds for the matrices $\boldsymbol{P}$, $\boldsymbol{Q}$ in (3.8) and also for the matrices $\boldsymbol{D}$, $\boldsymbol{G}$, $\boldsymbol{K}$, $\boldsymbol{N}$ in (3.10). Further, one obtains

$$\hat{\boldsymbol{h}}(t)=\boldsymbol{U}^T\boldsymbol{h}(t). \tag{3.62}$$

The equations of motion (3.8) are carried over by the transformation (3.58) into the transformed equations of motion

$$\hat{\boldsymbol{M}}(t)\ddot{\hat{\boldsymbol{y}}}(t)+\hat{\boldsymbol{P}}(t)\dot{\hat{\boldsymbol{y}}}(t)+\hat{\boldsymbol{Q}}(t)\hat{\boldsymbol{y}}(t)=\hat{\boldsymbol{h}}(t). \tag{3.63}$$

If an ordinary mechanical system is represented by its state equation (3.37), then the special structure (3.38) of the system matrix $\boldsymbol{A}$ is certainly preserved under a similarity transformation when the transformation matrix $\boldsymbol{T}$ is of the form

$$\boldsymbol{T}=\left[\begin{array}{c|c}\boldsymbol{U} & \mathbf{0} \\ \hline \mathbf{0} & \boldsymbol{U}\end{array}\right]. \tag{3.64}$$

An arbitrary transformation matrix $\boldsymbol{T}$ may destroy the structure of the system matrix $\boldsymbol{A}$.

Example 3.6. Transformation of the state equations of the double pendulum

In Examples 3.1 and 3.2 the derivation of the state equation for the double pendulum was given in terms of the position vector

$$\boldsymbol{y}(t)=[\alpha_1 \quad \alpha_2]^T. \tag{3.65}$$

Here the motion of the double pendulum can just as easily be described by the position vector

$$\hat{\boldsymbol{y}}(t)=[y_1 \quad y_2]^T \tag{3.66}$$

where y_1 and y_2 denote the small horizontal displacements. According to the kinematical relations in Fig. 3.3, the transformation matrix is

$$\boldsymbol{U}=\frac{1}{L}\begin{bmatrix} 1 & 0 \\ -1 & 1\end{bmatrix}. \tag{3.67}$$

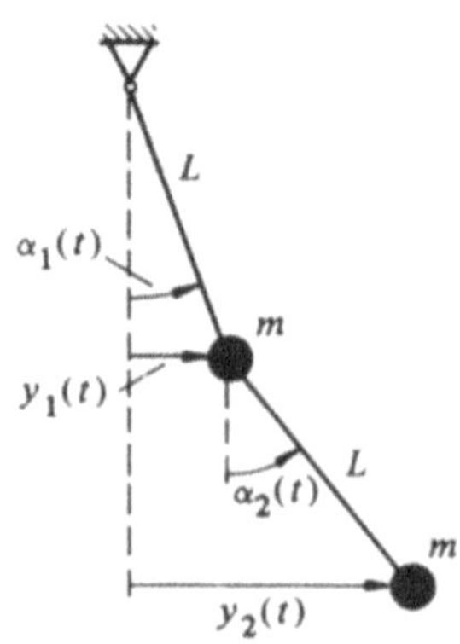

Fig. 3.3. Position coordinates of a double pendulum

Subjecting the matrices $\boldsymbol{M}$ and $\boldsymbol{K}$ in (3.33) to the congruence transformation (3.61), one gets

$$\hat{\boldsymbol{M}} = \frac{1}{L^2}\begin{bmatrix} 1 & -1 \\ 0 & 1 \end{bmatrix}\begin{bmatrix} 2 & 1 \\ 1 & 1 \end{bmatrix}\begin{bmatrix} 1 & 0 \\ -1 & 1 \end{bmatrix} = \frac{1}{L^2}\begin{bmatrix} 1 & 0 \\ 0 & 1 \end{bmatrix} \tag{3.68}$$

and

$$\hat{\boldsymbol{K}} = \frac{g}{L^3}\begin{bmatrix} 1 & -1 \\ 0 & 1 \end{bmatrix}\begin{bmatrix} 2 & 0 \\ 0 & 1 \end{bmatrix}\begin{bmatrix} 1 & 0 \\ -1 & 1 \end{bmatrix} = \frac{g}{L^3}\begin{bmatrix} 3 & -1 \\ -1 & 1 \end{bmatrix}. \tag{3.69}$$

Thus one obtains the transformed equations of motion

$$\hat{\boldsymbol{M}}\ddot{\hat{\boldsymbol{y}}}(t) + \hat{\boldsymbol{K}}\hat{\boldsymbol{y}}(t) = \boldsymbol{0}. \tag{3.70}$$

The system matrix of the corresponding state equation

$$\dot{\hat{\boldsymbol{x}}}(t) = \hat{\boldsymbol{A}}\hat{\boldsymbol{x}}(t) \tag{3.71}$$

is determined from (3.68) and (3.69),

$$\hat{\boldsymbol{A}} = \left[\begin{array}{cc|cc} 0 & 0 & 1 & 0 \\ 0 & 0 & 0 & 1 \\ \hline -3\dfrac{g}{L} & \dfrac{g}{L} & 0 & 0 \\ \dfrac{g}{L} & -\dfrac{g}{L} & 0 & 0 \end{array}\right]. \tag{3.72}$$

The matrix $\hat{\boldsymbol{A}}$ can also be obtained from (3.41) using the similarity transformation (3.56) and by taking into account (3.64) and (3.67).

Problem 3.1. Linearized equations of motion of the centrifuge

Consider the task of setting up the linear equations of motion for the vibrations of the rigid centrifuge discussed in Problem 2.1. Let x_1, y_1, z_1 be a Cartesian body-fixed coordinate system of the centrifuge. The position of the centrifuge is described by relating this coordinate system to an inertial system x_I, y_I, z_I, as shown on Fig. 3.4. The holonomic constraints on the centrifuge are provided by the ball bearing.

1. How many degrees of freedom are there?

☐ $f = 6$, ☐ $f = 3$, ☐ $f = 2$, ☐ $f = 1$.

The translation vector $\boldsymbol{r}_1$ and the rotation tensor $\boldsymbol{S}_1$ can be expressed in terms of the Cardan angles as follows:

$$\boldsymbol{r}_1 = \begin{bmatrix} \sin\beta \\ -\sin\alpha\cos\beta \\ \cos\alpha\cos\beta \end{bmatrix} L, \tag{3.73}$$

$$S_1 = \begin{bmatrix} \cos\beta\cos\gamma & -\cos\beta\sin\gamma & \sin\beta \\ \cos\alpha\sin\gamma & \cos\alpha\cos\gamma & \sin\alpha\cos\beta \\ +\sin\alpha\sin\beta\cos\gamma & -\sin\alpha\sin\beta\sin\gamma & \\ \sin\alpha\sin\gamma & \sin\alpha\cos\gamma & \cos\alpha\cos\beta \\ -\cos\alpha\sin\beta\cos\gamma & +\cos\alpha\sin\beta\sin\gamma & \end{bmatrix}, \qquad (3.74)$$

where L is the length of the translation vector.

2. Which of the following is the position vector for the centrifuge?

$\square\ z = \begin{bmatrix} r_x \\ r_y \\ r_z \\ \alpha \\ \beta \\ \gamma \end{bmatrix}$, $\square\ z = \begin{bmatrix} \alpha \\ \beta \\ \gamma \end{bmatrix}$, $\square\ z = \begin{bmatrix} \alpha \\ \beta \end{bmatrix}$, $\square\ z = \gamma$.

The reference motion of the centrifuge is the rotation about the vertical axis z_I with constant angular velocity Ω. Thus the position vector of the reference motion is $z_S = [0\ \ 0\ \ \Omega t]^T$. Vibrations in the neighborhood of this reference motion are characterized by the conditions $\alpha \ll 1$, $\beta \ll 1$ and $\gamma = \Omega t + \tilde{\gamma}$, $\tilde{\gamma} \ll 1$.

3. What is the Taylor expansion up to the second term of the translation vector and rotation tensor?

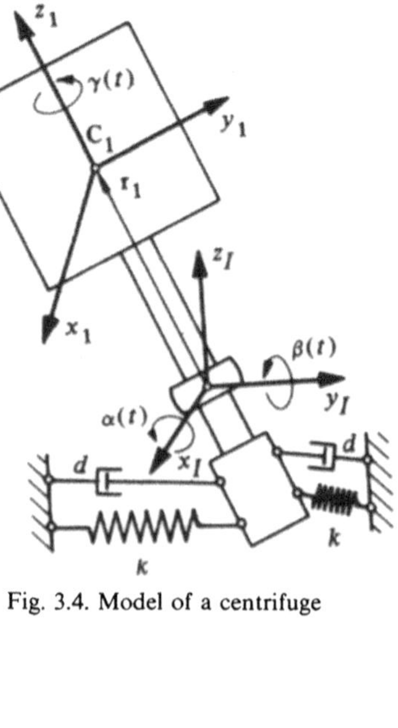

Fig. 3.4. Model of a centrifuge

□ $r_1(z)=\begin{bmatrix}0\\0\\1\end{bmatrix}L,$

□ $S_1(z)=\begin{bmatrix}(1-\frac{1}{2}\beta^2)\cos\gamma & -(1-\frac{1}{2}\beta^2)\sin\gamma & \beta\\ (1-\frac{1}{2}\alpha^2)\sin\gamma + \alpha\beta\cos\gamma & (1-\frac{1}{2}\alpha^2)\cos\gamma - \alpha\beta\sin\gamma & -\alpha\\ \alpha\sin\gamma - \beta\cos\gamma & \alpha\cos\gamma + \beta\sin\gamma & 1-\frac{1}{2}\alpha^2-\frac{1}{2}\beta^2\end{bmatrix}$

□ $r_1(z)=\begin{bmatrix}\beta\\-\alpha\\1\end{bmatrix}L,$ □ $S_1(z)=\begin{bmatrix}\cos\gamma & -\sin\gamma & \beta\\ \sin\gamma & \cos\gamma & -\alpha\\ \alpha\sin\gamma - \beta\cos\gamma & \alpha\cos\gamma + \beta\sin\gamma & 1\end{bmatrix},$

□ $r_1(z)=\begin{bmatrix}\beta\\-\alpha\\1-\frac{1}{2}\alpha^2-\frac{1}{2}\beta^2\end{bmatrix}L,$ □ $S_1(z)=\begin{bmatrix}\cos\gamma & -\sin\gamma & 0\\ \sin\gamma & \cos\gamma & 0\\ 0 & 0 & 1\end{bmatrix}.$

4. Which of the following equalities hold for the Jacobian matrices of translation and rotation?

□ $J_{T1}=\mathbf{0},$ □ $J_{T1}=\begin{bmatrix}0&1&0\\-1&0&0\\0&0&0\end{bmatrix}L,$ □ $J_{T1}(y)=\begin{bmatrix}0&1&0\\-1&0&0\\-\alpha&-\beta&0\end{bmatrix}L,$

□ $J_{R1}=\mathbf{0},$ □ $J_{R1}(y)=\begin{bmatrix}1&0&\beta\\0&1&-\alpha\\0&\alpha&1\end{bmatrix},$ $J_{R1}(y)=\begin{bmatrix}1&0&0\\0&1&0\\0&\alpha&0\end{bmatrix}.$

5. What are the local velocities?

□ $\bar{v}_1=\mathbf{0},$ □ $\bar{v}_1(\dot{y})=\begin{bmatrix}\dot{\beta}\\-\dot{\alpha}\\0\end{bmatrix}L,$

□ $\bar{\omega}_1=\begin{bmatrix}0\\0\\\Omega\end{bmatrix},$ □ $\bar{\omega}_1(y)=\begin{bmatrix}\beta\Omega\\-\alpha\Omega\\\Omega\end{bmatrix}.$

Remark: In calculating the local velocities one may omit the second terms in the Taylor expansion of the translation vector and the rotation tensor.

6. Which absolute velocities are obtained after linearization?

$$\square\ \boldsymbol{v}_1(\dot{y}) = \begin{bmatrix} \dot{\beta} \\ -\dot{\alpha} \\ 0 \end{bmatrix} L, \qquad \square\ \boldsymbol{v}_1(\dot{y}) = \begin{bmatrix} 2\dot{\beta} \\ -2\dot{\alpha} \\ 0 \end{bmatrix} L,$$

$$\square\ \boldsymbol{\omega}_1(\dot{y}) = \begin{bmatrix} \dot{\alpha} \\ \dot{\beta} \\ \dot{\gamma} \end{bmatrix}, \qquad \square\ \boldsymbol{\omega}_1(y, \dot{y}) = \begin{bmatrix} \dot{\alpha} + \beta\Omega \\ \dot{\beta} - \alpha\Omega \\ \dot{\gamma} + \Omega \end{bmatrix}.$$

Suppose that the centrifuge has the mass $m_1 = m$ and the tensor of inertia

$$\boldsymbol{I}_1(y, t) = \begin{bmatrix} I_x & 0 & (I_z - I_x)\beta - I_{yz}\sin\gamma \\ 0 & I_x & -(I_z - I_x)\alpha + I_{yz}\cos\gamma \\ (I_z - I_x)\beta - I_{yz}\sin\gamma & -(I_z - I_x)\alpha + I_{yz}\cos\gamma & I_z \end{bmatrix} \tag{3.75}$$

in the inertial system relative the point C_1. Here I_x and I_z are the moments of inertia of the symmetrical centrifuge and I_{yz} denotes the small dynamic unbalance. The external forces are gravity and the constraint forces,

$$\boldsymbol{f}_1 = [0 \quad 0 \quad -mg]^T + \boldsymbol{f}_Z. \tag{3.76}$$

The acting external torques are the radially symmetric linear spring and damping torques, the small driving torque $a(t)$, and the constraint torques,

$$\boldsymbol{l}_1 = \begin{bmatrix} -k\alpha - d\dot{\alpha} \\ -k\beta - d\dot{\beta} \\ a(t) \end{bmatrix} + \boldsymbol{l}_Z, \tag{3.77}$$

where k and d are the spring and damping constants, respectively. If one substitutes mass and tensor of inertia, velocity and angular velocity, and the external forces and torques into the Newton-Euler equations (2.57) and (2.58), one gets

$$\overline{\boldsymbol{M}}\ddot{y}(t) + \overline{\boldsymbol{P}}\dot{y}(t) + \overline{\boldsymbol{Q}}y(t) = \overline{\boldsymbol{h}}(t) \tag{3.78}$$

with the matrices

$$\overline{\boldsymbol{M}} = \begin{bmatrix} 0 & mL & 0 \\ -mL & 0 & 0 \\ 0 & 0 & 0 \\ \hdashline I_x & 0 & 0 \\ 0 & I_x & 0 \\ 0 & 0 & I_z \end{bmatrix}, \quad \overline{\boldsymbol{P}} = \begin{bmatrix} 0 & 0 & 0 \\ 0 & 0 & 0 \\ 0 & 0 & 0 \\ \hdashline d & I_z\Omega & 0 \\ -I_z\Omega & d & 0 \\ 0 & 0 & 0 \end{bmatrix},$$

$$\overline{\boldsymbol{Q}} = \begin{bmatrix} 0 & 0 & 0 \\ 0 & 0 & 0 \\ 0 & 0 & 0 \\ \hdashline k & 0 & 0 \\ 0 & k & 0 \\ 0 & 0 & 0 \end{bmatrix}, \quad \bar{\boldsymbol{h}}(t) = \begin{bmatrix} f_{Zx} \\ f_{Zy} \\ f_{Zz} - mg \\ \hdashline I_{yz}\Omega^2 \cos \Omega t + l_{Zx} \\ I_{yz}\Omega^2 \sin \Omega t + l_{Zy} \\ a(t) + l_{Zz} \end{bmatrix} \tag{3.79}$$

7. What is the form of the transposed Jacobian matrix?

$$\bar{\boldsymbol{J}}^T(y) = \left[\begin{array}{ccc:ccc} & & & & & \\ \hline & & & & & \\ \hline & & & & & \end{array}\right].$$

If (3.78) is multiplied from the left by $\bar{\boldsymbol{J}}^T(y)$, one obtains the time-invariant equations of motion

$$\boldsymbol{M}\ddot{y}(t) + (\boldsymbol{D} + \boldsymbol{G})\dot{y}(t) + (\boldsymbol{K} + \boldsymbol{N})y(t) = \boldsymbol{h}(t), \tag{3.80}$$

provided that the vanishing of the virtual work done by the constraint forces and torques, $\boldsymbol{J}_{T1}^T f_Z + \boldsymbol{J}_{R1}^T \boldsymbol{l}_Z = \boldsymbol{0}$, is taken into account.

8. Determine the 3×3 matrices

$$\boldsymbol{M} = \left[\begin{array}{c|c|c} & & \\ \hline & & \\ \hline & & \end{array}\right],$$

$$\boldsymbol{D} = \left[\begin{array}{c|c|c} & & \\ \hline & & \\ \hline & & \end{array}\right], \quad \boldsymbol{G} = \left[\begin{array}{c|c|c} & & \\ \hline & & \\ \hline & & \end{array}\right],$$

$$K = \begin{bmatrix} \quad & \quad & \quad \\ \quad & \quad & \quad \\ \quad & \quad & \quad \end{bmatrix}, \qquad N = \begin{bmatrix} \quad & \quad & \quad \\ \quad & \quad & \quad \\ \quad & \quad & \quad \end{bmatrix},$$

and the 3×1 excitation vector

$$h(t) = \begin{bmatrix} \quad \\ \quad \\ \quad \end{bmatrix}.$$

9. What kind of forces are acting on the centrifuge?
   - ☐ inertial forces
   - ☐ damping forces
   - ☐ gyroscopic forces
   - ☐ conservative position forces
   - ☐ nonconservative position forces

10. The spatial vibrations of the centrifuge are determined by the angles α, β, and the variations in angular velocity are described by the angle $\tilde{\gamma}$. Are the spatial vibrations influenced by variations in angular velocity?

☐ yes ☐ no

11. What are the normalized equations of motion for the spatial vibrations of the centrifuge with the position vector $y_1 = [\alpha \ \beta]^T$?

☐ $\ddot{y}_1(t) + (\vartheta E + g\Omega S)\dot{y}_1(t) + \kappa y_1(t) = h_1(t)$,

☐ $\ddot{y}_1(t) + (\vartheta S + g\Omega E)\dot{y}_1(t) + \kappa y_1(t) = h_1(t)$,

if one introduces the notation

$$\vartheta = \frac{d}{I_x + mL^2}, \qquad g = \frac{I_z}{I_x + mL^2}, \qquad \kappa = \frac{k - mgL}{I_x + mL^2}, \tag{3.81}$$

the 2×2 unit matrix E, the skewsymmetric 2×2 matrix

$$S = \begin{bmatrix} 0 & 1 \\ -1 & 0 \end{bmatrix}, \tag{3.82}$$

and the 2×1 excitation vector

$$h_1(t) = e\Omega^2 \begin{bmatrix} \cos \Omega t \\ \sin \Omega t \end{bmatrix}, \qquad e = \frac{I_{yz}}{I_x + mL^2}. \tag{3.83}$$

12. Under which conditions do there occur free or forced vibrations?

	Free	Forced
	vibrations	
$I_{yz} > 0$	□	□
$I_{yz} = 0$	□	□
$I_{yz} < 0$	□	□

◆

Problem 3.2. State equation of the vertical vibrations of a motor car

The equations of motion of the vertical vibrations of a motor car, Fig. 2.12, was determined in Problem 2.2,

$$\begin{bmatrix} m_1 & 0 \\ 0 & m_2 \end{bmatrix}\begin{bmatrix} \ddot{y}_1 \\ \ddot{y}_2 \end{bmatrix} + \begin{bmatrix} d_1 & -d_1 \\ -d_1 & d_1 + d_2 \end{bmatrix}\begin{bmatrix} \dot{y}_1 \\ \dot{y}_2 \end{bmatrix} + \begin{bmatrix} k_1 & -k_1 \\ -k_1 & k_1 + k_2 \end{bmatrix}\begin{bmatrix} y_1 \\ y_2 \end{bmatrix}$$

$$= \begin{bmatrix} 0 \\ k_2 y_e + d_2 \dot{y}_e \end{bmatrix}. \tag{3.84}$$

In order to set up the state equation one needs the state vector $\boldsymbol{x}(t)$, the system matrix $\boldsymbol{A}$, and the excitation vector $\boldsymbol{b}(t)$.

1. Which of the following is the state vector?

□ $\boldsymbol{x}(t) = [y_1 \quad y_2 \quad \dot{y}_1 \quad \dot{y}_2]^T$,

□ $\boldsymbol{x}(t) = [y_1 \quad \dot{y}_1 \quad y_2 \quad \dot{y}_2]^T$,

□ $\boldsymbol{x}(t) = [y_1 \quad \dot{y}_2 \quad y_2 \quad \dot{y}_1]^T$,

□ $\boldsymbol{x}(t) = [y_1 \quad y_2 \quad y_e \quad \dot{y}_e]^T$.

2. Which submatrices appear in the matrix $\boldsymbol{A}$ according to (3.38)?

□ $\boldsymbol{A} = \left[\begin{array}{c|c} 0 & \boldsymbol{E} \\ \hline \boldsymbol{M}^{-1}\boldsymbol{K} & \boldsymbol{M}^{-1}\boldsymbol{D} \end{array}\right]$,

□ $\boldsymbol{A} = \left[\begin{array}{c|c} \boldsymbol{0} & \boldsymbol{E} \\ \hline -\boldsymbol{M}^{-1}\boldsymbol{K} & -\boldsymbol{M}^{-1}\boldsymbol{D} \end{array}\right]$,

□ $\boldsymbol{A} = \left[\begin{array}{c|c} \boldsymbol{0} & \boldsymbol{E} \\ \hline -\boldsymbol{M}^{-1}\boldsymbol{D} & -\boldsymbol{M}^{-1}\boldsymbol{K} \end{array}\right]$.

3. Which is the inverse inertia matrix?

□ $\boldsymbol{M}^{-1} = \begin{bmatrix} \frac{1}{m_1} & 0 \\ 0 & \frac{1}{m_2} \end{bmatrix}$,

□ $\boldsymbol{M}^{-1} = \frac{1}{\det \boldsymbol{M}} \begin{bmatrix} m_1 & 0 \\ 0 & m_2 \end{bmatrix}$,

□ $\boldsymbol{M}^{-1} = (\det \boldsymbol{M})^{-1} \begin{bmatrix} m_2 & 0 \\ 0 & m_1 \end{bmatrix}$.

4. Determine the system matrix $\boldsymbol{A}$ and the excitation vector $\boldsymbol{b}$ using the abbreviations

$$\vartheta_{ij} = \frac{d_i}{m_j}, \quad \kappa_{ij} = \frac{k_i}{m_j}, \qquad i, j = 1, 2. \tag{3.85}$$

$$\boldsymbol{A} = \left[\begin{array}{cc:cc} \quad & \quad & \quad & \quad \\ \quad & \quad & \quad & \quad \\ \hdashline \quad & \quad & \quad & \quad \\ \quad & \quad & \quad & \quad \end{array}\right], \qquad \boldsymbol{b} = \left[\begin{array}{c} \quad \\ \quad \\ \hdashline \quad \\ \quad \end{array}\right]. \blacklozenge$$

Part II

Time-invariant vibrating systems

The mathematical description of vibrating systems given in Part I shows that both time-invariant and time-variant coefficients may appear in linear systems. Part II of this book will be devoted to an extensive treatment of the time-invariant vibrating systems since they appear frequently in practice and the methods of solution are well developed. In Chapter 4 the general solution will be considered as well as the closely connected fundamental matrix. Chapter 5 is devoted to the analysis of stability and boundedness of vibrating systems. In Chapters 6 and 7 the free and forced vibrations will be dealt with. In Chapter 8 the phenomena of resonance, pseudoresonance, and absorption will be considered. In Chapter 9 we will look at random vibrations, their spectral analysis in the frequency domain, and their covariance analysis in the time domain. The following have been chosen as accompanying examples: the vibrations of a double pendulum, the vertical vibrations of a motor car, the vibrations of a symmetrical centrifuge, and the vibrations of a magnetically levitated vehicle.

4. The general solution of time-invariant vibrating systems

It has been shown in Chapter 3 that linear time-invariant vibrating systems can be described by the system (3.10) of f second order differential equations

$$\boldsymbol{M}\ddot{\boldsymbol{y}}(t)+(\boldsymbol{D}+\boldsymbol{G})\dot{\boldsymbol{y}}(t)+(\boldsymbol{K}+\boldsymbol{N})\boldsymbol{y}(t)=\boldsymbol{h}(t),$$
$$\boldsymbol{y}(t_0)=\boldsymbol{y}_0,\quad \dot{\boldsymbol{y}}(t_0)=\dot{\boldsymbol{y}}_0 \tag{4.1}$$

or, more generally, by the system (3.43) of n first order differential equations

$$\dot{\boldsymbol{x}}(t)=\boldsymbol{A}\boldsymbol{x}(t)+\boldsymbol{b}(t),\qquad \boldsymbol{x}(t_0)=\boldsymbol{x}_0. \tag{4.2}$$

The general solution

$$\boldsymbol{y}(t)=\boldsymbol{y}(t;\ \boldsymbol{y}_0,\ \dot{\boldsymbol{y}}_0,\ t_0;\ \boldsymbol{h}(t)),\qquad t\geqslant t_0, \tag{4.3}$$

respectively

$$\boldsymbol{x}(t)=\boldsymbol{x}(t;\ \boldsymbol{x}_0,\ t_0;\ \boldsymbol{b}(t)),\qquad t\geqslant t_0, \tag{4.4}$$

of these systems of differential equations yields information about the motion of the vibrating system when it is perturbed from its equilibrium position, or if it is subject to continuing excitation. The general solution is in principle determined by the *fundamental matrix* of the system of differential equations. Therefore, this chapter will be devoted to discussing the properties of the fundamental matrix and the possibilities of its calculation.

The *time-invariance* of the coefficients in (4.1) and (4.2) implies that the general homogeneous solutions of the systems (4.1) and (4.2) do not depend on the choice of the initial instant t_0,

$$\boldsymbol{y}(t;\ \boldsymbol{y}_0,\ \dot{\boldsymbol{y}}_0,\ t_0;\ \boldsymbol{0})\equiv\boldsymbol{y}(t+T;\ \boldsymbol{y}_0,\ \dot{\boldsymbol{y}}_0,\ t_0+T;\ \boldsymbol{0})\text{ for all }T, \tag{4.5}$$

$$\boldsymbol{x}(t;\ \boldsymbol{x}_0,\ t_0;\ \boldsymbol{0})\equiv\boldsymbol{x}(t+T;\ \boldsymbol{x}_0,\ t_0+T;\ \boldsymbol{0})\text{ for all }T. \tag{4.6}$$

The solution is only a function of the time difference between the running time and the initial time, provided the same initial condition $(y_0, \dot{y}_0)$, respectively $\boldsymbol{x}_0$, is chosen for $t = t_0$ or $t = t_0 + T$. Because of this, one can write, without loss of generality,

$$t_0 = 0 \tag{4.7}$$

for time-invariant systems (4.1), (4.2).

4.1. Fundamental matrix

Given a homogeneous linear system of differential equations of order n,

$$\dot{\boldsymbol{x}}(t) = \boldsymbol{A}\boldsymbol{x}(t), \tag{4.8}$$

there always exists a fundamental system of n linearly independent solution vectors $\boldsymbol{\varphi}_i(t)$, $i = 1(1)n$, which coincide for $t = 0$ with the n unit vectors $\boldsymbol{e}_i$,

$$\boldsymbol{\varphi}_i(0) = \boldsymbol{e}_i, \qquad i = 1(1)n, \tag{4.9}$$

and satisfy condition (4.8),

$$\dot{\boldsymbol{\varphi}}_i(t) = \boldsymbol{A}\boldsymbol{\varphi}_i(t). \tag{4.10}$$

These solution vectors form the columns of a regular $n \times n$ matrix

$$\boldsymbol{\Phi}(t) = \left[\boldsymbol{\varphi}_1(t) \vdots \boldsymbol{\varphi}_2(t) \vdots \ldots \vdots \boldsymbol{\varphi}_n(t)\right]. \tag{4.11}$$

which is called the *fundamental matrix* *. The matrix (4.11) is also known as state transition matrix. The fundamental matrix exhibits certain characteristic properties. Thus it follows from (4.9) and (4.10) that the differential equation (4.8) is satisfied by the matrix $\boldsymbol{\Phi}(t)$ itself,

$$\dot{\boldsymbol{\Phi}}(t) = \boldsymbol{A}\boldsymbol{\Phi}(t), \qquad \boldsymbol{\Phi}(0) = \boldsymbol{E}. \tag{4.12}$$

Observing that $x(t) = \mathrm{e}^{at}x_0$ is a solution of the scalar differential equation $\dot{x}(t) = ax(t)$, one can represent the fundamental matrix (4.11) as a solution of (4.12) by the *matrix exponential function*

$$\boldsymbol{\Phi}(t) = \mathrm{e}^{\boldsymbol{A}t}. \tag{4.13}$$

* In general, any regular matrix $\tilde{\boldsymbol{\Phi}}(t)$ satisfying the matrix differential equation $\dot{\tilde{\boldsymbol{\Phi}}}(t) = \boldsymbol{A}\tilde{\boldsymbol{\Phi}}(t)$ is called a fundamental matrix. The fundamental matrix defined by (4.11) has the additional property that $\boldsymbol{\Phi}(0) = \boldsymbol{E}$. The relation between $\tilde{\boldsymbol{\Phi}}(t)$ and $\boldsymbol{\Phi}(t)$ is $\tilde{\boldsymbol{\Phi}}(t) = \boldsymbol{\Phi}(t)\tilde{\boldsymbol{\Phi}}(0)$.

This matrix exponential function is defined for every system matrix $\boldsymbol{A}$ by the converging infinite series

$$\mathrm{e}^{\boldsymbol{A}t}=\boldsymbol{E}+\boldsymbol{A}t+\frac{1}{2!}(\boldsymbol{A}t)^2+\frac{1}{3!}(\boldsymbol{A}t)^3+\ldots=\sum_{k=0}^{\infty}\frac{(\boldsymbol{A}t)^k}{k!}. \tag{4.14}$$

It is an obvious consequence of (4.13) and (4.14) that

$$\dot{\boldsymbol{\Phi}}(t)=\boldsymbol{A}\boldsymbol{\Phi}(t)=\boldsymbol{\Phi}(t)\boldsymbol{A}, \tag{4.15}$$

i.e. it is indeed true that (4.13) satisfies (4.12), and moreover $\boldsymbol{\Phi}(t)$ and $\boldsymbol{A}$ commute.

Furthermore, (4.13) and (4.14) imply the *transition property* of the fundamental matrix $\boldsymbol{\Phi}(t)$. Indeed, it follows from

$$\begin{aligned}
\mathrm{e}^{\boldsymbol{A}(t_1+t_2)}&=\sum_{k=0}^{\infty}\frac{\boldsymbol{A}^k(t_1+t_2)^k}{k!}\\
&=\sum_{k=0}^{\infty}\frac{1}{k!}\left[\sum_{j=0}^{k}\binom{k}{j}(\boldsymbol{A}t_1)^j(\boldsymbol{A}t_2)^{k-j}\right]\\
&=\sum_{k=0}^{\infty}\sum_{j=0}^{k}\frac{1}{j!}(\boldsymbol{A}t_1)^j\frac{1}{(k-j)!}(\boldsymbol{A}t_2)^{k-1}\\
&=\sum_{j=0}^{\infty}\frac{1}{j!}(\boldsymbol{A}t_1)^j\cdot\sum_{k=0}^{\infty}\frac{1}{k!}(\boldsymbol{A}t_2)^k\\
&=\mathrm{e}^{\boldsymbol{A}t_1}\cdot\mathrm{e}^{\boldsymbol{A}t_2}=\mathrm{e}^{\boldsymbol{A}t_2}\cdot\mathrm{e}^{\boldsymbol{A}t_1}
\end{aligned} \tag{4.16}$$

that

$$\boldsymbol{\Phi}(t_1+t_2)=\boldsymbol{\Phi}(t_1)\boldsymbol{\Phi}(t_2)=\boldsymbol{\Phi}(t_2)\boldsymbol{\Phi}(t_1). \tag{4.17}$$

A further illustration of the transition property (4.17) will be given later by (4.35).

If $t_1=-t_2=t$ in (4.17), one gets $\boldsymbol{\Phi}(0)=\boldsymbol{\Phi}(t)\boldsymbol{\Phi}(-t)$. Since $\boldsymbol{\Phi}(0)=\boldsymbol{E}$, this implies

$$\boldsymbol{\Phi}^{-1}(t)=\boldsymbol{\Phi}(-t), \tag{4.18}$$

i.e., the reversal of the time argument in the fundamental matrix yields the inverse fundamental matrix. The property (4.18) is also an immediate consequence of (4.13):

$$\boldsymbol{\Phi}^{-1}(t)=(\mathrm{e}^{\boldsymbol{A}t})^{-1}=\mathrm{e}^{-\boldsymbol{A}t}=\boldsymbol{\Phi}(-t).$$

The last important property of $\boldsymbol{\Phi}(t)$ to be derived here is the Jacobi-Liouville formula for the Wronski determinant, i.e. for the determinant of the fundamental matrix. Using (4.10) one gets, according to the rules for differentiating determinants

$$\begin{aligned}\frac{\mathrm{d}}{\mathrm{d}t}\det \boldsymbol{\Phi}(t) &= \det[\dot{\boldsymbol{\varphi}}_1(t) \vdots \boldsymbol{\varphi}_2(t) \vdots \ldots \vdots \boldsymbol{\varphi}_n(t)] \\ &\quad + \det[\boldsymbol{\varphi}_1(t) \vdots \dot{\boldsymbol{\varphi}}_2(t) \vdots \ldots \vdots \boldsymbol{\varphi}_n(t)] + \ldots \\ &\quad + \det[\boldsymbol{\varphi}_1(t) \vdots \boldsymbol{\varphi}_2(t) \vdots \ldots \vdots \dot{\boldsymbol{\varphi}}_n(t)] \\ &= \det[\boldsymbol{A}\boldsymbol{\varphi}_1(t) \vdots \boldsymbol{\varphi}_2(t) \vdots \ldots \vdots \boldsymbol{\varphi}_n(t)] + \ldots \\ &\quad + \det[\boldsymbol{\varphi}_1(t) \vdots \boldsymbol{\varphi}_2(t) \vdots \ldots \vdots \boldsymbol{A}\boldsymbol{\varphi}_n(t)],\end{aligned}$$

which in combination with (4.15), i.e., with

$$\boldsymbol{A}\boldsymbol{\varphi}_k(t) = \boldsymbol{\Phi}(t)\boldsymbol{A}\boldsymbol{e}_k, \qquad k = 1(1)n, \tag{4.19}$$

implies the auxiliary result

$$\begin{aligned}\frac{\mathrm{d}}{\mathrm{d}t}\det \boldsymbol{\Phi}(t) &= \det[a_{11}\boldsymbol{\varphi}_1(t) \vdots \boldsymbol{\varphi}_2(t) \vdots \ldots \vdots \boldsymbol{\varphi}_n(t)] \\ &\quad + \det[\boldsymbol{\varphi}_1(t) \vdots a_{22}\boldsymbol{\varphi}_2(t) \vdots \ldots \vdots \boldsymbol{\varphi}_n(t)] \\ &\quad + \ldots + \det[\boldsymbol{\varphi}_1(t) \vdots \boldsymbol{\varphi}_2(t) \vdots \ldots \vdots a_{nn}\boldsymbol{\varphi}_n(t)] \\ &= \sum_{i=1}^{n} a_{ii} \det \boldsymbol{\Phi}(t).\end{aligned} \tag{4.20}$$

The sum of the diagonal elements a_{ii} of the system matrix $\boldsymbol{A}$ is called the trace of $\boldsymbol{A}$ and it will be denoted here by $\operatorname{tr} \boldsymbol{A}$:

$$\sum_{i=1}^{n} a_{ii} = \operatorname{tr} \boldsymbol{A}. \tag{4.21}$$

Thus the Wronski determinant satisfies the scalar differential equation

$$\frac{\mathrm{d}}{\mathrm{d}t}\det \boldsymbol{\Phi}(t) = \operatorname{tr} \boldsymbol{A} \det \boldsymbol{\Phi}(t). \tag{4.22}$$

Integrating (4.22) with the initial condition $\det \boldsymbol{\Phi}(0) = \det \boldsymbol{E} = 1$, one obtains the formula of Jacobi and Liouville

$$\det \boldsymbol{\Phi}(t) = \mathrm{e}^{t \operatorname{tr} \boldsymbol{A}}. \tag{4.23}$$

This relation again shows that the fundamental matrix (4.11) whose columns form a fundamental system of solution vectors (4.9, 4.10), is regular for all finite values of t, and its determinant is positive.

To complete the above list of properties of the fundamental matrix it should be mentioned that the following useful relation can be derived from (4.12) and (4.13)

$$\int_0^t \boldsymbol{\Phi}(\tau)\mathrm{d}\tau = \boldsymbol{A}^{-1}(\boldsymbol{\Phi}(t)-\boldsymbol{E}) = (\boldsymbol{\Phi}(t)-\boldsymbol{E})\boldsymbol{A}^{-1}, \tag{4.24}$$

whenever $\det \boldsymbol{A} \neq 0$.

If the vibrating system is described by (4.1) instead of (4.2), then by analogy with (4.8), one may also associate to the homogeneous system of differential equations

$$\boldsymbol{M}\ddot{\boldsymbol{y}}(t) + (\boldsymbol{D}+\boldsymbol{G})\dot{\boldsymbol{y}}(t) + (\boldsymbol{K}+\boldsymbol{N})\boldsymbol{y}(t) = \boldsymbol{0} \tag{4.25}$$

a fundamental system of $2f$ linearly independent vector functions $\boldsymbol{\psi}_i(t)$, $i = 1(1)2f$, where $\boldsymbol{\psi}_i$ satisfies (4.25) and the initial conditions

$$\left.\begin{aligned} &\boldsymbol{\psi}_i(0) = \boldsymbol{e}_i, && \dot{\boldsymbol{\psi}}_i(0) = \boldsymbol{0}, && i = 1(1)f,\\ &\boldsymbol{\psi}_{f+i}(0) = \boldsymbol{0}, && \dot{\boldsymbol{\psi}}_{f+i}(0) = \boldsymbol{e}_i, && i = 1(1)f. \end{aligned}\right\} \tag{4.26}$$

These vector solutions form the columns of two matrices

$$\left.\begin{aligned} \boldsymbol{\Psi}_1(t) &= [\boldsymbol{\psi}_1(t) \mid \dots \mid \boldsymbol{\psi}_f(t)],\\ \boldsymbol{\Psi}_2(t) &= [\boldsymbol{\psi}_{f+1}(t) \mid \dots \mid \boldsymbol{\psi}_{2f}(t)]. \end{aligned}\right\} \tag{4.27}$$

These matrices (4.27) also satisfy the differential equation (4.25), but with different initial conditions:

$$\left.\begin{aligned} &\boldsymbol{M}\ddot{\boldsymbol{\Psi}}_i(t) + (\boldsymbol{D}+\boldsymbol{G})\dot{\boldsymbol{\Psi}}_i(t) + (\boldsymbol{K}+\boldsymbol{N})\boldsymbol{\Psi}_i(t) = \boldsymbol{0},\ i = 1, 2,\\ &\boldsymbol{\Psi}_1(0) = \boldsymbol{E}, \quad \dot{\boldsymbol{\Psi}}_1(0) = \boldsymbol{0},\\ &\boldsymbol{\Psi}_2(0) = \boldsymbol{0}, \quad \dot{\boldsymbol{\Psi}}_2(0) = \boldsymbol{E}. \end{aligned}\right\} \tag{4.28}$$

At the transition from the equations of motion (4.25) to the equivalent state equation (4.8), according to (3.35, 3.38), one obtains the following relation between the fundamental matrix (4.11) and the solution matrices (4.27):

$$\boldsymbol{\Phi}(t) = \left[\begin{array}{c|c} \boldsymbol{\Psi}_1(t) & \boldsymbol{\Psi}_2(t) \\ \hline \dot{\boldsymbol{\Psi}}_1(t) & \dot{\boldsymbol{\Psi}}_2(t) \end{array}\right]. \tag{4.29}$$

Substituting (3.38) for $\boldsymbol{A}$ in the exponential series (4.14),

$$\boldsymbol{A} = \left[\begin{array}{c|c} 0 & \boldsymbol{E} \\ \hline -\boldsymbol{M}^{-1}(\boldsymbol{K}+\boldsymbol{N}) & -\boldsymbol{M}^{-1}(\boldsymbol{D}+\boldsymbol{G}) \end{array}\right] \tag{4.30}$$

one can calculate the first few terms of the convergent power series for $\boldsymbol{\Psi}_1(t)$ and $\boldsymbol{\Psi}_2(t)$:

$$\left.\begin{aligned} \boldsymbol{\Psi}_1(t) &= \boldsymbol{E} - \frac{t^2}{2!}\boldsymbol{M}^{-1}(\boldsymbol{K}+\boldsymbol{N}) + \frac{t^3}{3!}\boldsymbol{M}^{-1}(\boldsymbol{D}+\boldsymbol{G})\boldsymbol{M}^{-1}(\boldsymbol{K}+\boldsymbol{N}) \\ &\quad + \frac{t^4}{4!}\left[\boldsymbol{M}^{-1}(\boldsymbol{K}+\boldsymbol{N}) - \left(\boldsymbol{M}^{-1}(\boldsymbol{D}+\boldsymbol{G})\right)^2\right]^{\boldsymbol{M}-1}(\boldsymbol{K}+\boldsymbol{N}) + \dots, \\ \boldsymbol{\Psi}_2(t) &= \frac{t}{1!}\boldsymbol{E} - \frac{t^2}{2!}\boldsymbol{M}^{-1}(\boldsymbol{D}+\boldsymbol{G}) + \frac{t^3}{3!}\left[\left(\boldsymbol{M}^{-1}(\boldsymbol{D}+\boldsymbol{G})\right)^2\right. \\ &\quad \left. -\boldsymbol{M}^{-1}(\boldsymbol{K}+\boldsymbol{N})\right] + \frac{t^4}{4!}\left[\boldsymbol{M}^{-1}(\boldsymbol{D}+\boldsymbol{G})\boldsymbol{M}^{-1}(\boldsymbol{K}+\boldsymbol{N})\right. \\ &\quad \left. +\boldsymbol{M}^{-1}(\boldsymbol{K}+\boldsymbol{N})\boldsymbol{M}^{-1}(\boldsymbol{D}+\boldsymbol{G}) - \left(\boldsymbol{M}^{-1}(\boldsymbol{D}+\boldsymbol{G})\right)^3\right] + \dots . \end{aligned}\right\} \tag{4.31}$$

In the next section the fundamental matrix will be used as a simple tool for finding the general solutions (4.3) and (4.4).

4.2. General solution

The *general solution* $\boldsymbol{x}(t)$ of a *homogeneous* system of differential equations (4.8) with initial condition $\boldsymbol{x}_0$ is obtained as a linear combination of the fundamental solutions $\boldsymbol{\varphi}_i(t)$,

$$\boldsymbol{x}(t) = \sum_{i=1}^{n} \boldsymbol{\varphi}_i(t) x_{i0}. \tag{4.32}$$

Using the fundamental matrix (4.11), one can also write (4.32) in matrix notation

$$\boldsymbol{x}(t) = \boldsymbol{\Phi}(t)\boldsymbol{x}_0. \tag{4.33}$$

In other words, the state transition matrix indicates how the initial state $\boldsymbol{x}_0$ transforms into the state $\boldsymbol{x}(t)$.

To illustrate the transition property (4.17) of the fundamental matrix, consider a trajectory $\boldsymbol{x}(t) = \boldsymbol{x}(t;\ \boldsymbol{x}_0, 0)$, Fig. 4.1. Looking at the solution for $t = t_1$ and $t = t_3 = t_1 + t_2$, one obtains

$$\boldsymbol{x}_1 = \boldsymbol{x}(t_1) = \boldsymbol{\Phi}(t_1)\boldsymbol{x}_0, \tag{4.34a}$$

$$\boldsymbol{x}_3 = \boldsymbol{x}(t_3) = \boldsymbol{\Phi}(t_3)\boldsymbol{x}_0. \tag{4.34b}$$

It is also possible to calculate $\boldsymbol{x}_3$ by using $\boldsymbol{x}_1 = \boldsymbol{x}(t_1)$ as an initial condition at time $t = t_1$:

$$\boldsymbol{x}_3 = \boldsymbol{x}(t_3) = \boldsymbol{\Phi}(t_2)\boldsymbol{x}_1. \tag{4.34c}$$

The three relations (4.34) can be collected to form

$$\boldsymbol{x}_3 = \boldsymbol{\Phi}(t_3)x_0 = \boldsymbol{\Phi}(t_2)\boldsymbol{\Phi}(t_1)\boldsymbol{x}_0. \tag{4.35}$$

Since (4.35) holds for every initial condition $\boldsymbol{x}_0$, and $t_3 = t_1 + t_2$, one again gets a proof of the transition property (4.17) of the fundamental matrix.

The *general solution* of the *inhomogeneous* system of differential equations (4.2) is obtained by adding a particular solution $\boldsymbol{x}_p(t)$ of (4.2) to the general solution (4.33) of the homogeneous system (4.8),

$$\boldsymbol{x}(t) = \boldsymbol{x}(t;\ \boldsymbol{x}_0, 0;\ \boldsymbol{b}(t)) = \boldsymbol{\Phi}(t)\boldsymbol{x}_0 + \boldsymbol{x}_p(t). \tag{4.36}$$

A particular solution can be found by the method of variation of constants. For this a particular solution of the form

$$\boldsymbol{x}_p(t) = \boldsymbol{\Phi}(t)\boldsymbol{c}(t) \tag{4.37}$$

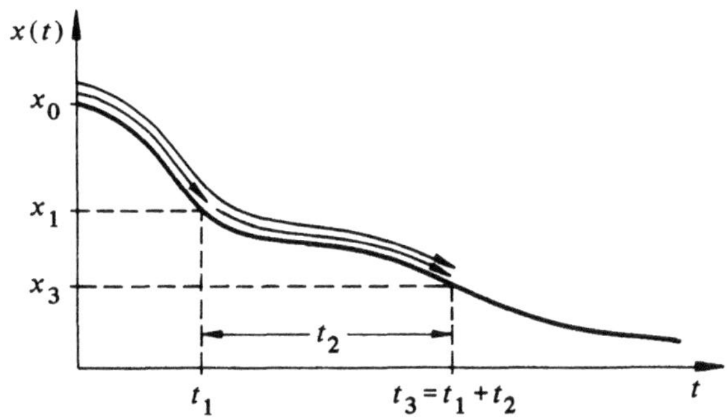

Fig. 4.1. The transition $\boldsymbol{x}_0 \to \boldsymbol{x}_1$, $\boldsymbol{x}_1 \to \boldsymbol{x}_3$, $\boldsymbol{x}_0 \to \boldsymbol{x}_3$

is introduced which has to satisfy the differential equation (4.2), i.e.,

$$\begin{aligned}\dot{\boldsymbol{x}}_p(t) &= \dot{\boldsymbol{\Phi}}(t)\boldsymbol{c}(t) + \boldsymbol{\Phi}(t)\dot{\boldsymbol{c}}(t)\\ &= \boldsymbol{A}\boldsymbol{\Phi}(t)\boldsymbol{c}(t) + \boldsymbol{\Phi}(t)\dot{\boldsymbol{c}}(t)\\ &\overset{!}{=} \boldsymbol{A}\boldsymbol{x}_p(t) + \boldsymbol{b}(t).\end{aligned}$$

Hence the determining differential equation for $\boldsymbol{c}(t)$ is

$$\boldsymbol{\Phi}(t)\dot{\boldsymbol{c}}(t) = \boldsymbol{b}(t). \tag{4.38}$$

Applying (4.18) one gets

$$\dot{\boldsymbol{c}}(t) = \boldsymbol{\Phi}(-t)\boldsymbol{b}(t)$$

and by integrating one finally obtains

$$\boldsymbol{c}(t) = \boldsymbol{c}(0) + \int_0^t \boldsymbol{\Phi}(-\tau)\boldsymbol{b}(\tau)\mathrm{d}\tau. \tag{4.39}$$

If $t = 0$ is taken in (4.36), one gets the particular solution with $\boldsymbol{x}_p(0) = \boldsymbol{0}$. Hence

$$\boldsymbol{c}(0) = \boldsymbol{0} \tag{4.40}$$

and this, together with (4.37) and (4.17), implies the particular solution

$$\left.\begin{aligned}\boldsymbol{x}_p(t) &= \boldsymbol{\Phi}(t)\int_0^t \boldsymbol{\Phi}(-\tau)\boldsymbol{b}(\tau)\mathrm{d}\tau\\ &= \int_0^t \boldsymbol{\Phi}(t-\tau)\boldsymbol{b}(\tau)\mathrm{d}\tau.\end{aligned}\right\} \tag{4.41}$$

Therefore the general solution of the inhomogeneous vector differential equation (4.2) is given by

$$\boldsymbol{x}(t) = \boldsymbol{\Phi}(t)\boldsymbol{x}_0 + \int_0^t \boldsymbol{\Phi}(t-\tau)\boldsymbol{b}(\tau)\mathrm{d}\tau. \tag{4.42}$$

In the case of ordinary mechanical systems with a finite number of degrees of freedom, the general solution (4.3) of the equations of motion

(4.1) can also be written with the use of the matrices $\boldsymbol{\Psi}_1(t)$, $\boldsymbol{\Psi}_2(t)$, according to (4.27). Utilizing (4.29), one gets from (4.42)

$$y(t) = \boldsymbol{\Psi}_1(t)\, y_0 + \boldsymbol{\Psi}_2(t)\, \dot{y}_0 + \int_0^t \boldsymbol{\Psi}_2(t-\tau)\boldsymbol{M}^{-1}\boldsymbol{h}(\tau)\mathrm{d}\tau. \tag{4.43}$$

Example 4.1. Spring-mass pendulum

According to Newton's equation (2.57) the equation of motion of the vertically vibrating spring-mass pendulum shown in Fig. 4.2 reads as

$$m\ddot{z}(t) = -kz(t) - mg. \tag{4.44}$$

The external forces taken into account here are the spring force $-kz$ and the gravity force $-mg$. The equilibrium position is

$$z_S = -\frac{mg}{k}, \tag{4.45}$$

so that, according to (3.4), the vibration about the equilibrium position is given by

$$\ddot{y}(t) + \kappa y(t) = 0, \qquad \kappa = \frac{k}{m} \tag{4.46}$$

with $z = z_S + y$.

The general solution of this scalar equation of a harmonic oscillator with eigenfrequency ω is known as

$$y(t) = y_0 \cos \omega t + \frac{\dot{y}_0}{\omega} \sin \omega t, \qquad \omega = \sqrt{\kappa}\,. \tag{4.47}$$

This result, however, will be confirmed here by means of the matrix

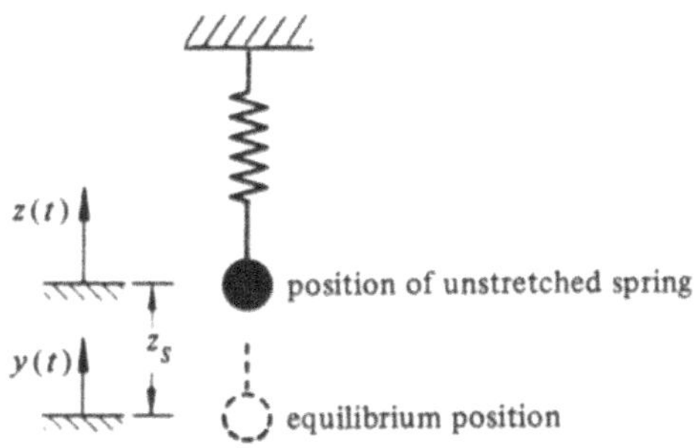

Fig. 4.2. Vertically vibrating spring-mass pendulum

exponential function (4.13, 4.14). The state equation equivalent to (4.46) is

$$\underbrace{\begin{bmatrix} \dot{y} \\ \ddot{y} \end{bmatrix}}_{\dot{\boldsymbol{x}}(t)} = \underbrace{\begin{bmatrix} 0 & 1 \\ -\omega^2 & 0 \end{bmatrix}}_{\boldsymbol{A}} \underbrace{\begin{bmatrix} y \\ \dot{y} \end{bmatrix}}_{\boldsymbol{x}(t)} . \tag{4.48}$$

The series (4.14) is built by of the powers $\boldsymbol{A}^k$. These satisfy

$$\boldsymbol{A}^{2k} = (-1)^k \omega^{2k} \boldsymbol{E}, \qquad \boldsymbol{A}^{2k+1} = (-1)^k \omega^{2k} \boldsymbol{A}, \quad k = 0(1)\infty. \tag{4.49}$$

Consequently

$$\boldsymbol{\Phi}(t) = \begin{bmatrix} \sum_{k=0}^{\infty} \frac{(-1)^k}{(2k)!} (\omega t)^{2k} & \frac{1}{\omega} \sum_{k=0}^{\infty} \frac{(-1)^k}{(2k+1)!} (\omega t)^{2k+1} \\ -\omega \sum_{k=0}^{\infty} \frac{(-1)^k}{(2k+1)!} (\omega t)^{2k+1} & \sum_{k=0}^{\infty} \frac{(-1)^k}{(2k)!} (\omega t)^{2k} \end{bmatrix}$$

$$= \begin{bmatrix} \cos \omega t & \frac{1}{\omega} \sin \omega t \\ -\omega \sin \omega t & \cos \omega t \end{bmatrix} \tag{4.50}$$

if one recalls the power series expansions of $\cos \omega t$ and $\sin \omega t$. The general solution (4.47) is now confirmed by (4.29) and (4.43). ◆

The transition matrix is the essential operator for the solution of linear systems of differential equations. Thus it is of great importance to provide methods for analytical or numerical calculation of this matrix. The first calculation possibility is offered by (4.13, 4.14). The analytic evaluation of (4.14) can be, however, in general quite difficult, as indicated by Example 4.1. Also, if the calculation is an approximation, the question arises after which term the series expansion should be terminated. Satisfactory results are obtained only for small values of t. This will be discussed in greater detail in connection with the numerical calculation of $\boldsymbol{\Phi}(t) = e^{\boldsymbol{A}t}$ in Section 14.5. In the following two more possibilities of analytic calculation will be derived which will offer new insights into the behavior of the solution. Firstly, the solution of linear time-invariant systems is characterized by eigenvalues and eigenvectors and their connection with the fundamental matrix is shown. Secondly, references to the theorem of Cayley and Hamilton will be presented.

4.3. Eigenvalues, eigenvectors

While the fundamental matrix (4.11) is given by special solution vectors $\boldsymbol{\varphi}_i(t)$, normed at the initial instant $t=0$ so that $\boldsymbol{\varphi}_i(0)=\boldsymbol{e}_i$, the *eigenvalue problem* characterizes the behavior of a dynamic system (4.8) by particular solutions with the property that the state vector $\boldsymbol{x}(t)$ and its derivative $\dot{\boldsymbol{x}}(t)$ have the same direction in the state space,

$$\dot{\boldsymbol{x}}(t)=\lambda \boldsymbol{x}(t). \tag{4.51}$$

If such particular solutions do exist, then they must satisfy not only equation (4.8), but also condition (4.51). This implies that trajectories keeping a constant direction in the state space must satisfy

$$\boldsymbol{x}(t)=\mathrm{e}^{\lambda t}\tilde{\boldsymbol{x}}. \tag{4.52}$$

This is in agreement with condition (4.51). Substituting (4.52) into (4.8),

$$\lambda\, \mathrm{e}^{\lambda t}\tilde{\boldsymbol{x}}=\boldsymbol{A}\ \mathrm{e}^{\lambda t}\tilde{\boldsymbol{x}},$$

one gets the homogeneous system of equations

$$(\lambda \boldsymbol{E}-\boldsymbol{A})\tilde{\boldsymbol{x}}=\boldsymbol{0}, \tag{4.53}$$

which is called the *eigenvalue problem* associated with the matrix $\boldsymbol{A}$.

The eigenvalue problem (4.53) possesses nontrivial vector solutions $\tilde{\boldsymbol{x}}$ if and only if $(\lambda\boldsymbol{E}-\boldsymbol{A})$ is a singular matrix. This is precisely the case when λ is a root of the characteristic polynomial

$$p(\lambda)\equiv \det(\lambda\boldsymbol{E}-\boldsymbol{A})\equiv \lambda^n+a_1\lambda^{n-1}+a_2\lambda^{n-2}+\ldots+a_{n-1}\lambda+a_n=0. \tag{4.54}$$

For every real $n\times n$ matrix $\boldsymbol{A}$ the *characteristic polynomial* $p(\lambda)$ is of degree n with real characteristic coefficients a_i, $i=1(1)n$, which are uniquely determined by $\boldsymbol{A}$ (see (13.53)). In particular, one always has

$$a_1=-\operatorname{tr}\boldsymbol{A}, \tag{4.55}$$

$$a_n=(-1)^n\det\boldsymbol{A}. \tag{4.56}$$

The characteristic equation (4.54) has precisely n roots $\lambda=\lambda_j$,

$$p(\lambda_j)=0, \qquad j=1(1)n, \tag{4.57}$$

where multiple roots are counted according to their *multiplicity* υ_j. If complex roots appear, $\lambda_j = \delta_j + i\omega_j$, then their conjugates $\bar{\lambda}_j = \delta_j - i\omega_j$ are also roots of the characteristic equation (4.54). Denoting λ_j, $j = 1(1)s \leqslant n$, the s distinct roots of (4.54) and v_j, $j = 1(1)s$, their multiplicities, one also has in addition to the relations (4.55, 4.56) the following two equations given by Vieta's theorem

$$a_1 = -\sum_{j=1}^{s} v_j \lambda_j, \tag{4.58}$$

$$a_n = (-1)^n \prod_{j=1}^{s} \lambda_j^{v_j}. \tag{4.59}$$

The roots $\lambda = \lambda_j$, $j = 1(1)s$, of (4.54) are called the *eigenvalues* of $\boldsymbol{A}$. The nontrivial solutions $\tilde{\boldsymbol{x}}_j = \boldsymbol{x}(\lambda_j)$ of (4.53) are called *eigenvectors*. To all conjugate complex eigenvalues there corresponds conjugate complex eigenvectors. While the eigenvectors corresponding to different eigenvalues are always linearly independent, one must in the case of multiple eigenvalues ($v_j > 1$) distinguish several possibilities. Besides the multiplicity v_j, one also has to consider the *rank decrease* or *nullity* of the matrix $(\lambda_j \boldsymbol{E} - \boldsymbol{A})$:

$$\left.\begin{aligned} &d_j = n - \operatorname{rank}(\lambda_j \boldsymbol{E} - \boldsymbol{A}), \\ &1 \leqslant d_j \leqslant v_j, \quad j = 1(1)s. \end{aligned}\right\} \tag{4.60}$$

There then exists for an eigenvalue λ_j of multiplicity v_j precisely d_j linearly independent eigenvectors.

Example 4.2. Eigenvectors for multiple eigenvalues

a) The 3×3 unit matrix $\boldsymbol{E}$ has the characteristic polynomial

$$p(\lambda) \equiv \det(\lambda \boldsymbol{E} - \boldsymbol{E}) \equiv (\lambda - 1)^3. \tag{4.61}$$

The value $\lambda = \lambda_1 = 1$ is an eigenvalue with multiplicity $v_1 = 3$. The nullity

$$d_1 = 3 - \operatorname{rank}(\lambda_1 \boldsymbol{E} - \boldsymbol{E}) = 3 - \operatorname{rank}(\boldsymbol{0}) = 3 \tag{4.62}$$

also has the value $d_1 = v_1 = 3$. Hence there exists three linearly independent eigenvectors, e.g. the three unit vectors

$$\tilde{\boldsymbol{x}}_1 = \boldsymbol{e}_1, \qquad \tilde{\boldsymbol{x}}_2 = \boldsymbol{e}_2, \qquad \tilde{\boldsymbol{x}}_3 = \boldsymbol{e}_3. \tag{4.63}$$

b) Also the matrix

$$A = \begin{bmatrix} 1 & 1 & 0 \\ 0 & 1 & 0 \\ 0 & 0 & 1 \end{bmatrix} \tag{4.64}$$

has the characteristic polynomial (4.61), and thus the triple eigenvalue $\lambda_1 = 1$. But due to

$$\operatorname{rank}(\lambda_1 E - A) = \operatorname{rank}\begin{bmatrix} 0 & 1 & 0 \\ 0 & 0 & 0 \\ 0 & 0 & 0 \end{bmatrix} = 1 \tag{4.65}$$

one has $d_1 = 2$. Hence there exists here only two linearly independent eigenvectors, e.g.

$$\tilde{x}_1 = e_1 \qquad \tilde{x}_2 = e_3. \tag{4.66}$$

c) Also for the matrix

$$B = \begin{bmatrix} 1 & 1 & 0 \\ 0 & 1 & 1 \\ 0 & 0 & 1 \end{bmatrix} \tag{4.67}$$

there exists only one triple eigenvalue $\lambda_1 = 1$, but its nullity is only $d_1 = 1$. Thus the equation (4.53) can be satisfied for only one direction in the state space, given for example by

$$\tilde{x}_1 = e_1. \; \blacklozenge \tag{4.68}$$

Case I: $d_j = v_j$ for all eigenvalues.

For each eigenvalue λ_j there exists $d_j = v_j$ linearly independent eigenvectors $\tilde{x}_j$, i.e., there exists exactly n linearly independent particular solutions (4.52). The linear combination of these is the general solution of (4.8):

$$x(t) = \sum_{j=1}^{n} \tilde{x}_j e^{\lambda_j t} c_j. \tag{4.69}$$

Here the constants c_j, $j = 1(1)n$, have to be determined from the initial condition

$$x_0 = \sum_{j=1}^{n} \tilde{x}_j c_j. \tag{4.70}$$

To write the last two relations in matrix notation, one has to introduce the *modal matrix*

$$X = [\tilde{x}_1 \vdots \tilde{x}_2 \vdots \dots \vdots \tilde{x}_n] \tag{4.71}$$

composed of the eigenvectors, the corresponding diagonal matrix of the eigenvalues

$$\Lambda = \operatorname{diag}\{\lambda_j\} = \begin{bmatrix} \lambda_1 & 0 & 0 & \dots & 0 \\ 0 & \lambda_2 & 0 & \dots & 0 \\ 0 & 0 & \lambda_3 & \ddots & \vdots \\ \vdots & \vdots & \vdots & \ddots & 0 \\ 0 & 0 & 0 & \dots & \lambda_n \end{bmatrix} \tag{4.72}$$

and the column vector

$$c = [c_1 \quad c_2 \ \dots \ c_n]^T. \tag{4.73}$$

Then (4.69) can be written as

$$x(t) = X\,e^{\Lambda t} c \tag{4.74}$$

and (4.70) as

$$x_0 = Xc. \tag{4.75}$$

Thus the general solution is

$$x(t) = X\,e^{\Lambda t} X^{-1} x_0, \tag{4.76}$$

where

$$e^{\Lambda t} = \operatorname{diag}\{e^{\lambda_j t}\} \tag{4.77}$$

is the diagonal matrix of the functions $e^{\lambda_j t}$. The comparison of (4.76) with (4.33) leads to a second possibility of calculating the *fundamental matrix*:

$$\Phi(t) = X\,e^{\Lambda t} X^{-1}. \tag{4.78}$$

On the other hand it holds

$$e^{\Lambda t} = X^{-1} \Phi(t) X, \tag{4.79}$$

i.e. e^{At} and $\Phi(t)$ are similar matrices (see Section 3.3). If one chooses a similarity transformation (3.53) with transformation matrix X,

$$x(t) = X\hat{x}(t), \tag{4.80}$$

then, due to

$$X^{-1}AX = \Lambda \tag{4.81}$$

the system (4.8) is transformed into the *uncoupled system*

$$\dot{\hat{x}}(t) = \Lambda\hat{x}(t), \qquad \hat{x}_0 = X^{-1}x_0. \tag{4.82}$$

The system (4.82) is also called the *normal form* of the vibrating system (4.8), and the state variables $\hat{x}_j$, $j = 1(1)n$, are called *normal or principal coordinates.*

This case is of considerable practical importance, for when there are n different eigenvalues, one has always $d_j = v_j = 1$. This will be discussed in greater detail in Chapter 6.

Case II: $d_j < v_j$ for at least one eigenvalue λ_j.

For the eigenvalue λ_j with $d_j < v_j$ there exists only d_j linearly independent eigenvectors

$$\tilde{x}_{j1}, \tilde{x}_{j2}, \ldots, \tilde{x}_j d_j.$$

To transform the given coordinates into normal coordinates, as in Case I, one has to obtain the remaining $(v_j - d_j)$ linearly independent vectors required for the construction of a modal matrix from the system of equations for the *generalized eigenvectors*

$$\left.\begin{aligned} &(\lambda_j E - A)\tilde{x}_{jk}^{(l)} = -\tilde{x}_{jk}^{(l-1)}, \quad \tilde{x}_{jk}^{(1)} = \tilde{x}_{jk},\\ &l = 2(1)r_{jk}, \quad \sum_{k=1}^{d_j}(r_{jk} - 1) = v_j - d_j. \end{aligned}\right\} \quad k = 1(1)d_j \tag{4.83}$$

Here $(r_{jk} - 1)$ is the number of generalized eigenvectors $\tilde{x}_{jk}^{(l)}$, $l = 2(1)r_{jk}$, belonging to the eigenvector $\tilde{x}_{jk}$, and it turns out that r_{jk} is the greatest number for which (4.83) is still solvable. In this process of determining the chains of generalized eigenvectors one has to take into consideration that in the calculation of the generalized eigenvectors of the l-th grade, the already known generalized eigenvectors of the $(l-1)$-st grade should undergo a "clearing" process, so that precisely those right-hand sides $-\tilde{x}_{jk}^{(l-1)}$ appear in (4.83) for which (4.83) is solvable. If a linear combination of the $(l-1)$-st grade generalized eigenvectors satisfying the above condition cannot be found, then the calculation algorithm terminates. Further details concerning the structure of generalized eigenvectors can be found in the book on matrices by Zurmühl (§ 19.5).

The generalized eigenvectors together with the eigenvectors make up a generalized, regular *modal matrix*

$$X = [X_1 \vdots X_2 \vdots \ldots \vdots X_s], \tag{4.84}$$

where the $n \times \upsilon_j$ submatrices X_j, $j = 1(1)s$, corresponding to the s different eigenvalues λ_j are composed of the eigenvectors and generalized eigenvectors in the following manner:

$$\begin{aligned} X_j &= \left[X_{j1} \ldots X_{jd_j} \right] \\ &= \left[\tilde{x}_{j1}^{(1)} \vdots \tilde{x}_1^{(2)} \vdots \ldots \vdots \tilde{x}^{(r_{j1})} \vdots \ldots \vdots \tilde{x}_{jd_j}^{(1)} \vdots \tilde{x}_{jd_j}^{(2)} \vdots \ldots \vdots \tilde{x}_{jd_j}^{(r_j d_j)} \right]. \end{aligned} \tag{4.85}$$

The corresponding similarity transformation leads to the *Jordan matrix*

$$J = X^{-1} A X = \operatorname{diag}\left\{ \ldots;\ J_{j1}, \ldots,\ J_{jd_j}; \ldots \right\}, \tag{4.86a}$$

in which there appears the $r_{jk} \times r_{jk}$ submatrices

$$J_{jk} = \begin{bmatrix} \lambda_j & 1 & 0 & \ldots & 0 \\ 0 & \lambda_j & 1 & \ldots & 0 \\ 0 & 0 & \lambda_j & \ddots & \vdots \\ \vdots & \vdots & \vdots & \ddots & 1 \\ 0 & 0 & 0 & \ldots & \lambda_j \end{bmatrix}. \tag{4.86b}$$

These matrices differ from the diagonal matrix discussed in Case I due to the presence of the elements 1 just above the diagonal whenever $r_{jk} > 1$.

The similarity transformation (4.80) together with the generalized modal matrix (4.84) yield the *normal form*

$$\dot{\hat{x}}(t) = J\hat{x}(t), \qquad \hat{x}_0 = X^{-1} x_0 \tag{4.87}$$

of the vibrating system (4.8) in Case II. This blockwise uncoupled system has the fundamental matrix

$$e^{Jt} = \operatorname{diag}\{ \ldots;\ e^{J_{j1}t}, \ldots, e^{J_{jd_j}t}; \ldots \} \tag{4.88}$$

with the $r_{jk} \times r_{jk}$ submatrices

$$e^{J_{jk}t} = e^{\lambda_j t} \begin{bmatrix} 1 & \frac{t}{1!} & \frac{t^2}{2!} & \cdots & \frac{t^{r_{jk}-1}}{(r_{jk}-1)!} \\ & 1 & \frac{t}{1!} & \ddots & \vdots \\ & & 1 & \ddots & \frac{t^2}{2!} \\ & & & \ddots & \frac{t}{1!} \\ & 0 & & & 1 \end{bmatrix}, \qquad k = 1(1)d_j. \tag{4.89}$$

Due to (4.78), the *fundamental matrix* $\boldsymbol{\Phi}(t)$ of the original system (4.8) is now given by

$$\boldsymbol{\Phi}(t) = \boldsymbol{X}\, e^{\boldsymbol{J}t} \boldsymbol{X}^{-1}. \tag{4.90}$$

In Case II the general solution differs essentially from the solution (4.69) in Case I by the presence of the submatrices (4.89). In place of (4.69) one can now write the general solution as a linear combination of the elementary solutions $\boldsymbol{X}_{jk} \exp(\boldsymbol{J}_{jk} t)$:

$$\boldsymbol{x}(t) = \sum_{j=1}^{s} \sum_{k=1}^{d_j} \boldsymbol{X}_{jk}\, e^{\boldsymbol{J}_{jk}t} \boldsymbol{c}_{jk}, \tag{4.91}$$

where c_{jk} are constant $r_{jk} \times 1$ vectors. Note that in the general solution (4.91) there appears not only the exponential functions but also exponential functions weighted with the power of time. This fact will play an essential role in the stability analysis in Chapter 5.

A coordinate transformation $\boldsymbol{x}(t) = \boldsymbol{T}\hat{\boldsymbol{x}}(t)$ has the following influence upon the eigenvalue problem (4.53). From (4.53) one gets, by (3.56),

$$(\lambda \boldsymbol{E} - \boldsymbol{T}\hat{\boldsymbol{A}}\boldsymbol{T}^{-1})\tilde{\boldsymbol{x}} = \boldsymbol{T}(\lambda \boldsymbol{E} - \hat{\boldsymbol{A}})\boldsymbol{T}^{-1}\tilde{\boldsymbol{x}} = \boldsymbol{0}, \tag{4.92}$$

i.e. due to the regularity of $\boldsymbol{T}$, the eigenvalue problem for the transformed system is

$$(\lambda \boldsymbol{E} - \hat{\boldsymbol{A}})\hat{\tilde{\boldsymbol{x}}} = \boldsymbol{0}, \quad \hat{\tilde{\boldsymbol{x}}} = \boldsymbol{T}^{-1}\tilde{\boldsymbol{x}}. \tag{4.93}$$

Relation (4.92) implies the invariance of the characteristic polynomial under coordinate transformation:

$$p(\lambda) \equiv \det(\lambda \boldsymbol{E} - \boldsymbol{A}) \equiv \det \boldsymbol{T} \cdot \det(\lambda \boldsymbol{E} - \boldsymbol{A}) \cdot \det \boldsymbol{T}^{-1}$$

$$\equiv \det \boldsymbol{T}(\lambda \boldsymbol{E} - \boldsymbol{A})\boldsymbol{T}^{-1} \equiv \det(\lambda \boldsymbol{E} - \hat{\boldsymbol{A}}) \equiv \hat{p}(\lambda). \tag{4.94}$$

Thus the characteristic coefficients a_i, $i = 1(1)n$, as well as the eigenvalues λ_j, $j = 1(1)n$, are the same for $\boldsymbol{A}$ and $\hat{\boldsymbol{A}}$. The corresponding eigenvectors and generalized eigenvectors, however, are subject to the transformation law (3.53),

$$\tilde{\boldsymbol{x}} = \boldsymbol{T}\tilde{\hat{\boldsymbol{x}}}. \tag{4.95}$$

Until now the eigenvalue problem was considered only for the state equations (4.8). But it can also be formulate for the equations of motion (4.25) of ordinary mechanical systems. Since (4.25) can be reduced to (4.8) by means of the system matrix (4.30), the above results can be simply carried over to the case of the equations of motion of ordinary mechanical systems. The eigenvectors satisfy

$$\tilde{\boldsymbol{x}}_j = \begin{bmatrix} \tilde{\boldsymbol{y}}_j \\ \hdashline \lambda_j \tilde{\boldsymbol{y}}_j \end{bmatrix}, \tag{4.96}$$

where $\tilde{\boldsymbol{y}}_j$ denote the linearly independent solutions of the characteristic matrix equation

$$\left[\boldsymbol{M}\lambda_j^2 + (\boldsymbol{D} + \boldsymbol{G})\lambda_j + (\boldsymbol{K} + \boldsymbol{N})\right] \tilde{\boldsymbol{y}}_j = \boldsymbol{0}. \tag{4.97}$$

The eigenvalues λ_j are the zeros of the characteristic polynomial

$$p(\lambda) \equiv \frac{1}{\det \boldsymbol{M}} \det\left[\boldsymbol{M}\lambda^2 + (\boldsymbol{D} + \boldsymbol{G})\lambda + (\boldsymbol{K} + \boldsymbol{N})\right], \tag{4.98}$$

which is equivalent to (4.54).

However, if generalized eigenvectors appear in case of multiple eigenvalues, then it is recommendable to calculate them from the state space representation. The solution of (4.53) is preferable when also the determination of eigenvalues and eigenvectors has to be done numerically, since there exist well established numerical procedures for solving (4.53) (see Section 14.3). The presentation (4.97) of the eigenvalue problem will be quite useful in discussing the vibrations of mechanical systems in Chapter 6.

Example 4.3. Vertical vibrations of a motor car

The vertical vibrations of a motor car were described in Problem 2.2 using a two-mass model (Fig. 2.12). If tyre damping is neglected, the corresponding equation of motion is

$$\underbrace{\begin{bmatrix} m_1 & 0 \\ 0 & m_2 \end{bmatrix}}_{M} \underbrace{\begin{bmatrix} \ddot{y}_1 \\ \ddot{y}_2 \end{bmatrix}}_{\ddot{y}(t)} + \underbrace{\begin{bmatrix} d_1 & -d_1 \\ -d_1 & d_1 \end{bmatrix}}_{D} \underbrace{\begin{bmatrix} \dot{y}_1 \\ \dot{y}_2 \end{bmatrix}}_{\dot{y}(t)} +$$

$$+ \underbrace{\begin{bmatrix} k_1 & -k_1 \\ -k_1 & k_1 + k_2 \end{bmatrix}}_{K} \underbrace{\begin{bmatrix} y_1 \\ y_2 \end{bmatrix}}_{y(t)=\mathbf{0}.} = \mathbf{0}, \tag{4.99}$$

The calculation of the eigenvalues, eigenvectors and the general solution of (4.99) will be carried out for the following numerical data:

$$m_1 = 1200 \text{ kg}, \qquad m_2 = 80 \text{ kg}, \qquad d_1 = 4800 \frac{\text{Ns}}{m},$$

$$k_1 = 300 \frac{\text{N}}{\text{cm}}, \qquad k_2 = 3200 \frac{\text{N}}{\text{cm}}.$$

Beginning with the characteristic polynomial, one has from (4.98)

$$p(\lambda) \equiv \frac{1}{m_1 m_2} \det \begin{bmatrix} m_1\lambda^2 + d_1\lambda + k_1 & -(d_1\lambda + k_1) \\ -(d_1\lambda + k_1) & m_2\lambda^2 + d_1\lambda + k_1 + k_2 \end{bmatrix}$$

$$\equiv \lambda^4 + (\vartheta_1 + \vartheta_2)\lambda^3 + (\kappa_{11} + \kappa_{22} + \kappa_{12})\lambda^2 + \vartheta_1\kappa_{22}\lambda + \kappa_{11}\kappa_{22} \tag{4.100}$$

with the abbreviations, following (3.85),

$$\vartheta_i = \frac{d_1}{m_i}, \quad i = 1, 2, \quad \kappa_{ij} = \frac{k_i}{m_j}, \quad i, j = 1, 2. \tag{4.101}$$

Substituting the numerical data, one gets

$$p(\lambda) \equiv \lambda^4 + 64\lambda^3 + 4400\lambda^2 + 16\,000\lambda + 100\,000. \tag{4.102}$$

This gives the eigenvalues

$$\left.\begin{aligned} \lambda_{1,2} &= -30{,}253 \pm 57{,}003 \text{ i}, \\ \lambda_{3,4} &= -\ 1{,}747 \pm\ 4{,}578 \text{ i}. \end{aligned}\right\} \tag{4.103}$$

The eigenvectors can, in this case, be calculated from (4.97):

$$\left[M\lambda_j^2 + D\lambda_j + K\right]\begin{bmatrix}\tilde{y}_{1j}\\ \tilde{y}_{2j}\end{bmatrix} = \mathbf{0}. \tag{4.104}$$

One obtains

$$\left.\begin{aligned} &\text{for } \lambda_{1,2}: && \tilde{y}_1 = -0{,}031 \pm 0{,}054\,\mathrm{i}, \\ & && \tilde{y}_2 = 1; \\ &\text{for } \lambda_{3,4}: && \tilde{y}_1 = 8{,}2 \mp 7{,}5\,\mathrm{i}, \\ & && \tilde{y}_2 = 1. \end{aligned}\right\} \tag{4.105}$$

From (4.96), the corresponding eigenvectors in the state space are

$$\left.\begin{aligned} \lambda_{1,2}:\quad \tilde{x}_{1,2} &= \begin{bmatrix} -0.031 \pm 0.054\,\mathrm{i} \\ 1 \\ -2.14 \mp 3.4\,\mathrm{i} \\ -30.25 \pm 57\,\mathrm{i} \end{bmatrix}, \\ \lambda_{3,4}:\quad \tilde{x}_{3,4} &= \begin{bmatrix} 8.2 \mp 7.5\,\mathrm{i} \\ 1 \\ 20 \pm 50.68\,\mathrm{i} \\ -1.75 \pm 4.58\,\mathrm{i} \end{bmatrix}. \end{aligned}\right\} \tag{4.106}$$

According to (4.69), the general solution is given by

$$x(t) = \begin{bmatrix} y_1(t) \\ y_2(t) \\ \dot{y}_1(t) \\ \dot{y}_2(t) \end{bmatrix} = \sum_{j=1}^{4} \tilde{x}_j\, e^{\lambda_j t} c_j. \tag{4.107}$$

To consider this solution for the initial condition $x(0) = x_0$, one has to determine the (complex) constants c_j, $j = 1(1)4$, according to (4.75):

$$c = X^{-1} x_0. \tag{4.108}$$

One finds

$$X^{-1} = \begin{bmatrix} 0.047+0.022\,\mathrm{i} & 0.53-0.28\,\mathrm{i} & -0.007+0.004\,\mathrm{i} & -0.009\,\mathrm{i} \\ -0.047-0.022\,\mathrm{i} & 0.53+0.28\,\mathrm{i} & -0.007-0.004\,\mathrm{i} & 0.009\,\mathrm{i} \\ 0.047+0.015\,\mathrm{i} & -0.03+0.03\,\mathrm{i} & 0.007-0.008\,\mathrm{i} & 0 \\ 0.047-0.015\,\mathrm{i} & -0.03-0.03\,\mathrm{i} & 0.007+0.008\,\mathrm{i} & 0 \end{bmatrix}. \tag{4.109}$$

The fundamental matrix is now determined by (4.78). Though $\boldsymbol{X}$ and $e^{\Lambda t}$ are complex matrices, the fundamental matrix is real. To illustrate this solution, the time functions of the coordinates $y_1(t)$, $y_2(t)$, $\dot{y}_1(t)$, $\dot{y}_2(t)$ for the initial vector

$$\boldsymbol{x}_0 = [y_{10} \quad 0 \quad 0 \quad 0]^T \tag{4.110}$$

has been plotted in Fig. 4.3. The initial condition was chosen such that the body vibration y_1 is dominating while the fast wheel vibration y_2 is quickly damped.

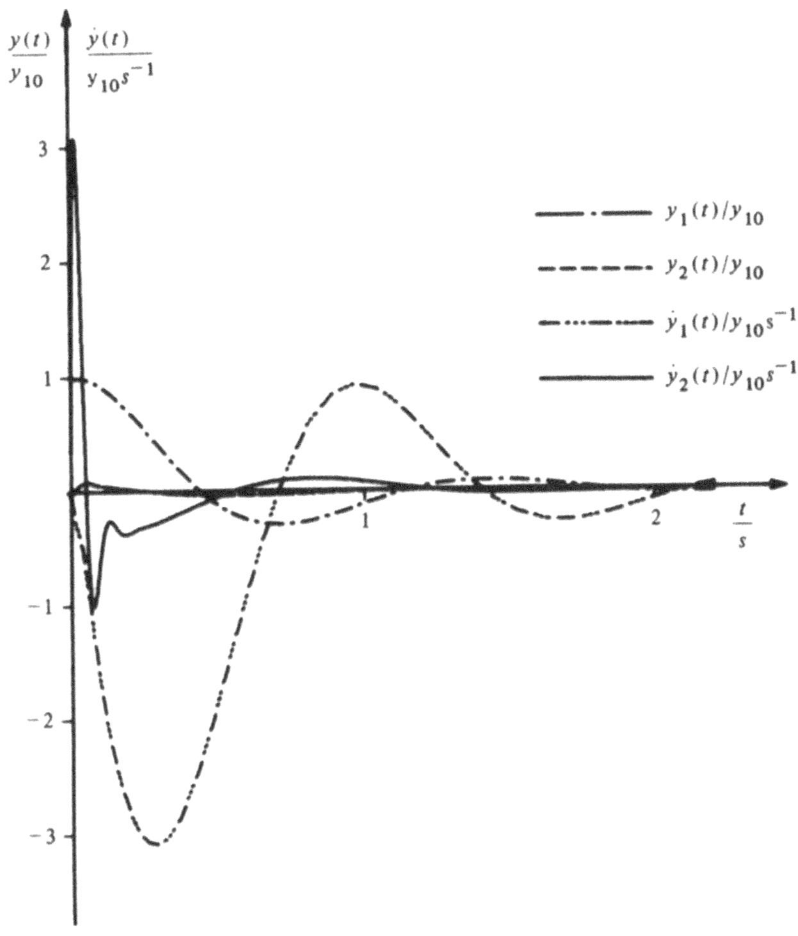

Fig. 4.3. Solution of the system (4.99) describing the vertical vibrations of a motor car

Example 4.4. Single-variable vibrating system

A homogeneous dynamic system described by an n-th order scalar differential equation

$$y^{(n)}(t)+a_1 y^{(n-1)}(t)+\ldots+a_{n-1}\dot{y}(t)+a_n y(t)=0, \tag{4.111}$$

allows a state space representation (4.8) with

$$\boldsymbol{x}=\begin{bmatrix} y \\ \dot{y} \\ \vdots \\ \vdots \\ y^{(n-1)} \end{bmatrix}, \quad \boldsymbol{A}=\boldsymbol{A}_{\mathrm{F}}=\begin{bmatrix} 0 & 1 & 0 & \ldots & 0 \\ 0 & 0 & 1 & \ldots & 0 \\ \vdots & \vdots & \vdots & \ddots & \vdots \\ 0 & 0 & 0 & \ldots & 1 \\ -a_n & -a_{n-1} & -a_{n-2} & \ldots & -a_1 \end{bmatrix} \tag{4.112}$$

The matrix $\boldsymbol{A}_{\mathrm{F}}$ is called the *Frobenius* or *companion matrix*. Its characteristic polynomial is

$$p(\lambda)\equiv\lambda^n+a_1\lambda^{n-1}+\ldots+a_{n-1}\lambda+a_n, \tag{4.113}$$

i.e. the coefficients of the differential equation (4.111) are the same as the coefficients of (4.54). Using the s different eigenvalues λ_j, $j=1(1)s$, which are the roots of (4.113), one can easily describe the eigenvectors and the generalized eigenvectors of $\boldsymbol{A}_{\mathrm{F}}$. Thus it can be stated that for every eigenvalue there exists precisely one eigenvector ($d_j=1\leqslant v_j$), and it is given by

$$\tilde{\boldsymbol{x}}_j=\left[1 \quad \lambda_j \quad \lambda_j^2 \ \ldots \ \lambda_j^{n-1}\right]^T. \tag{4.114}$$

The v_j-d_j generalized eigenvectors are given by

$$\tilde{\boldsymbol{x}}_j^{(l)}=\frac{1}{(l-1)!}\frac{d^{l-1}}{d\lambda^{l-1}}\left[1 \quad \lambda \quad \lambda^2\ldots\lambda^{n-1}\right]^T\bigg|_{\lambda=\lambda_j}, \quad l=2(1)v_j. \tag{4.115}$$

The corresponding modal matrix (4.84) is a (generalized) Vandermonde matrix. If all the eigenvalues λ_j, $j=1(1)n$, are distinct, this matrix is

$$\boldsymbol{X}=\begin{bmatrix} 1 & 1 & \ldots & 1 \\ \lambda_1 & \lambda_2 & \ldots & \lambda_n \\ \lambda_1^2 & \lambda_2^2 & \ldots & \lambda_n^2 \\ \vdots & \vdots & & \vdots \\ \lambda_1^{n-1} & \lambda_2^{n-1} & \ldots & \lambda_n^{n-1} \end{bmatrix}. \quad \blacklozenge \tag{4.116}$$

Example 4.5. Magnetically levitated vehicle

The state equation for a simple model (Fig. 3.2) of a magnetically levitated vehicle was derived in Example 3.5. If one neglects the roughness of the guideway, one gets according to (3.51), with $\boldsymbol{b}(t)=\mathbf{0}$,

$$\dot{\boldsymbol{x}}(t)=\begin{bmatrix} 0 & 1 & 0 \\ \frac{c}{m} & 0 & \frac{d}{m} \\ \frac{a}{T} & \frac{b-k}{T} & -\frac{1}{T} \end{bmatrix}\boldsymbol{x}(t). \tag{4.117}$$

The transition from the state coordinates y, $\dot{y}$, i (Example 3.5) to the new state coordinates y, $\dot{y}$, $\ddot{y}$ with the coordinate transformation

$$\boldsymbol{x}\equiv\begin{bmatrix} y \\ \dot{y} \\ i \end{bmatrix}=\begin{bmatrix} 1 & 0 & 0 \\ 0 & 1 & 0 \\ -\frac{c}{d} & 0 & \frac{m}{d} \end{bmatrix}\begin{bmatrix} y \\ \dot{y} \\ \ddot{y} \end{bmatrix}=\boldsymbol{T}\hat{\boldsymbol{x}} \tag{4.118}$$

yields, according to (3.53), the transformed state equation

$$\dot{\hat{\boldsymbol{x}}}(t)=\begin{bmatrix} 0 & 1 & 0 \\ 0 & 0 & 1 \\ \frac{c+ad}{mT} & \frac{c}{m}+\frac{d}{m}\frac{b-k}{T} & -\frac{1}{T} \end{bmatrix}\hat{\boldsymbol{x}}(t). \tag{4.119}$$

This presentation allows an application of the results found in Example 4.4. For further discussion the coefficients a, b of the linear voltage control (see (3.49)) will be prescribed as follows:

$$a=-\frac{1}{d}\left(c+\frac{m}{32T^2}\right),\qquad b=k-\frac{cT}{d}-\frac{5m}{16dT}. \tag{4.120}$$

This simplifies the Frobenius matrix appearing in (4.119) to

$$\boldsymbol{A}_{\mathrm{F}}=\begin{bmatrix} 0 & 1 & 0 \\ 0 & 0 & 1 \\ -\frac{1}{32T^3} & -\frac{5}{16T^2} & -\frac{1}{T} \end{bmatrix}. \tag{4.121}$$

The corresponding characteristic polynomial (4.113) is

$$\begin{aligned} p(\lambda)&=\lambda^3+\frac{1}{T}\lambda^2+\frac{5}{16T^2}\lambda+\frac{1}{32T^3} \\ &=\left(\lambda+\frac{1}{2T}\right)\left(\lambda+\frac{1}{4T}\right)^2 \end{aligned} \tag{4.122}$$

and its roots are

$$\lambda_1 = -\frac{1}{2T}, \qquad \lambda_{2,3} = -\frac{1}{4T}. \tag{4.123}$$

According to (4.114), the eigenvector belonging to the eigenvalue λ_1 is

$$\tilde{\boldsymbol{x}}_1 = \begin{bmatrix} 1 & -\frac{1}{2T} & \frac{1}{4T^2} \end{bmatrix}^T; \tag{4.124a}$$

to $\lambda_{2,3}$ there corresponds an eigenvector and a generalized eigenvector (4.115):

$$\tilde{\boldsymbol{x}}_2 = \begin{bmatrix} 1 & -\frac{1}{4T} & \frac{1}{16T^2} \end{bmatrix}^T, \tag{4.124b}$$

$$\tilde{\boldsymbol{x}}_2^{(2)} = \begin{bmatrix} 0 & 1 & -\frac{1}{2T} \end{bmatrix}^T. \tag{4.124c}$$

Hence the modal matrix $\hat{\boldsymbol{X}}$ (4.116) and its inverse $\hat{\boldsymbol{X}}^{-1}$ are

$$\hat{\boldsymbol{X}} = \begin{bmatrix} 1 & 1 & 0 \\ -\frac{1}{2T} & -\frac{1}{4T} & 1 \\ \frac{1}{4T^2} & \frac{1}{16T^2} & -\frac{1}{2T} \end{bmatrix}, \tag{4.125}$$

$$\hat{\boldsymbol{X}}^{-1} = \begin{bmatrix} 1 & 8T & 16T^2 \\ 0 & -8T & -16T^2 \\ \frac{1}{2T} & 3 & 4T \end{bmatrix}. \tag{4.126}$$

It is now easy to verify that

$$\hat{\boldsymbol{X}}^{-1}\boldsymbol{A}_{\mathrm{F}}\hat{\boldsymbol{X}} = \boldsymbol{J} = \begin{bmatrix} -\frac{1}{2T} & 0 & 0 \\ 0 & -\frac{1}{4T} & 1 \\ 0 & 0 & -\frac{1}{4T} \end{bmatrix}. \tag{4.127}$$

$$\hat{\boldsymbol{\Phi}}(t) = \hat{\boldsymbol{X}}\,\mathrm{e}^{Jt}\hat{\boldsymbol{X}}^{-1}$$

$$= \begin{bmatrix} \mathrm{e}^{-t/2T} + \dfrac{t}{2T}\mathrm{e}^{-t/4T} & 8T\left[\mathrm{e}^{-t/2T} + \left(\dfrac{3t}{8T} - 1\right)\mathrm{e}^{-t/4T}\right] & 16T^2\left[\mathrm{e}^{-t/2T} + \left(\dfrac{t}{4T} - 1\right)\mathrm{e}^{-t/4T}\right] \\ -\dfrac{1}{2T}\left[\mathrm{e}^{-t/2T} + \left(\dfrac{t}{4T} - 1\right)\mathrm{e}^{-t/4T}\right] & -4\left[\mathrm{e}^{-t/2T} + \left(\dfrac{3t}{16T} - \dfrac{5}{4}\right)\mathrm{e}^{-t/4T}\right] & -8T\left[\mathrm{e}^{-t/2T} + \left(\dfrac{t}{8T} - 1\right)\mathrm{e}^{-t/4T}\right] \\ -\dfrac{1}{4T^2}\left[\mathrm{e}^{-t/2T} + \left(\dfrac{t}{8T} - 1\right)\mathrm{e}^{-t/4T}\right] & \dfrac{2}{T}\left[\mathrm{e}^{-t/2T} + \left(\dfrac{3t}{32T} - 1\right)\mathrm{e}^{-t/4T}\right] & 4\left[\mathrm{e}^{-t/2T} + \left(\dfrac{t}{16T} - \dfrac{3}{4}\right)\mathrm{e}^{-t/4T}\right] \end{bmatrix}. \quad (4.129)$$

Because of

$$e^{Jt} = \begin{bmatrix} e^{-t/2T} & 0 & 0 \\ 0 & e^{-1/4T} & t\,e^{-t/4T} \\ 0 & 0 & e^{-t/4T} \end{bmatrix} \tag{4.128}$$

the fundamental matrix $\hat{\boldsymbol{\Phi}}(t) = e^{\boldsymbol{A}_F t}$ for the Frobenius matrix (4.121) has the form of equation (4.129) on the following page. The general solution $\hat{\boldsymbol{x}}(t) = \hat{\boldsymbol{\Phi}}(t)\hat{\boldsymbol{x}}_0$ is a linear combination of elementary solutions whose time behavior is determined by the functions $e^{-t/2T}$, $e^{-t/4T}$, $t\,e^{-t/4T}$. ◆

4.4. Theorem of Cayley and Hamilton

The theorem of Cayley and Hamilton states that every $n \times n$ matrix $\boldsymbol{A}$ satisfies the related characteristic equation (4.54):

$$p(\boldsymbol{A}) \equiv \boldsymbol{A}^n + a_1\boldsymbol{A}^{n-1} + \ldots + a_{n-1}\boldsymbol{A} + a_n\boldsymbol{E} = \boldsymbol{0}. \tag{4.130}$$

Hence every matrix power $\boldsymbol{A}^k$ with integer $k \geqslant n$ can be represented by a matrix polynomial of degree, at most, $(n-1)$. Consequently, there exists for the infinite matrix series (4.14) a *finite polynomial*

$$\boldsymbol{\Phi}(t) = e^{\boldsymbol{A}t} \equiv \alpha_0(t)\boldsymbol{E} + \alpha_1(t)\boldsymbol{A} + \ldots + \alpha_{n-1}(t)\boldsymbol{A}^{n-1}. \tag{4.131}$$

The unknown coefficients $\alpha_i(t)$, $i = 0(1)n-1$, can be determined by solving the inhomogeneous, linear n-th order system of equations

$$\frac{d^s}{d\lambda^s}\left[\alpha_0(t) + \alpha_1(t)\lambda + \cdots + \alpha_{n-1}(t)\lambda^{n-1}\right]\Big|_{\lambda=\lambda_j} = t^s\, e^{\lambda_j t},$$

$$s = 0(1)v_j - 1, \qquad j = 1(1)s, \qquad \sum_{j=1}^{s} v_j = n. \tag{4.132}$$

Its system matrix is a generalized Vandermonde matrix belonging to the s different eigenvalues, hence it is always regular. The coefficient functions $\alpha_i(t)$ are uniquely determined by (4.132). If the n eigenvalues are all distinct, then one gets

$$\begin{bmatrix} 1 & \lambda_1 & \lambda_1^2 & \ldots & \lambda_1^{n-1} \\ 1 & \lambda_2 & \lambda_2^2 & \ldots & \lambda_2^{n-1} \\ \vdots & \vdots & \vdots & \ldots & \vdots \\ 1 & \lambda_n & \lambda_n^2 & \ldots & \lambda_n^{n-1} \end{bmatrix} \begin{bmatrix} \alpha_0(t) \\ \alpha_1(t) \\ \vdots \\ \alpha_{n-1}(t) \end{bmatrix} = \begin{bmatrix} e^{\lambda_1 t} \\ e^{\lambda_2 t} \\ \vdots \\ e^{\lambda_n t} \end{bmatrix}. \tag{4.133}$$

The system matrix of (4.133) coincides with the transposed modal matrix (4.116) of a Frobenius matrix. In the case of multiple eigenvalues, the individual rows are formed according to the prescription (4.115) while, according to (4.132), corresponding right-hand sides are exponential functions weighted with time powers. A distinction between Case I and II as in Section 4.3 is not needed here.

Example 4.6. Double pendulum

The equations of motion of a double pendulum (Fig. 3.3) were derived in Example 3.6,

$$\ddot{\boldsymbol{y}}(t)+\boldsymbol{K}\boldsymbol{y}(t)=\boldsymbol{0}, \qquad \boldsymbol{K}=\frac{g}{L}\begin{bmatrix} 3 & -1 \\ -1 & 1 \end{bmatrix} \tag{4.134}$$

or

$$\dot{\boldsymbol{x}}(t)=\boldsymbol{A}\boldsymbol{x}(t), \quad \boldsymbol{A}=\left[\begin{array}{c|c} \boldsymbol{0} & \boldsymbol{E} \\ \hline -\boldsymbol{K} & \boldsymbol{0} \end{array}\right]=\begin{bmatrix} 0 & 0 & 1 & 0 \\ 0 & 0 & 0 & 1 \\ -3\frac{g}{L} & \frac{g}{L} & 0 & 0 \\ \frac{g}{L} & -\frac{g}{L} & 0 & 0 \end{bmatrix}. \tag{4.135}$$

The characteristic equation is

$$p(\lambda)\equiv\det(\lambda\boldsymbol{E}-\boldsymbol{A})\equiv\det(\lambda^2\boldsymbol{E}+\boldsymbol{K})$$

$$\equiv\lambda^4+4\frac{g}{L}\lambda^2+2\left(\frac{g}{L}\right)^2=0. \tag{4.136}$$

This leads to the eigenvalues

$$\lambda_{1,2,3,4}=\pm\mathrm{i}\omega_{1,2}, \qquad \omega_{1,2}=\sqrt{\frac{g}{L}(2\pm\sqrt{2})}\,. \tag{4.137}$$

To calculate the fundamental matrix, one can use the finite polynomial (4.131):

$$\boldsymbol{\Phi}(t)=\mathrm{e}^{\boldsymbol{A}t}\equiv\alpha_0(t)\boldsymbol{E}+\alpha_1(t)\boldsymbol{A}+\alpha_2(t)\boldsymbol{A}^2+\alpha_3(t)\boldsymbol{A}^3. \tag{4.138}$$

The coefficient functions $\alpha_i(t)$ can be determined, according to (4.133), from the system of equations

$$\begin{bmatrix} 1 & \mathrm{i}\omega_1 & -\omega_1^2 & -\mathrm{i}\omega_1^3 \\ 1 & -\mathrm{i}\omega_1 & -\omega_1^2 & \mathrm{i}\omega_1^3 \\ 1 & \mathrm{i}\omega_2 & -\omega_2^2 & -\mathrm{i}\omega_2^3 \\ 1 & -\mathrm{i}\omega_2 & -\omega_2^2 & \mathrm{i}\omega_2^3 \end{bmatrix}\begin{bmatrix} \alpha_0(t) \\ \alpha_1(t) \\ \alpha_2(t) \\ \alpha_3(t) \end{bmatrix}=\begin{bmatrix} \mathrm{e}^{\mathrm{i}\omega_1 t} \\ \mathrm{e}^{-\mathrm{i}\omega_1 t} \\ \mathrm{e}^{\mathrm{i}\omega_2 t} \\ \mathrm{e}^{-\mathrm{i}\omega_2 t} \end{bmatrix}. \tag{4.139}$$

By adding and subtracting the first and last two equations (4.139) one gets the four real equations

$$\left.\begin{aligned}
\alpha_0(t)-\omega_1^2\alpha_2(t)&=\cos\omega_1 t,\\
\alpha_0(t)-\omega_2^2\alpha_2(t)&=\cos\omega_2 t,\\
\alpha_1(t)-\omega_1^2\alpha_3(t)&=\frac{1}{\omega_1}\sin\omega_1 t,\\
\alpha_1(t)-\omega_2^2\alpha_3(t)&=\frac{1}{\omega_2}\sin\omega_2 t,
\end{aligned}\right\}\tag{4.140}$$

which imply the following formula for the $\alpha_i(t)$, $i=0(1)3$:

$$\left.\begin{aligned}
\alpha_0(t)&=\frac{1}{\omega_1^2-\omega_2^2}\left[-\omega_2^2\cos\omega_1 t+\omega_1^2\cos\omega_2 t\right],\\
\alpha_1(t)&=\frac{1}{\omega_1^2-\omega_2^2}\left[-\frac{\omega_2^2}{\omega_1}\sin\omega_1 t+\frac{\omega_1^2}{\omega_2}\sin\omega_2 t\right],\\
\alpha_2(t)&=\frac{1}{\omega_1^2-\omega_2^2}\left[-\cos\omega_1 t+\cos\omega_2 t\right],\\
\alpha_3(t)&=\frac{1}{\omega_1^2-\omega_2^2}\left[-\frac{1}{\omega_1}\sin\omega_1 t+\frac{1}{\omega_2}\sin\omega_2 t\right].
\end{aligned}\right\}\tag{4.141}$$

Consequently, the fundamental matrix of the system (4.135) is

$$\boldsymbol{\Phi}(t)=\left[\begin{array}{c|c}
\alpha_0(t)\boldsymbol{E}-\alpha_2(t)\boldsymbol{K} & \alpha_1(t)\boldsymbol{E}-\alpha_3(t)\boldsymbol{K}\\ \hline
-\alpha_1(t)\boldsymbol{K}+\alpha_3(t)\boldsymbol{K}^2 & \alpha_0(t)\boldsymbol{E}-\alpha_2(t)\boldsymbol{K}
\end{array}\right].\tag{4.142}$$

If this result is compared with (4.29), one finds the general solution (4.43) of the equations of motion (4.134), to be

$$\boldsymbol{y}(t)=(\alpha_0(t)\boldsymbol{E}-\alpha_2(t)\boldsymbol{K})\boldsymbol{y}_0+(\alpha_1(t)\boldsymbol{E}-\alpha_3(t)\boldsymbol{K})\dot{\boldsymbol{y}}_0.\tag{4.143}$$

Fig. 4.4 shows the time behavior of $y_1(t)$ and $y_2(t)$, i.e. the displacements from the vertical of the upper and lower mass, for the initial conditions $\boldsymbol{y}_0=[y_{10}\quad 0]^T$, $\dot{\boldsymbol{y}}_0=\boldsymbol{0}$. ◆

Example 4.7. Vertical vibrations of a motor car

The fundamental matrix $\boldsymbol{\Phi}(t)$ for the system (4.99) describing the vertical vibrations of a motor car was established in Example 4.3 by means of eigenvalues and eigenvectors. This calculation will be confirmed

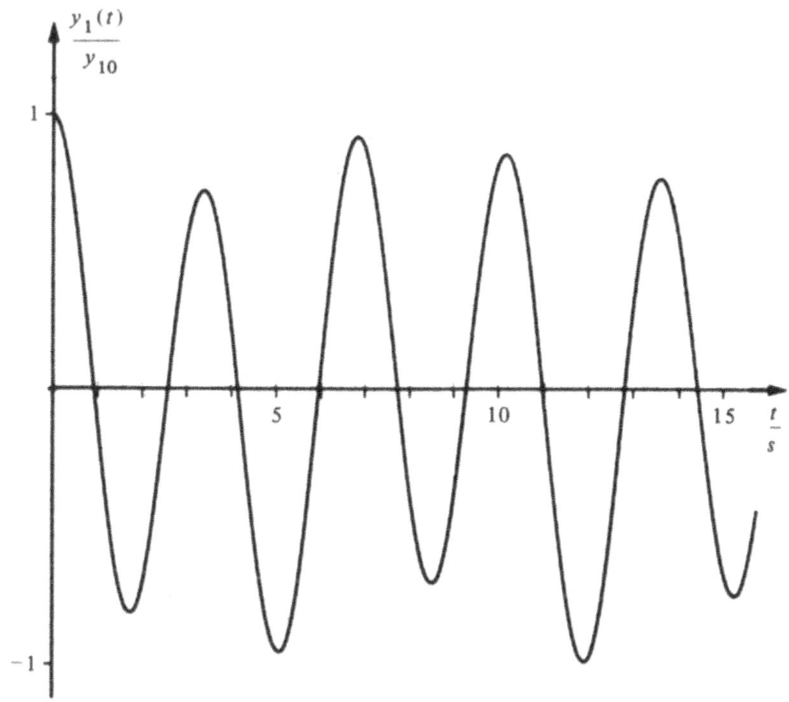

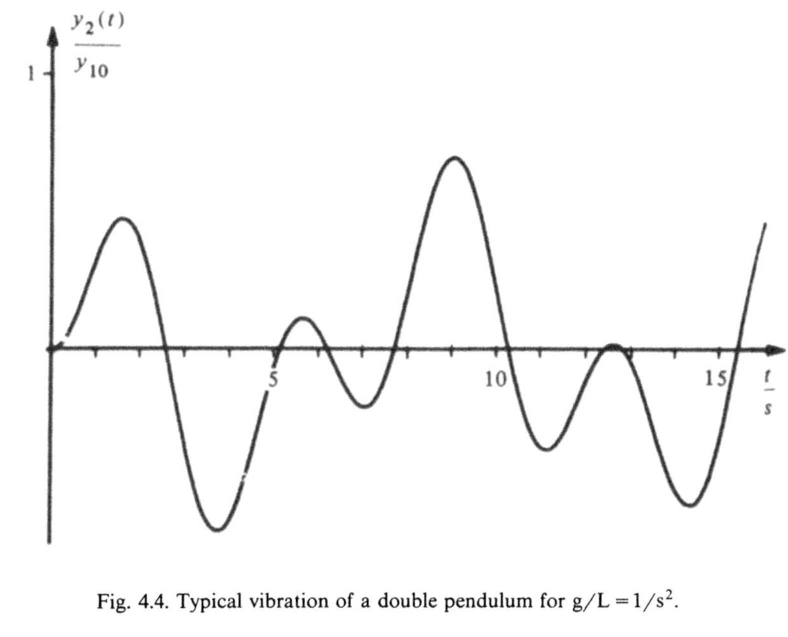

Fig. 4.4. Typical vibration of a double pendulum for $g/L = 1/s^2$.

according to (4.131). The eigenvalues (4.103),

$$\left.\begin{aligned}\lambda_{1,2} &= \delta_1 \pm \mathrm{i}\omega_1 = -30.253 \pm 57.003\,\mathrm{i}\\ \lambda_{3,4} &= \delta_2 \pm \mathrm{i}\omega_2 = -\ \ 1.747 \pm \ \ 4.578\,\mathrm{i}\end{aligned}\right\} \tag{4.144}$$

have to be substituted in the system of equations (4.133) for the functions $\alpha_i(t)$, $i = 0(1)3$,

$$\begin{bmatrix} 1 & \lambda_1 & \lambda_1^2 & \lambda_1^3 \\ 1 & \lambda_2 & \lambda_2^2 & \lambda_2^3 \\ 1 & \lambda_3 & \lambda_3^2 & \lambda_3^3 \\ 1 & \lambda_4 & \lambda_4^2 & \lambda_4^3 \end{bmatrix} \begin{bmatrix} \alpha_0(t) \\ \alpha_1(t) \\ \alpha_2(t) \\ \alpha_3(t) \end{bmatrix} = \begin{bmatrix} \mathrm{e}^{\lambda_1 t} \\ \mathrm{e}^{\lambda_2 t} \\ \mathrm{e}^{\lambda_3 t} \\ \mathrm{e}^{\lambda_4 t} \end{bmatrix}. \tag{4.145}$$

By adding and subtracting the first and last two equations in (4.145) one obtains the equivalent real system

$$\begin{bmatrix} 1 & \delta_1 & \delta_1^2 - \omega_1^2 & \delta_1(\delta_1^2 - 3\omega_1^2) \\ 0 & \omega_1 & 2\omega_1\delta_1 & \omega_1(3\delta_1^2 - \omega_1^2) \\ 1 & \delta_2 & \delta_2^2 - \omega_2^2 & \delta_2(\delta_2^2 - 3\omega_2^2) \\ 0 & \omega_2 & 2\omega_2\delta_2 & \omega_2(3\delta_2^2 - \omega_2^2) \end{bmatrix} \begin{bmatrix} \alpha_0(t) \\ \alpha_1(t) \\ \alpha_2(t) \\ \alpha_3(t) \end{bmatrix} = \begin{bmatrix} \mathrm{e}^{\delta_1 t} \cos \omega_1 t \\ \mathrm{e}^{\delta_1 t} \sin \omega_1 t \\ \mathrm{e}^{\delta_2 t} \cos \omega_2 t \\ \mathrm{e}^{\delta_2 t} \sin \omega_2 t \end{bmatrix}. \tag{4.146}$$

Inserting the numerical values (4.144), one derives

$$\left.\begin{aligned}
\alpha_0(t) &= \left(-1.0 \cdot 10^{-3} \cos \omega_1 t + 6.6 \cdot 10^{-3} \sin \omega_1 t\right) \mathrm{e}^{\delta_1 t} \\
&\qquad + (1.001 \cos \omega_2 t + 0.458 \sin \omega_2 t)\, \mathrm{e}^{\delta_2 t}, \\
\alpha_1(t) &= \left(-6.4 \cdot 10^{-5} \cos \omega_1 t + 1.0 \cdot 10^{-3} \sin \omega_1 t\right) \mathrm{e}^{\delta_1 t} \\
&\qquad + \left(6.4 \cdot 10^{-5} \cos \omega_2 t + 0.231 \sin \omega_2 t\right) \mathrm{e}^{\delta_2 t}, \\
\alpha_2(t) &= \left(-3.0 \cdot 10^{-5} \cos \omega_1 t + 2.9 \cdot 10^{-4} \sin \omega_1 t\right) \mathrm{e}^{\delta_1 t} \\
&\qquad + \left(3.0 \cdot 10^{-5} \cos \omega_2 t + 0.003 \sin \omega_2 t\right) \mathrm{e}^{\delta_2 t}, \\
\alpha_3(t) &= \left(3.5 \cdot 10^{-6} \cos \omega_1 t + 2.6 \cdot 10^{-6} \sin \omega_1 t\right) \mathrm{e}^{\delta_1 t} \\
&\qquad + \left(-3.5 \cdot 10^{-6} \cos \omega_2 t + 5.4 \cdot 10^{-5} \sin \omega_2 t\right) \mathrm{e}^{\delta_2 t}.
\end{aligned}\right\} \tag{4.147}$$

Hence the fundamental matrix is

$$\Phi(t) = \alpha_0(t)E + \alpha_1(t)A + \alpha_2(t)A^2 + \alpha_3(t)A^3$$

$$= \alpha_0(t)\left[\begin{array}{c|c} E & 0 \\ \hline 0 & E \end{array}\right] + \alpha_1(t)\left[\begin{array}{c|c} 0 & E \\ \hline -M^{-1}K & -M^{-1}D \end{array}\right]$$

$$+\alpha_2(t)\left[\begin{array}{c|c} -M^{-1}K & -M^{-1}D \\ \hline M^{-1}DM^{-1}K & -M^{-1}K + (M^{-1}D)^2 \end{array}\right]$$

$$+\alpha_3(t)\left[\begin{array}{c|c} M^{-1}DM^{-1}K & -M^{-1}K + (M^{-1}D)^2 \\ \hline (M^{-1}K)^2 - (M^{-1}D)^2 M^{-1}K & \begin{array}{c} M^{-1}DM^{-1}K + M^{-1}KM^{-1}D - \\ -(M^{-1}D)^3 \end{array} \end{array}\right].$$

(4.148) ◆

Problem 4.1. Centrifuge

The equations of motion for a centrifuge were derived in Problem 3.1. The normalized description of the linear, unperturbed system is given by

$$\ddot{y}(t) + (\vartheta E + g\Omega S)\dot{y}(t) + \kappa y(t) = 0 \tag{4.149}$$

with the position vector $y = [\alpha \quad \beta]^T$, the skewsymmetric matrix $S = \begin{bmatrix} 0 & 1 \\ -1 & 0 \end{bmatrix}$, and the abbreviations

$$\vartheta = \frac{d}{I_x + mL^2}, \qquad g = \frac{I_z}{I_x + mL^2}, \qquad \kappa = \frac{k - mgL}{I_x + mL^2}. \tag{4.150}$$

a) What is the characteristic polynomial $p(\lambda)$?

- ☐ $p(\lambda) \equiv \lambda^4 + \vartheta\lambda^3 + 2\kappa\lambda^2 + \vartheta\kappa\lambda + \kappa^2$;
- ☐ $p(\lambda) \equiv \lambda^4 + 2\vartheta\lambda^3 + (2\kappa + \vartheta^2 + g^2\Omega^2)\lambda^2 + 2\vartheta\kappa\lambda + \kappa^2$;
- ☐ $p(\lambda) \equiv \lambda^4 + 2\vartheta\lambda^3 + (\vartheta^2 + g^2\Omega^2)\lambda^2 + 2\vartheta\kappa\lambda + \kappa^2$;
- ☐ $p(\lambda) \equiv \lambda^4 + \vartheta\lambda^3 + (\vartheta^2 + g^2\Omega^2)\lambda^2 + \vartheta\kappa\lambda + \kappa^2$.

b) What are the eigenvalues of the system (4.149) for $\kappa = 0$?

☐ $\lambda_{1,2,3} = 0, \qquad \lambda_4 = -\vartheta;$

☐ $\lambda_{1,2} = 0, \qquad \lambda_{3,4} = -\vartheta;$

☐ $\lambda_{1,2} = 0, \qquad \lambda_{3,4} = -(\vartheta \pm \mathrm{i}g\Omega);$

☐ $\lambda_{1,2} = 0, \qquad \lambda_{3,4} = -\frac{\vartheta}{2} \pm \mathrm{i}\sqrt{\frac{3}{4}\vartheta^2 + g^2\Omega^2}\,.$

c) For $\kappa = 0$ the state equation of the centrifuge is

$$\underbrace{\frac{\mathrm{d}}{\mathrm{d}t}\begin{bmatrix} \alpha(t) \\ \beta(t) \\ \dot{\alpha}(t) \\ \dot{\beta}(t) \end{bmatrix}}_{\dot{\boldsymbol{x}}(t)} = \underbrace{\begin{bmatrix} 0 & 0 & 1 & 0 \\ 0 & 0 & 0 & 1 \\ 0 & 0 & -\vartheta & -g\Omega \\ 0 & 0 & g\Omega & -\vartheta \end{bmatrix}}_{\boldsymbol{A}} \underbrace{\begin{bmatrix} \alpha(t) \\ \beta(t) \\ \dot{\alpha}(t) \\ \dot{\beta}(t) \end{bmatrix}}_{\boldsymbol{x}(t).} \tag{4.151}$$

Which multiplicity and nullity correspond to each of the eigenvalues found in b)?

☐ $v_1 = 2, \qquad v_{3,4} = 1,$
$d_1 = 1, \qquad d_{3,4} = 1;$

☐ $v_1 = 2, \qquad v_{3,4} = 1,$
$d_1 = 2, \qquad d_{3,4} = 1;$

☐ $v_{1,2} = 1, \qquad v_{3,4} = 1,$
$d_{1,2} = 1, \qquad d_{3,4} = 1;$

☐ $v_1 = 3, \qquad v_4 = 1,$
$d_1 = 1, \qquad d_4 = 1.$

d) What are the eigenvectors of the system (4.151)?

☐ $\tilde{\boldsymbol{x}}_1 = \begin{bmatrix} 1 \\ 0 \\ 0 \\ 0 \end{bmatrix}, \qquad \tilde{\boldsymbol{x}}_2 = \begin{bmatrix} 0 \\ 1 \\ 0 \\ 0 \end{bmatrix}, \qquad \tilde{\boldsymbol{x}}_3 = \begin{bmatrix} 1 \\ i \\ -(\vartheta + \mathrm{i}g\Omega) \\ -i(\vartheta + \mathrm{i}g\Omega) \end{bmatrix},$

$\tilde{\boldsymbol{x}}_4 = \begin{bmatrix} 1 \\ -i \\ -(\vartheta - \mathrm{i}g\Omega) \\ i(\vartheta - \mathrm{i}g\Omega) \end{bmatrix};$

$\square$ $\tilde{x}_1 = \begin{bmatrix} 1 \\ 0 \\ 0 \\ 0 \end{bmatrix}, \quad \tilde{x}_2 = \begin{bmatrix} 1 \\ 1 \\ 0 \\ 0 \end{bmatrix}, \quad \tilde{x}_3 = \begin{bmatrix} i \\ 1 \\ -i(\vartheta + \mathrm{i}g\Omega) \\ (\vartheta + \mathrm{i}g\Omega) \end{bmatrix},$

$$\tilde{x}_4 = \begin{bmatrix} i \\ -1 \\ -i(\vartheta - \mathrm{i}g\Omega) \\ -(\vartheta - \mathrm{i}g\Omega) \end{bmatrix};$$

$\square$ $\tilde{x}_1 = \begin{bmatrix} 0 \\ 0 \\ 1 \\ 0 \end{bmatrix}, \quad \tilde{x}_2 = \begin{bmatrix} 0 \\ 0 \\ 0 \\ 1 \end{bmatrix}, \quad \tilde{x}_3 = \begin{bmatrix} 1 \\ i \\ -(\vartheta + \mathrm{i}g\Omega) \\ i(\vartheta + \mathrm{i}g\Omega) \end{bmatrix},$

$$\tilde{x}_4 = \begin{bmatrix} 1 \\ -i \\ -(\vartheta - \mathrm{i}g\Omega) \\ -i(\vartheta - \mathrm{i}g\Omega) \end{bmatrix};$$

$\square$ $\tilde{x}_1 = \begin{bmatrix} 1 \\ 0 \\ 0 \\ 0 \end{bmatrix}, \quad \tilde{x}_2 = \begin{bmatrix} 0 \\ 1 \\ 0 \\ 0 \end{bmatrix}, \quad \tilde{x}_3 = \begin{bmatrix} 0 \\ 0 \\ -(\vartheta + \mathrm{i}g\Omega) \\ i(\vartheta + \mathrm{i}g\Omega) \end{bmatrix},$

$$\tilde{x}_4 = \begin{bmatrix} 0 \\ 0 \\ -(\vartheta - \mathrm{i}g\Omega) \\ -i(\vartheta - \mathrm{i}g\Omega) \end{bmatrix}.$$

e) According to the theorem of Cayley and Hamilton the fundamental matrix $\boldsymbol{\Phi}(t)$ of the state equation (4.151) has the form

$$\begin{aligned}
\boldsymbol{\Phi}(t) &= \alpha_0(t)\boldsymbol{E} + \alpha_1(t)\boldsymbol{A} + \alpha_2(t)\boldsymbol{A}^2 + \alpha_3(t)\boldsymbol{A}^3 \\
&= \alpha_0(t)\left[\begin{array}{c|c} \boldsymbol{E} & \boldsymbol{0} \\ \hline \boldsymbol{0} & \boldsymbol{E} \end{array}\right] + \alpha_1(t)\left[\begin{array}{c|c} \boldsymbol{0} & \boldsymbol{E} \\ \hline \boldsymbol{0} & -\vartheta\boldsymbol{E} - g\Omega\boldsymbol{S} \end{array}\right] \\
&\quad + \alpha_2(t)\left[\begin{array}{c|c} \boldsymbol{0} & -\vartheta\boldsymbol{E} - g\Omega\boldsymbol{S} \\ \hline \boldsymbol{0} & (\vartheta^2 - g^2\Omega^2)\boldsymbol{E} + 2\vartheta g\Omega\boldsymbol{S} \end{array}\right] \\
&\quad + \alpha_3(t)\left[\begin{array}{c|c} \boldsymbol{0} & (\vartheta^2 - g^2\Omega^2)\boldsymbol{E} + 2\vartheta g\Omega\boldsymbol{S} \\ \hline \boldsymbol{0} & -\vartheta(\vartheta^2 - 3g^2\Omega^2)\boldsymbol{E} + g\Omega(-3\vartheta^2 + g^2\Omega^2)\boldsymbol{S} \end{array}\right].
\end{aligned} \tag{4.152}$$

What are the coefficient functions $\alpha_i(t)$, $i = 0(1)3$, in (4.152)?

☐ $\alpha_0(t) = 1, \qquad \alpha_1(t) = t,$

$(\vartheta^2 - g^2\Omega^2)\alpha_2(t) + \vartheta(-\vartheta^2 + 3g^2\Omega^2)\alpha_3(t) = e^{-\vartheta t}\cos g\Omega t + \vartheta t - 1,$

$-2\vartheta g\Omega\alpha_2(t) + g\Omega(3\vartheta^2 - g^2\Omega^2)\alpha_3(t) = e^{-\vartheta t}\sin g\Omega t - g\Omega t;$

☐ $\alpha_0(t) = 1, \qquad \alpha_1(t) = e^{-\vartheta t},$

$2\vartheta g\Omega\alpha_2(t) + g\Omega(g^2\Omega^2 - 3\vartheta^2)\alpha_3(t) = g\Omega t - e^{-\vartheta t}\sin g\Omega t,$

$(g^2\Omega^2 - \vartheta^2)\alpha_2(t) + \vartheta(3g^2\Omega^2 - \vartheta^2)\alpha_3(t) = 1 - \vartheta t - e^{-\vartheta t}\cos g\Omega t;$

☐ $\alpha_0(t) + (\vartheta^2 - g^2\Omega^2)\alpha_2(t) = e^{-\vartheta t}\cos g\Omega t,$

$\alpha_0(t) + 2\vartheta g\Omega\alpha_2(t) = e^{-\vartheta t}(\vartheta\cos g\Omega t - g\Omega\sin g\Omega t),$

$\alpha_1(t) + (\vartheta^2 - g^2\Omega^2)\alpha_3(t) = \frac{1}{g\Omega}e^{-\vartheta t}\sin g\Omega t,$

$\alpha_1(t) + 2\vartheta g\Omega\alpha_3(t) = e^{-\vartheta t}\left(\sin g\Omega t + \frac{\vartheta}{g\Omega}\cos g\Omega t\right);$

☐ $\alpha_0(t) = 1, \qquad \alpha_1(t) = t,$

$(\vartheta^2 - g^2\Omega^2)\alpha_2(t) + \vartheta(-\vartheta^2 + 3g^2\Omega^2)\alpha_3(t) = e^{-\vartheta t}\cos g\Omega t,$

$2\vartheta g\Omega\alpha_2(t) + g\Omega(-3\vartheta^2 + g^2\Omega^2)\alpha_3(t) = -e^{-\vartheta t}\sin g\Omega t.$

f) What is the general solution

$$\boldsymbol{y}(t) = \boldsymbol{\Psi}_1(t)\,\boldsymbol{y}_0 + \boldsymbol{\Psi}_2(t)\,\dot{\boldsymbol{y}}_0$$

of the system (4.149) for $\kappa = 0$?

☐ $\boldsymbol{y}(t) = e^{-\vartheta t}\boldsymbol{y}_0 + \left\{\left[t - \frac{\vartheta}{\vartheta^2 + g^2\Omega^2}(e^{-\vartheta t}\cos g\Omega t + \vartheta t - 1)\right.\right.$

$\left. + \frac{g\Omega}{\vartheta^2 + g^2\Omega^2}(e^{-\vartheta t}\sin g\Omega t - g\Omega t)\right]\boldsymbol{E}$

$+\left[\frac{g\Omega}{\vartheta^2 + g^2\Omega^2}(e^{-\vartheta t}\cos g\Omega t + \vartheta t - 1)\right.$

$$+\frac{\vartheta}{\vartheta^2+g^2\Omega^2}(\mathrm{e}^{-\vartheta t}\sin g\Omega t - g\Omega t)\bigg]\boldsymbol{S}\bigg\}\dot{y}_0;$$

□ $$y(t)=y_0+\bigg\{\bigg[-\frac{\vartheta}{\vartheta^2+g^2\Omega^2}(\mathrm{e}^{-\vartheta t}\cos g\Omega t-1)$$

$$+\frac{g\Omega}{\vartheta^2+g^2\Omega^2}\mathrm{e}^{-\vartheta t}\sin g\Omega t\bigg]\boldsymbol{E}$$

$$+\bigg[\frac{g\Omega}{\vartheta^2+g^2\Omega^2}(\mathrm{e}^{-\vartheta t}\cos g\Omega t-1)$$

$$+\frac{\vartheta}{\vartheta^2+g^2\Omega^2}\mathrm{e}^{-\vartheta t}\sin g\Omega t\bigg]\boldsymbol{S}\bigg\}\dot{y}_0;$$

□ $$y(t)=(1+t)y_0+\bigg\{\bigg[-\frac{\vartheta}{\vartheta^2+g^2\Omega^2}(\mathrm{e}^{-\vartheta t}\cos g\Omega t-1)$$

$$+\frac{g\Omega}{\vartheta^2+g^2\Omega^2}\mathrm{e}^{-\vartheta t}\sin g\Omega t\bigg]\boldsymbol{E}$$

$$+\bigg[\frac{g\Omega}{\vartheta^2+g^2\Omega^2}(\mathrm{e}^{-\vartheta t}\cos g\Omega t-1)$$

$$+\frac{\vartheta}{\vartheta^2+g^2\Omega^2}\mathrm{e}^{-\vartheta t}\sin g\Omega t\bigg]\boldsymbol{S}\bigg\}\dot{y}_0.$$

g) Supplementary question: What are the eigenvalues of the system (4.149) for $\kappa \neq 0$?

□ $\lambda_{1,2}=-\frac{1}{2}\left(\frac{\vartheta}{2}+\mathrm{i}g\Omega\right)\pm\sqrt{\frac{1}{4}\left(\frac{\vartheta^2}{4}-g^2\Omega^2-\kappa\right)+\frac{i}{4}\vartheta g\Omega},\ \lambda_{3,4}=\bar{\lambda}_{1,2};$

□ $\lambda_{1,2}=\pm\sqrt{-\kappa}, \qquad \lambda_{3,4}=-\frac{\vartheta}{2}\pm\sqrt{\frac{\vartheta^2}{4}-\kappa};$

□ $\lambda_{1,2}=-\frac{\vartheta}{4}\pm\sqrt{\frac{1}{4}\left(\frac{\vartheta^2}{4}-g^2\Omega^2\right)-\kappa}\ , \qquad \lambda_{3,4}=\bar{\lambda}_{1,2};$

□ $\lambda_{1,2}=-\tfrac{1}{2}(\vartheta+\mathrm{i}g\Omega)\pm\sqrt{\tfrac{1}{4}(\vartheta^2-g^2\Omega^2-4\kappa)+\frac{i}{2}\vartheta g\Omega},\ \lambda_{3,4}=\bar{\lambda}_{1,2}.$ ◆

Problem 4.2. Magnetically levitated vehicle

The general solution of the system (4.119) was calculated in Example 4.5 by means of eigenvectors. For comparison, the fundamental matrix $\hat{\boldsymbol{\Phi}}(t)$ (4.129) will be calculated here by applying the theorem of Cayley and Hamilton. The state equations are given by (4.119 to 121),

$$\dot{\hat{x}}(t)=\begin{bmatrix} 0 & 1 & 0 \\ 0 & 0 & 1 \\ -\dfrac{1}{32T^3} & -\dfrac{5}{16T^2} & -\dfrac{1}{T} \end{bmatrix}\hat{x}(t)=A_F\hat{x}(t). \tag{4.153}$$

a) Which system of equations serves for the determination of the coefficient functions $\alpha_i(t)$, $i=0, 1, 2$, in

$$\hat{\boldsymbol{\Phi}}(t)=\alpha_0(t)E+\alpha_1(t)A_F+\alpha_2(t)A_F^2? \tag{4.154}$$

☐ $$\begin{bmatrix} 1 & -\dfrac{1}{2T} & \dfrac{1}{4T^2} \\ 1 & -\dfrac{1}{4T} & \dfrac{1}{16T^2} \\ 1 & -\dfrac{1}{4T} & \dfrac{1}{16T^2} \end{bmatrix}\begin{bmatrix} \alpha_0(t) \\ \alpha_1(t) \\ \alpha_2(t) \end{bmatrix}=\begin{bmatrix} e^{-t/2T} \\ e^{-t/4T} \\ e^{-t/4T} \end{bmatrix};$$

☐ $$\begin{bmatrix} 1 & -\dfrac{1}{2T} & -\dfrac{1}{4T^2} \\ 0 & 1 & -\dfrac{1}{T} \\ 1 & -\dfrac{1}{4T} & \dfrac{1}{16T^2} \end{bmatrix}\begin{bmatrix} \alpha_0(t) \\ \alpha_1(t) \\ \alpha_2(t) \end{bmatrix}=\begin{bmatrix} e^{-t/2T} \\ t\,e^{-t/2T} \\ e^{-t/4T} \end{bmatrix};$$

☐ $$\begin{bmatrix} 1 & -\dfrac{1}{2T} & \dfrac{1}{4T^2} \\ 1 & -\dfrac{1}{4T} & \dfrac{1}{16T^2} \\ 0 & 1 & -\dfrac{1}{2T} \end{bmatrix}\begin{bmatrix} \alpha_0(t) \\ \alpha_1(t) \\ \alpha_2(t) \end{bmatrix}=\begin{bmatrix} e^{-t/2T} \\ e^{-t/4T} \\ t\,e^{-t/4T} \end{bmatrix};$$

☐ $$\begin{bmatrix} 1 & -\dfrac{1}{4T} & \dfrac{1}{16T^2} \\ 0 & 1 & -\dfrac{1}{2T} \\ 0 & 0 & 2 \end{bmatrix}\begin{bmatrix} \alpha_0(t) \\ \alpha_1(t) \\ \alpha_2(t) \end{bmatrix}=\begin{bmatrix} e^{-t/4T} \\ t\,e^{-t/4T} \\ t^2\,e^{-t/4T} \end{bmatrix}.$$

b) What are the coefficient functions $\alpha_i(t)$, $i = 0, 1, 2$?

☐ $\alpha_0(t) = \left(1 + \frac{t}{4T} + \frac{t^3}{32T^2}\right) e^{-t/4T}$,

$\alpha_1(t) = \left(t + \frac{t^2}{4T}\right) e^{-t/4T}$,

$\alpha_2(t) = \frac{t^2}{2} e^{-t/4T}$;

☐ $\alpha_0(t) = 4\, e^{-t/4T} + e^{-t/2T}\left(-3 + \frac{9t}{4T}\right)$,

$\alpha_1(t) = 16T\, e^{-t/4T} + e^{-t/2T}(-16T + 5t)$,

$\alpha_2(t) = 16T^2\, e^{-t/4T} + e^{-t/2T}(-16T^2 + 4Tt)$;

☐ $\alpha_0(t) = e^{-t/4T} - e^{-t/2T}$,

$\alpha_1(t) = -2T\, e^{-t/4T} - 4T\, e^{-t/2T}$;

$\alpha_2(t) = -8T^2\, e^{-t/4T}$;

☐ $\alpha_0(t) = \frac{t}{2T} e^{-t/4T} + e^{-t/2T}$

$\alpha_1(t) = (3t - 8T)\, e^{-t/4T} + 8T\, e^{-t/2T}$

$\alpha_2(t) = (4Tt - 16T^2)\, e^{-t/4T} + 16T^2\, e^{-t/2T}$.

c) Which of the following is $\boldsymbol{A}_F^2$?

☐ $\boldsymbol{A}_F^2 = \begin{bmatrix} 0 & 1 & 0 \\ 0 & 0 & 1 \\ \dfrac{1}{32T^4} & \dfrac{9}{32T^3} & \dfrac{11}{16T^2} \end{bmatrix}$,

☐ $\boldsymbol{A}_F^2 = \begin{bmatrix} 0 & 0 & 1 \\ -\dfrac{1}{32T^3} & -\dfrac{5}{16T^2} & -\dfrac{1}{T} \\ 0 & 1 & 0 \end{bmatrix}$,

☐ $\boldsymbol{A}_F^2 = \begin{bmatrix} 0 & 0 & 1 \\ -\dfrac{1}{32T^3} & -\dfrac{5}{16T^2} & -\dfrac{1}{T} \\ \dfrac{1}{32T^4} & \dfrac{9}{32T^3} & \dfrac{11}{16T^2} \end{bmatrix}$,

$$\square \quad A_F^2 = \begin{bmatrix} 0 & 0 & 1 \\ \dfrac{1}{32T^4} & \dfrac{9}{32T^3} & \dfrac{11}{16T^2} \\ -\dfrac{1}{32T^3} & -\dfrac{5}{16T^2} & -\dfrac{1}{T} \end{bmatrix}.$$

d) Using (4.154) confirm the result (4.129). ◆

Problem 4.3. Vibrating system with elastic damper

The state equation (3.47) for a vibrating system with elastic damper shown on Fig. 3.1 was derived in Example 3.4,

$$\dot{x}(t) = \begin{bmatrix} 0 & 1 & 0 \\ -\dfrac{b+k}{m} & 0 & \dfrac{b}{m} \\ \dfrac{b}{d} & 0 & -\dfrac{b}{d} \end{bmatrix} x(t). \tag{4.155}$$

Under the coordinate transformation (3.53) of the state vector (3.46) given by

$$x = \begin{bmatrix} y_1 \\ \dot{y}_1 \\ y_2 \end{bmatrix} = \begin{bmatrix} 1 & 0 & 0 \\ 0 & 1 & 0 \\ \dfrac{b+k}{b} & 0 & \dfrac{m}{b} \end{bmatrix} \begin{bmatrix} y_1 \\ \dot{y}_1 \\ \ddot{y}_1 \end{bmatrix} = T\hat{x} \tag{4.156}$$

the system (4.155) becomes

$$\dot{\hat{x}}(t) = \begin{bmatrix} 0 & 1 & 0 \\ 0 & 0 & 1 \\ -\dfrac{bk}{md} & -\dfrac{b+k}{m} & -\dfrac{b}{d} \end{bmatrix} \hat{x}(t). \tag{4.157}$$

a) What is the characteristic polynomial $p(\lambda)$?

$\square \quad p(\lambda) \equiv \lambda^3 + \dfrac{b}{d}\lambda^2 + \dfrac{b+k}{m}\lambda + \dfrac{b}{m};$

$\square \quad p(\lambda) \equiv \lambda^3 + \dfrac{m}{b}\lambda^2 + \dfrac{b+k}{b};$

$\square \quad p(\lambda) \equiv \lambda^3 + \dfrac{bk}{md}\lambda^2 + \dfrac{b+k}{m}\lambda + \dfrac{b}{d};$

$\square \quad p(\lambda) \equiv \lambda^3 + \dfrac{b}{d}\lambda^2 + \dfrac{b+k}{m}\lambda + \dfrac{bk}{md}.$

b) What are the eigenvalues of the system (4.155) for $b = k + d^2/m$?

☐ $\lambda_1 = -\frac{d}{m}, \qquad \lambda_{2,3} = -\frac{1}{2}\frac{d}{m} \pm i\sqrt{\frac{k}{m} + \frac{3}{4}\frac{d^2}{m^2}}\,;$

☐ $\lambda_1 = -\frac{k}{d}, \qquad \lambda_{2,3} = -\frac{1}{2}\frac{d}{m} \pm i\sqrt{\frac{k}{m} + \frac{3}{4}\frac{d^2}{m^2}}\,;$

☐ $\lambda_1 = -\frac{k}{d}, \qquad \lambda_{2,3} = -\frac{1}{2}\frac{k}{d} \pm i\sqrt{\frac{1}{4}\left(\frac{k}{d}\right)^2 + \frac{k}{m}}\,;$

☐ $\lambda_1 = -\frac{d}{m}, \qquad \lambda_{2,3} = -\frac{1}{2}\frac{k}{d} \pm i\sqrt{\frac{1}{4}\left(\frac{k}{d}\right)^2 + \frac{k}{m}}\,.$

c) The system (4.157) has the eigenvectors $\tilde{\hat{\boldsymbol{x}}}_i = [1 \quad \lambda_i \quad \lambda_i^2]^T$, $i = 1, 2, 3$. What are the eigenvectors of the system (4.155)? (With λ_i, $i = 1, 2, 3$, and $b = k + d^2/m$).

☐ $\tilde{\boldsymbol{x}}_i = \begin{bmatrix} 1 \\ \lambda_i \\ \lambda_i^2 \end{bmatrix}. \qquad i = 1, 2, 3;$

☐ $\tilde{\boldsymbol{x}}_i = \begin{bmatrix} 1 \\ \lambda_i \\ \frac{mk + d^2}{m^2}(\lambda_i^2 - 1) - \frac{k}{m} \end{bmatrix}, \qquad i = 1, 2, 3;$

☐ $\tilde{\boldsymbol{x}}_i = \begin{bmatrix} 1 \\ \lambda_i \\ \frac{1}{mk + d^2}(2mk + d^2 + m^2\lambda_i^2) \end{bmatrix}, \qquad i = 1, 2, 3;$

☐ $\tilde{\boldsymbol{x}}_i = \begin{bmatrix} 1 \\ \lambda_i \\ 1 + \frac{mk}{mk + d^2} + \frac{m^2}{mk + d^2}\lambda_i^2 \end{bmatrix}, \qquad i = 1, 2, 3.$ ◆

5. Stability and boundedness

The general solution of linear time-invariant vibrating systems with given initial conditions and excitation function was determined in Chapter 4. However, in the engineering practice the initial conditions and the excitation function are frequently only approximately known, and because of this the general solution cannot be exactly calculated. Thus the notions of *stability* and *boundedness* which describe qualitatively the general behavior of the solution are of fundamental importance for analysing dynamic systems. Stable and bounded systems satisfy the requirement that from small disturbances only small changes in the behavior of the system occur. Such disturbances include here externally applied excitations or deviations in the initial conditions. The theoretical presentation of the stability problem is preceded by the definitions of the concepts of stability and boundedness in Section 5.1. Further, in Section 5.2, the stability of homogeneous systems and their criteria for the verification of stability behavior are investigated. The corresponding treatment of the boundedness problem for inhomogeneous systems follows in Section 5.3.

5.1. Definitions

The stability and boundedness of the motions of a vibrating system will be defined relative to a reference motion which, according to (3.4), is always given by $\boldsymbol{y}(t) \equiv \boldsymbol{0}$ or $\boldsymbol{x}(t) \equiv \boldsymbol{0}$. Stability concerns the behavior of homogeneous systems (natural vibrations) while boundedness describes the qualitative behavior of inhomogeneous systems (forced vibrations).

For the definitions which follow, it will be convenient to introduce a generalization of the concept of the length of a vector $\boldsymbol{x}$, called the *norm* $\|\boldsymbol{x}\|$ of the vector. The norm has the following properties:

$$\|\boldsymbol{x}\| = 0 \quad \Leftrightarrow \quad \boldsymbol{x} = \boldsymbol{0}, \tag{5.1a}$$

$$\|\boldsymbol{x}\| > 0 \quad \Leftrightarrow \quad \boldsymbol{x} \neq \boldsymbol{0}, \tag{5.1b}$$

$$\|\lambda \boldsymbol{x}\| = |\lambda|\,\|\boldsymbol{x}\|, \tag{5.1c}$$

$$\|\boldsymbol{x}+\boldsymbol{y}\| \leqslant \|\boldsymbol{x}\| + \|\boldsymbol{y}\|. \tag{5.1d}$$

Besides the usual Euclidean length

$$\|\boldsymbol{x}\|_E = \sqrt{x_1^2 + x_2^2 + \cdots + x_n^2} = \sqrt{\boldsymbol{x}^{\mathrm{T}}\boldsymbol{x}}\,, \tag{5.2}$$

the following quantities may also serve as examples of such generalized lengths:

$$\|\boldsymbol{x}\|_R = \sqrt{\boldsymbol{x}^{\mathrm{T}}\boldsymbol{R}\boldsymbol{x}}\,, \qquad \boldsymbol{R} = \boldsymbol{R}^{\mathrm{T}} > \boldsymbol{0}, \tag{5.3}$$

where $\boldsymbol{R}$ is a positive definite symmetric $n \times n$ weighting matrix,

$$\|\boldsymbol{x}\|_m = \max_{i=1,\ldots,n} |x_i|, \tag{5.4}$$

or

$$\|\boldsymbol{x}\|_b = |x_1| + |x_2| + \ldots + |x_n| = \sum_{i=1}^{n} |x_i|. \tag{5.5}$$

Using this concept of norm, the following definitions of stability and boundedness will apply: The first is the definition of stability (5.7, 5.8) given in 1892 by Ljapunov pertaining not only to linear time-invariant systems, but also to particular solutions of nonlinear time-variant systems in general. Next in (5.10, 5.11) a definition of stability will be presented especially suited to linear systems which will facilitate the transition to the definitions of boundedness.

The linear time-invariant homogeneous vibrating system

$$\dot{\boldsymbol{x}}(t) = \boldsymbol{A}\boldsymbol{x}(t), \qquad \boldsymbol{x}(0) = \boldsymbol{x}_0 \tag{5.6}$$

is called *stable* (*in the sense of Ljapunov*) if for every positive $\epsilon > 0$ there exists a positive number $\delta = \delta(\epsilon) > 0$ such that for all initial conditions bounded by

$$\|\boldsymbol{x}_0\| < \delta = \delta(\epsilon) \tag{5.7}$$

the corresponding trajectories $\boldsymbol{x}(t)$ remain bounded for all $t \geqslant 0$ by ϵ:

$$\|\boldsymbol{x}(t)\| < \epsilon, \qquad t \geqslant 0. \tag{5.8}$$

The vibrating system (5.6) is called *asymptotically stable* if it is stable and if, for every bounded initial condition (5.7), the corresponding trajectory tends to zero as $t \to \infty$,

$$\lim_{t \to \infty} \boldsymbol{x}(t) = \mathbf{0}. \tag{5.9}$$

If the vibrating system (5.6) is not stable, it will be called *unstable*.

Since an asymptotically stable system could also be termed stable, it is desirable to adopt the following convention. A vibrating system will be called *marginally stable* if it is stable, but not asymptotically stable.

The various concepts of stability are illustrated in Fig. 5.1 for a two dimensional vibrating system. Here it is again shown geometrically that the system is called stable when small perturbations of the initial conditions imply only small deviations from the reference trajectory $\boldsymbol{x}(t) \equiv \mathbf{0}$. For linear systems the restriction to small initial conditions is not necessary, since (4.33) implies that if the system (5.6) is at all stable, then it is so for arbitrary initial conditions. Thus the stability definition (5.7, 5.8) can be simplified in case of linear systems: The vibrating system (5.6) is called *stable* if, for every bounded initial condition

$$\|\boldsymbol{x}_0\| < \delta \tag{5.10}$$

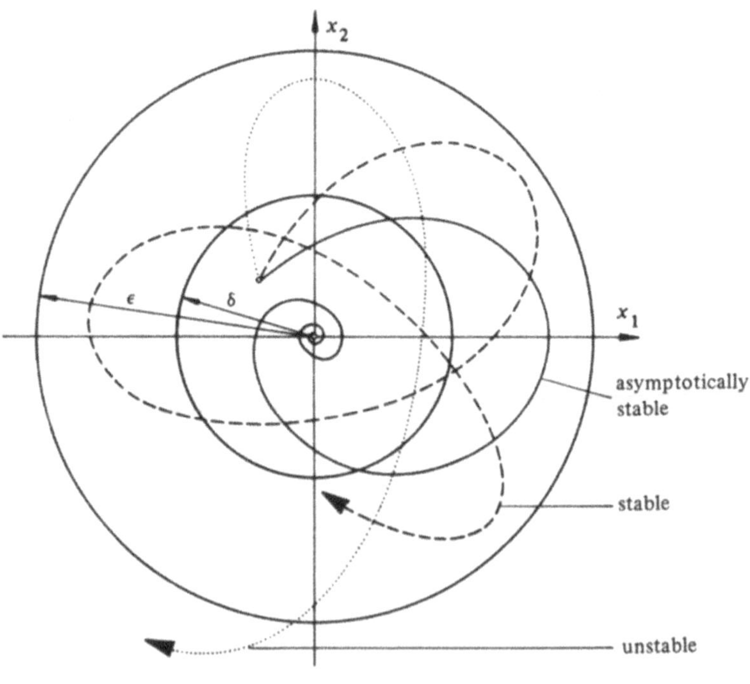

Fig. 5.1. Illustration of the stability concepts

the corresponding solution is bounded for all $t \geqslant 0$,

$$\|x(t)\| < \epsilon = \epsilon(\delta), \qquad t \geqslant 0. \tag{5.11}$$

If additional excitation forces act on the system (5.6),

$$\dot{x}(t) = Ax(t) + b(t), \qquad x(0) = x_0, \tag{5.12}$$

then the qualitative behavior of the system (5.12) cannot be characterized by means of the above definitions of stability. Whereas for (5.6) one needs to consider only the deviations x_0 from $x(t) \equiv \mathbf{0}$, for (5.12) it is essential to investigate the influence of the excitation function $b(t)$. This is done by means of the definitions of boundedness.

The externally excited system (5.12) is called *generally bounded* if, for every bounded initial condition

$$\|x_0\| < \delta \tag{5.13}$$

and every bounded excitation function

$$\|b(t)\| < \gamma \tag{5.14}$$

the corresponding solution $x(t)$ is bounded for all $t \geqslant 0$,

$$\|x(t)\| \leqslant \epsilon = \epsilon(\delta, \gamma), \qquad t \geqslant 0. \tag{5.15}$$

The vibrating system (5.12) is called *bounded relative to an excitation function* $b(t)$ if, for this function $b(t)$ and any bounded initial condition $\|x_0\| < \delta$, the corresponding solution $x(t)$ is bounded for all time values $t \geqslant 0$,

$$\|x(t)\| \leqslant \epsilon = \epsilon(\delta, b(t)). \tag{5.16}$$

If the vibrating system (5.12) is not bounded (in the sense of (5.15) or (5.16)), then it is called *generally unbounded* or *unbounded relative to an excitation function.*

Obviously the requirements in the definition of general boundedness of a system (5.12) are more rigorous than those in the definition of boundedness relative to a special excitation function. This difference will be also seen during the discussion of the boundedness criteria in Section 5.3.

In engineering applications only the bounded, stable, or asymptotically stable systems are worth to be considered. Unbounded or unstable behavior jeopardizes the functioning of the vibrating system and leads mostly to its destruction. Thus it is essential to have at one's disposal criteria allowing verification of stability and boundedness, without having to calculate the whole variety of trajectories underlying the definitions

5.2. Stability

5.2.1. Stability criteria by eigenvalues

Applying the stability definitions (5.7 to 5.9) to a linear system (5.6) one can characterize its stability behavior simply by means of the fundamental matrix (4.11). Since the solution has the form $\boldsymbol{x}(t)=\boldsymbol{\Phi}(t)\boldsymbol{x}_0$, the requirements on $\boldsymbol{x}_0$ and $\boldsymbol{x}(t)$ can be expressed by properties of $\boldsymbol{\Phi}(t)$. For this purpose it is expedient to define the norm $\|\boldsymbol{\Phi}(t)\|$ of the matrix $\boldsymbol{\Phi}(t)$ by analogy with (5.1 to 5.5). Here care must be taken to assure compatibility of the chosen matrix and vector norms, so that for all $\boldsymbol{x}_0$

$$\|\boldsymbol{\Phi}(t)\boldsymbol{x}_0\| \leqslant \|\boldsymbol{\Phi}(t)\| \cdot \|\boldsymbol{x}_0\| \tag{5.17}$$

holds. Corresponding to (5.2 to 5.5) such compatible matrix norms are given by

$$\|\boldsymbol{\Phi}(t)\|_E = \sqrt{\operatorname{tr} \boldsymbol{\Phi}^T(t)\boldsymbol{\Phi}(t)}\,, \tag{5.18}$$

$$\|\boldsymbol{\Phi}(t)\|_R = \sqrt{\operatorname{tr} \boldsymbol{\Phi}^T(t)\boldsymbol{R}\boldsymbol{\Phi}(t)}\,, \qquad \boldsymbol{R}=\boldsymbol{R}^T>\boldsymbol{0}, \tag{5.19}$$

$$\|\boldsymbol{\Phi}(t)\|_m = n \max_{i,j=1,\dots,n} |\Phi_{ij}(t)|, \tag{5.20}$$

$$\|\boldsymbol{\Phi}(t)\|_b = \max_{j=1,\dots,n} \sum_{i=1}^{n} |\Phi_{ij}(t)|. \tag{5.21}$$

By any of these norms one gets for a bounded fundamental matrix

$$\|\boldsymbol{\Phi}(t)\| \leqslant c, \qquad 0<c<\infty, \qquad t\geqslant 0, \tag{5.22}$$

that $\|\boldsymbol{x}_0\|<\delta$ implies in view of (4.33), (5.17), and (5.22) the estimate $\|\boldsymbol{x}((t)\|<\epsilon=\delta c$; the system (5.6) is then stable. Conversely, if $\boldsymbol{\Phi}(t)$ is not bounded, then at least one of the elements, say $\boldsymbol{\Phi}_{ij}(t)$ is not bounded, and the solution $\boldsymbol{x}(t)$ with $\boldsymbol{x}_0=x_{0j}\boldsymbol{e}_j$ is also unbounded. This can be shown by (5.5):

$$\|\boldsymbol{x}(t)\|_b = \sum_{k=1}^{n} |x_k(t)| = \sum_{k=1}^{n} \|\Phi_{kj}(t)x_{0j}| = \sum_{k=1}^{n} |\Phi_{kj}(t)\|x_{0j}.$$

$$\geqslant |\Phi_{ij}(t)|\,|x_{0j}|.$$

It follows that the system (5.6) is then unstable.

This shows that the boundedness (5.22) of the fundamental matrix is a necessary and sufficient condition for the stability of the system (5.6). Moreover, the system (5.6) is asymptotically stable if additionally

$$\lim_{t \to \infty} \boldsymbol{\Phi}(t) = \mathbf{0} \tag{5.23}$$

holds. This second condition follows immediately from (4.33) and (5.9).

After having uniquely characterized the stability behavior of (5.6) by the fundamental matrix $\boldsymbol{\Phi}(t)$, the investigation of the stability problem can be concentrated on the modal fundamental matrix $e^{\boldsymbol{J}t}$ due to (4.90). The modal matrix $\boldsymbol{X}$ is a constant regular matrix and thus does not essentially influence the properties (5.22, 5.23). According to Section 4.3, $e^{\boldsymbol{J}t}$ equals in Case I ($d_i = v_i$ for all eigenvalues λ_i) the diagonal matrix (4.77), while in Case II ($d_i < v_i$ for at least one eigenvalue λ_i) $e^{\boldsymbol{J}t}$ contains, by (4.88), at least one submatrix (4.89) with exponential functions weighted by time powers. The relation (4.88) allows a characterization of the stability behavior of the system (5.6) by means of the eigenvalues λ_i:

The linear time-invariant homogeneous system $\dot{\boldsymbol{x}}(t) = \boldsymbol{A}\boldsymbol{x}(t)$ is

asymptotically stable if and only if all eigenvalues of $\boldsymbol{A}$ have negative real parts,

$$\text{Re}\,\lambda_i < 0, \qquad i = 1(1)n; \tag{5.24}$$

marginally stable if there are no eigenvalues of $\boldsymbol{A}$ with positive real parts and there exists at least one eigenvalue with a vanishing real part, moreover the multiplicity v_j of each eigenvalue with a vanishing real part equals the nullity d_j of the corresponding characteristic matrix $(\lambda_j \boldsymbol{E} - \boldsymbol{A})$:

$$\text{Re}\,\lambda_i \leqslant 0,\ i = 1(1)n,\ \text{Re}\,\lambda_j = 0 : d_j = v_j; \tag{5.25}$$

unstable if at least one eigenvalue of $\boldsymbol{A}$ has a positive real part or there exists at least one multiple eigenvalue with a vanishing real part whose multiplicity v_j exceeds the nullity d_j of its characteristic matrix $(\lambda_j \boldsymbol{E} - \boldsymbol{A})$:

$$\text{Re}\,\lambda_i > 0 \text{ for one } i \text{ or } \text{Re}\,\lambda_j = 0,\ v_j > d_j \text{ for one } j. \tag{5.26}$$

These stability conditions follow from (4.88, 4.89). The dependence of the stability behavior on eigenvalues λ_i with $\text{Re}\,\lambda_i < 0$ or $\text{Re}\,\lambda_i > 0$ is clearly related to the growing or decaying of $|e^{\lambda_i t}| = e^{\text{Re}\,\lambda_i t}$ in the solution (4.88, 4.89). Different cases have to be discussed only for the critical eigenvalues λ_j with $\text{Re}\,\lambda_j = 0$. If in Case I for some λ_j with $\text{Re}\,\lambda_j = 0$ one has $d_j = v_j$ then the corresponding Jordan block in (4.86) is a diagonal matrix and no secular terms (i.e. terms of the form $e^{\lambda_j t} t^i$, $i = 1(1)r_{jk}$) appear in (4.89); the natural vibrations remain bounded and thus stable. On the other hand, if in Case II one has $d_j < v_j$, then there exists a Jordan submatrix (4.86b) with nonvanishing elements above the diagonal and the corresponding solutions according to (4.89). Due to the presence of the secular terms $e^{\lambda_j t} t^i (\text{Re}\,\lambda_j = 0)$ these solutions are not bounded and thus unstable. Hence the dynamic problem of stability has been reduced to the algebraic problem of the eigenvalue distribution of a matrix. This problem can be solved either by a numerical calculation of the eigenvalues (Section 14.3) or by consideration of special stability criteria. They will be treated in the subsequent sections. In the discussion, two classes of stability criteria will be distinguished: The first class contains the criteria based on the characteristic polynomial (Section 5.2.2), while the criteria of the second class are derived from the Ljapunov matrix equation (Section 5.2.3). The second class allows the derivation of simple stability criteria for mechanical vibrating systems (3.10) (Section 5.2.4).

Example 5.1. Vertical vibrations of a motor car

In Example 4.3 the eigenvalues (4.103) of the system (4.99) were calculated:

$$\lambda_{1,2} = -30.253 \pm 57.003\,\mathrm{i},$$

$$\lambda_{3,4} = -1.747 \pm 4.578\,\mathrm{i}.$$

Since $\text{Re}\,\lambda_i < 0$, $i = 1(1)4$, it follows from (5.24) that the system is asymptotically stable. ◆

Example 5.2. Magnetically levitated vehicle

In Example 4.5 the eigenvalue problem was solved for a special design (4.120) of the linear voltage control (3.49) of the electromagnet carrying a magnetically levitated vehicle (Fig. 3.2). The system (4.119) is asymptotically stable since its eigenvalues (4.123) satisfy (5.24). ◆

Example 5.3. Double pendulum

The eigenvalues of the linearized equations of motion of a double pendulum were calculated in Example 4.6. Recalling (4.137),

$$\lambda_{1,2,3,4} = \pm \mathrm{i}\omega_{1,2}, \qquad \omega_{1,2} = \sqrt{\frac{g}{L}(2 \pm \sqrt{2})},$$

one gets

$$\text{Re}\,\lambda_i = 0, \qquad d_i = v_i = 1, \qquad i = 1(1)4.$$

Hence, from (5.25), the motions of a double pendulum are marginally stable. ◆

Example 5.4. Centrifuge

The eigenvalues for a centrifuge with $\kappa = 0$ were determined in Problem 4.1:

$$\lambda_{1,2} = 0, \qquad \lambda_{3,4} = -(\vartheta \pm \mathrm{i} g\Omega). \tag{5.27}$$

Since $\vartheta > 0$, one has Re $\lambda_{3,4} < 0$, however Re $\lambda_{1,2} = 0$ yields a critical case. It is necessary to determine the nullity $d_{1,2}$ of $(\lambda_{1,2}\boldsymbol{E} - \boldsymbol{A})$ for the system matrix $\boldsymbol{A}$ according to (4.151):

$$\begin{aligned} d_{1,2} &= n - \text{rank}(\lambda_{1,2}\boldsymbol{E} - \boldsymbol{A}) \\ &= 4 - \text{rank}(-\boldsymbol{A}) = 4 - \text{rank}\left[\begin{array}{c|c} \mathbf{0} & -\boldsymbol{E} \\ \hline \mathbf{0} & \vartheta\boldsymbol{E} + g\Omega\boldsymbol{S} \end{array}\right] \\ &= 4 - 2 = 2. \end{aligned} \tag{5.28}$$

Thus the characteristic matrix corresponding to the double eigenvalue $\lambda_{1,2}$ $(v = 2)$ has nullity two $(d = 2)$. According to (5.25) the system is marginally stable. The stable behavior can be read from the general solution $y(t)$ which is the second answer to question f) in Problem 4.1.

Moreover, if $\lambda_{3,4} = 0$, one gets an eigenvalue with multiplicity four and, by (5.28), the corresponding nullity is two. According to (5.26), the centrifuge is then unstable. ◆

5.2.2. Stability criteria by characteristic coefficients

It was shown is Section 5.2.1 that the stability problem for a dynamic system (5.6) is totally answered by the distribution of the eigenvalues of the system matrix $\boldsymbol{A}$. These eigenvalues in turn are uniquely defined as the roots of the characteristic polynomial

$$p(\lambda) \equiv \lambda^n + a_1\lambda^{n-1} + a_2\lambda^{n-2} + \cdots + a_{n-1}\lambda + a_n. \tag{5.29}$$

If the characteristic coefficients a_i, $i = 1(1)n$, of (5.29) are known, then indirectly the distribution of the eigenvalues is also known. Thus, the relations between the eigenvalues and the characteristic coefficients are used in order to derive conditions on stability from the knowledge of the coefficients a_i without calculating explicitly the eigenvalues.

As a first step, one can set up necessary conditions for stability from Stodola which dates back to the years 1893/94. A polynomial (5.29) with eigenvalues λ_i, $i = 1(1)n$, allows a factorization

$$p(\lambda) = (\lambda - \lambda_1)(\lambda - \lambda_2)\ldots(\lambda - \lambda_n). \tag{5.30}$$

If $\lambda_i = \delta_i$ is real, then $(\lambda - \lambda_i) = (\lambda - \delta_i)$ is a real linear factor. But, if $\lambda_i = \delta_i + \mathrm{i}\omega_i$ is complex, then the corresponding complex linear factor taken together with its complex conjugate yields a real quadratic factor,

$$\begin{aligned}(\lambda - \lambda_i)(\lambda - \lambda_{i+1}) &= (\lambda - \lambda_i)(\lambda - \bar{\lambda}_i) \\ &= (\lambda - \delta_i - \mathrm{i}\omega_i)(\lambda - \delta_i + \mathrm{i}\omega_i) \\ &= \lambda^2 - 2\delta_i\lambda + \left(\delta_i^2 + \omega_i^2\right). \end{aligned} \tag{5.31}$$

Thus one arrives at a real factorization of $p(\lambda)$,

$$p(\lambda) = \prod_{i=1}^{k} (\lambda - \delta_i) \prod_{i=k+1}^{k+(n-k)/2} \left(\lambda - 2\delta_i\lambda + \delta_i^2 + \omega_i^2\right), \tag{5.32}$$

where k is the number of real and $(n-k)$ the number of complex conjugate eigenvalues. If the system (5.6) is asymptotically stable, i.e. $\delta_i < 0$, then it is seen from (5.32) that $p(\lambda)$ is a product of real polynomials with positive coefficients. Hence in this case all the coefficients a_i, $i = 1(1)n$, must also be positive. This yields the

Stodola criterion:

A necessary condition for the asymptotic stability of (5.6) is that all coefficients of the characteristic polynomial (5.29) are positive,

$$a_i > 0, \qquad i = 1(1)n. \tag{5.33}$$

If $n = 2$, then this criterion is also sufficient, as can be seen by analysing the roots

$$\lambda_{1,2} = -\tfrac{1}{2}a_1 \pm \sqrt{\tfrac{1}{4}a_1^2 - a_2} \tag{5.34a}$$

of the equation

$$\lambda^2 + a_1\lambda + a_2 = 0. \tag{5.34b}$$

However for systems of higher orders, the condition (5.33) is only necessary as shown by the example

$$\left.\begin{aligned} p(\lambda) &= \lambda^3 + 8\lambda^2 + 81\lambda + 1010 \\ &= (\lambda + 10)(\lambda - 1 - 10\,\mathrm{i})(\lambda - 1 + 10\,\mathrm{i}), \\ \lambda_1 &= -10, \qquad \lambda_{2,3} = 1 \pm 10\,\mathrm{i}. \end{aligned}\right\} \tag{5.35}$$

Thus it is necessary to look for further conditions on the coefficients a_i in order to obtain *sufficient* conditions on stability.

The first result in this field is from Hermite (1856), however the later proposed criteria by Routh (1877) and Hurwitz (1895) are more frequently used. These last two criteria which are both *necessary and sufficient* are given here without proof.

Routh criterion:

A necessary and sufficient condition for the asymptotic stability of (5.6) is that all Routh numbers R_i are positive:

$$R_i > 0, \qquad i = 1(1)n. \tag{5.36}$$

Here the Routh numbers R_i are calculated from the Routh scheme

$$\begin{array}{c|l|llll}
 & 1 & a_2 & a_4 & a_6 & \cdots \\
 & R_1 = a_1 & a_3 & a_5 & a_7 & \cdots \\
\hline
r_2 = \dfrac{1}{R_1} & R_2 = c_{21} = a_2 - r_2 a_3 & c_{22} = a_4 - r_2 a_5 & c_{23} = a_6 - r_2 a_7 & \cdots & \\
r_3 = \dfrac{R_1}{R_2} & R_3 = c_{31} = a_3 - r_3 c_{22} & c_{32} = a_5 - r_3 c_{23} & \cdots & \cdots & \\
\vdots & \quad\vdots & \quad\vdots & & & \\
r_j = \dfrac{R_{j-2}}{R_{j-1}} & R_j = c_{j1} & c_{jk} = c_{j-2,k+1} - r_j c_{j+1,k+1} & & & \\
\vdots & \quad\vdots & & & & \\
r_n = \dfrac{R_{n-2}}{R_{n-1}} & R_n = c_{n1} = a_n & & & &
\end{array} \tag{5.37}$$

The first two rows are given by the coefficients of the characteristic polynomial (5.29). The other quantities of this scheme are calculated line by line, so that the elements of the j-th row are obtained from the elements of the (j-1)-st and (j-2)-nd row in the next right adjoining column, by means of the recursion formula using the auxiliary quantity r_j given in (5.37). This computational scheme terminates either at the j-th row if $R_j = c_{j1} = 0$, or it continues until $R_{n-1} = c_{n-1,1}$, since $R_n = c_{n1} =$

a_n. For proving asymptotic stability one can stop the calculation as soon as one gets a negative value for $R_j = c_{j1}$.

Independently of Routh, Hurwitz gave another solution to the stability problem.

Hurwitz criterion:

A necessary and sufficient condition for the asymptotic stability of the system (5.6) is that all Hurwitz determinants H_i are positive:

$$H_i > 0, \qquad i = 1(1)n. \tag{5.38}$$

Here the numbers H_i are the main principal minors of the $n \times n$ Hurwitz matrix

$$\boldsymbol{H} = \begin{bmatrix} a_1 & 1 & 0 & 0 & 0 & 0 & \dots & 0 \\ a_3 & a_2 & a_1 & 1 & 0 & 0 & \dots & 0 \\ a_5 & a_4 & a_3 & a_2 & a_1 & 1 & \dots & 0 \\ \vdots & & & & & & & \vdots \\ 0 & \dots & \dots & \dots & \dots & 0 & & a_n \end{bmatrix}, \tag{5.39}$$

i.e.

$$\left.\begin{aligned} &H_1 = a_1, \qquad H_2 = \det\begin{bmatrix} a_1 & 1 \\ a_3 & a_2 \end{bmatrix} = a_1 a_2 - a_3, \\ &H_3 = \det\begin{bmatrix} a_1 & 1 & 0 \\ a_3 & a_2 & a_1 \\ a_5 & a_4 & a_3 \end{bmatrix} = a_3 H_2 - a_1(a_1 a_4 - a_5), \\ &\dots \\ &H_n = a_n H_{n-1} = \det \boldsymbol{H}. \end{aligned}\right\} \tag{5.40}$$

The first two columns of the Hurwitz matrix (5.39) are formed from the coefficients of the characteristic polynomial (5.29). The further columns are found according to the following rule: The k-th columns is obtained from the $(k-2)$-nd column by lowering all the elements by one position and filling the top vacancy with the element 0.

If the Hurwitz determinants do not vanish, the relation between the Routh and Hurwitz criteria is given by

$$R_i = \frac{H_i}{H_{i-1}}, \qquad H_0 = 1, \qquad i = 1(1)n. \tag{5.41}$$

It is characteristic for both criteria (5.36) and (5.38) that they make no use of the necessary conditions (5.33). Since for systems of large order the calculation of the quantities R_i or H_i is quite laborious, any stability criterion issuing from (5.33) and avoiding the calculation of certain R_i or H_i offers a great simplification of stability proofs. The first results in this direction are due to Liénard and Chipart (1914) whose criterion, with due regard to later developments, can be stated as follows.

Liénard-Chipart criterion:

A necessary and sufficient condition for the asymptotic stability of (5.6) is that the following conditions are satisfied:

$$a_n > 0,\ H_{n-1} > 0,\ a_{n-2} > 0,\ H_{n-3} > 0,\ \ldots,\ H_1 = a_1 > 0. \tag{5.42}$$

Using this criterion, one needs to calculate only every second Hurwitz determinant.

For systems of lower order, the n inequalities (5.36) of the Routh criterion as well as the n inequalities (5.38) or (5.42) of the Hurwitz or the Liénard-Chipart criterion can be conveniently used for establishing the analytical dependence of the stability region from the system parameters. For higher order systems one generally encounters numerical difficulties. These can be avoided if one formulates the stability criteria by means of the "inners" of Jury. The definitions of the "inners", the corresponding stability criteria, and algorithms for their calculation are given in the book by Jury.

In the estimation of stability regions the idea of a *critical stability condition* is of great importance. Varying the parameters of a system matrix to establish the boundaries of the region of asymptotic stability in the parameter space, one moves from the region of asymptotic stability to the stability boundary. Here at least one of the stability conditions (5.36) or (5.38) must fail to hold. This condition is characterized by at least one eigenvalue with a vanishing real part. Since $H_n = a_n H_{n-1}$ (5.40) and $a_n = (-1)^n \prod_{i=1}^{n} \lambda_i$ (4.59) the formula from Orlando relating the eigenvalues to H_{n-1} can be used to obtain

$$H_n = (-1)^{n(n+1)/2} \frac{1}{2^n} \prod_{i \leqslant k}^{1,\ldots,n} (\lambda_i + \lambda_k). \tag{5.43}$$

Thus the n-th Hurwitz determinant vanishes precisely when for some eigenvalue λ_i with $p(\lambda_i) = 0$ also $p(-\lambda_i) = 0$ holds. This fact is an exact characterization for the eigenvalues on the stability boundary. Combining these observations with (5.43) one can make the following statement.

Critical stability boundary:

If the parameters of a system (5.6) are varied beginning within the region of asymptotic stability, then reaching the stability boundary is equivalent to the condition

$$H_n = a_n H_{n-1} = 0. \tag{5.44}$$

Here $a_n = 0$ characterizes the monotone marginal stability ($\lambda_i = 0$) and $H_{n-1} = 0$ the oscillatory marginal stability ($\lambda_j = i\omega$, $\omega \neq 0$).

Example 5.5. Magnetically levitated vehicle

The magnetically levitated vehicle with the voltage control (3.49, 4.120) can be described by the system of differential equations (4.119, 4.120). The system matrix (4.121) has the characteristic polynomial (4.122) with the coefficients

$$a_1 = \frac{1}{T}, \qquad a_2 = \frac{5}{16T^2}, \qquad a_3 = \frac{1}{32T^3}. \tag{5.45}$$

The necessary conditions of Stodola (5.33) for asymptotically stable behavior are satisfied. The stability test according to the Routh criterion yields:

	1	$\frac{5}{16T^2}$
	$R_1 = \frac{1}{T}$	$\frac{1}{32T^3}$
$r_2 = T$	$R_2 = \frac{9}{16T^2}$	0
$r_3 = \frac{16T}{9}$	$R_3 = \frac{1}{32T^3}$	0

Routh-scheme

The Routh numbers R_i, $i = 1, 2, 3$, are all positive, hence the system is asymptotically stable. Of course with the Hurwitz criterion (5.38) the same result is obtained. By (5.40), one has

$$\left.\begin{aligned} &H_1 = a_1 = \frac{1}{T} > 0, \quad H_2 = a_1 a_2 - a_3 = \frac{9}{16T^3} > 0, \\ &H_3 = a_3 H_2 = \frac{9}{512T^6} > 0. \end{aligned}\right\} \tag{5.46}$$

It is also easy to confirm the relations (5.41):

$$R_1 = H_1 = a_1 = \frac{1}{T}, \qquad R_2 = \frac{H_2}{H_1} = \frac{9}{16T^2},$$

$$R_3 = \frac{H_3}{H_2} = a_3 = \frac{1}{32T^3}.$$

The conditions of Liénard-Chipart are also satisfied:

$$a_3 > 0, \qquad H_2 > 0, \qquad a_1 > 0. \; \blacklozenge$$

Example 5.6. Double pendulum

For the double pendulum the characteristic equation (4.136) was derived with the coefficients

$$a_1 = 0, \quad a_2 = 4\frac{g}{L}, \quad a_3 = 0, \quad a_4 = 2\left(\frac{g}{L}\right)^2. \tag{5.47}$$

Here the Stodola criterion (5.33) shows that because of $a_1 = a_3 = 0$ the motions of a double pendulum do not have an asymptotically stable behavior. Further, the Routh and Hurwitz criteria confirm this result with

$R_1 = 0$ (then the Routh scheme terminates),

$$H_1 = 0, \quad H_2 = 0, \quad H_3 = 0, \quad H_4 = 0. \tag{5.48}$$

However with only those criteria it cannot be determined whether marginal stability or instability takes place. ◆

Example 5.7. Vibrating system with elastic damper

The eigenvalue problem for the vibrating system with elastic damper shown in Fig. 3.1 was solved in Problem 4.3. The characteristic polynomial was of degree three and it had the coefficients

$$a_1 = \frac{b}{d}, \quad a_2 = \frac{b+k}{m}, \quad a_3 = \frac{bk}{md}. \tag{5.49}$$

If $b > 0$, $k > 0$, $d > 0$ and $m > 0$ as required by the technical problem then the necessary conditions (5.33) for asymptotical stability are satisfied. By the Liénard-Chipart criterion, to complete the verification of asymptotic stability, the condition

$$H_2 = a_1 a_2 - a_3 = \frac{b}{d}\frac{b+k}{m} - \frac{bk}{md} = \frac{b^2}{md} > 0$$

has to be checked. The example will be concluded by an analysis of the influence of the two spring constants b and k upon the critical stability boundary. By (5.44), the values

$$a_3 = 0 \quad \text{and} \quad H_2 = 0$$

are critical. If $k \to 0$, then $a_3 \to 0$, i.e. $k \to 0$ yields one eigenvalue $\lambda_1 \to 0$. But the other two eigenvalues retain their negative real parts. If $b \to 0$, then $a_3 \to 0$ as well as $H_2 \to 0$, i.e., in this case one has not only $\lambda_1 \to 0$, but in addition $\lambda_{2,3} \to \pm i\omega$. This means if $b \to 0$, then all three eigenvalues approach the stability boundary. ◆

Example 5.8. Centrifuge

The characteristic polynomial for the equations of motion (4.149) of a centrifuge (Fig. 3.4) was obtained in Problem 4.1,

$$p(\lambda) = \lambda^4 + 2\vartheta\lambda^3 + (2\kappa + \vartheta^2 + g^2\Omega^2)\lambda^2 + 2\vartheta\kappa\lambda + \kappa^2. \tag{5.50}$$

Thus the characteristic coefficients are

$$a_1 = 2\vartheta, \quad a_2 = 2\kappa + \vartheta^2 + g^2\Omega^2, \quad a_3 = 2\vartheta\kappa, \quad a_4 = \kappa^2. \tag{5.51}$$

The necessary stability conditions (5.33) are satisfies for $\vartheta > 0$, $\kappa > 0$. The Routh criterion (5.36) yields with the scheme

	1	$2\kappa + \vartheta^2 + g^2\Omega^2$	κ^2
	$R_1 = 2\vartheta$	$2\vartheta\kappa$	0
$r_2 = \dfrac{1}{2\vartheta}$	$R_2 = \kappa + \vartheta^2 + g^2\Omega^2$	κ^2	0
$r_3 = \dfrac{2\vartheta}{\kappa + \vartheta^2 + g^2\Omega^2}$	$R_3 = \dfrac{2\vartheta\kappa(\vartheta^2 + g^2\Omega^2)}{\kappa + \vartheta^2 + g^2\Omega^2}$	0	0
$r_4 = \dfrac{(\kappa + \vartheta^2 + g^2\Omega^2)^2}{2\vartheta\kappa(\vartheta^2 + g^2\Omega^2)}$	$R_4 = \kappa^2$	0	0

the conditions

$$\left.\begin{aligned} &R_1 = 2\vartheta > 0, \qquad && R_2 = \kappa + \vartheta^2 + g^2\Omega^2 > 0, \\ &R_3 = \frac{2\vartheta\kappa(\vartheta^2 + g^2\Omega^2)}{\kappa + \vartheta^2 + g^2\Omega^2} > 0, \qquad && R_4 = \kappa^2 > 0. \end{aligned}\right\} \tag{5.52}$$

The Hurwitz criterion (5.38) yields with the Hurwitz matrix (5.39)

$$\boldsymbol{H}=\begin{bmatrix} 2\vartheta & 1 & 0 & 0 \\ 2\vartheta\kappa & 2\kappa+\vartheta^2+g^2\Omega^2 & 2\vartheta & 1 \\ 0 & \kappa^2 & 2\vartheta\kappa & 2\kappa+\vartheta^2+g^2\Omega^2 \\ 0 & 0 & 0 & \kappa^2 \end{bmatrix} \tag{5.53}$$

the necessary and sufficient conditions

$$\left.\begin{aligned} &H_1=2\vartheta>0, && H_2=2\vartheta\left(\kappa+\vartheta^2+g^2\Omega^2\right)>0, \\ &H_3=4\vartheta^2\kappa\left(\vartheta^2+g^2\Omega^2\right)>0, && H_4=4\vartheta^2\kappa^3\left(\vartheta^2+g^2\Omega^2\right)>0. \end{aligned}\right\} \tag{5.54}$$

Evidently the least computational effort is required by the Liénard-Chipart criterion (5.42); it yields the following necessary and sufficient conditions for stability:

$$\left.\begin{aligned} &a_4=\kappa^2>0, && H_3=4\vartheta^2\kappa\left(\vartheta^2+g^2\Omega^2\right)>0, \\ &a_2=2\kappa+\vartheta^2+g^2\Omega^2>0, && H_1=a_1=2\vartheta>0. \end{aligned}\right\} \tag{5.55}$$

New requirements don't follow from all these different, but equivalent stability conditions (5.52), (5.54) and (5.55). The essential requirements

$$\vartheta>0. \qquad \kappa>0. \tag{5.56}$$

are retained. If the centrifuge is subject to damping ($\vartheta>0$) proportional to velocity and to a statically stable attachment ($\kappa>0$) to its equilibrium position, then its motion is asymptotically stable.

If $\kappa\to 0$, one gets for the critical stability boundary $a_4\to 0$ and $H_3\to 0$. This implies (because of $a_1>0$, $a_2>0$) that for $\kappa\to 0$ there appears at least one double eigenvalue. Thus, by (5.25, 5.26), it is possible that instability and monotonous drift occur in the centrifuge system already on the stability boundary $\kappa=0$. But the investigation in Example 5.4 shows that, in fact, one has here marginal stability. The centrifuge axis can deviate from the vertical by its initial value $\boldsymbol{y}_0=[\alpha_0 \quad \beta_0]^T$.

If $\vartheta\to 0$, then $H_3\to 0$. The centrifuge becomes marginally stable with the eigenvalues

$$\lambda_{1,2}=\mathrm{i}\left(-\tfrac{1}{2}g\Omega\pm\sqrt{\tfrac{1}{4}g^2\Omega^2+\kappa}\right), \quad \lambda_{3,4}=\bar{\lambda}_{1,2} \tag{5.57}$$

(see Problem 4.1). For $\vartheta=0$ and $\kappa>-\frac{1}{4}g^2\Omega^2$ two distinct pairs of imaginary eigenvalues are obtained.

If at the same time $\vartheta=0$ and $\kappa=0$, then the two previously obtained boundary cases are superimposed one upon another. For $g\Omega \neq 0$ one gets with

$$\lambda_{1,2} = 0, \quad \lambda_{3,4} = \pm \mathrm{i} g\Omega \tag{5.58}$$

according to Example 5.4, a system which is marginally stable and vibrating with frequency $\omega = g\Omega$ about the shifted equilibrium position $y_0 = [\alpha_0 \quad \beta_0]^T$. Only when $g\Omega = 0$, i.e. if there is no rotation, a monotonous drift of the centrifuge axis can occur and the system is unstable.

The discussion in this example shows that the stability criteria can serve to determine the stability conditions and the stability boundaries, but more precise investigations are always required to determine the behavior of a dynamic system on the stability boundary. E.g., the eigenvalues for $\kappa = 0$ and for $\vartheta = 0$ can serve this purpose. ◆

5.2.3. Stability criteria by the Ljapunov matrix equation

The distribution of the eigenvalues of a matrix can be characterized not only by the characteristic coefficients a_i, $i = 1(1)n$, but also by the properties of the solutions of the *algebraic Ljapunov matrix equation*

$$\boldsymbol{A}^T\boldsymbol{R} + \boldsymbol{R}\boldsymbol{A} = -\boldsymbol{S}. \tag{5.59}$$

To show the importance of (5.59), the stability problem as defined in (5.7 to 5.9) will be treated by means of the vector norm (5.3):

$$\|\boldsymbol{x}(t)\|_{\boldsymbol{R}}^2 = \boldsymbol{x}^T(t)\boldsymbol{R}\boldsymbol{x}(t), \quad \boldsymbol{R} = \boldsymbol{R}^T > \boldsymbol{0}. \tag{5.60}$$

If the system (5.6) is stable, then there must exist a positive definite symmetric matrix $\boldsymbol{R}$ such that

$$\|\boldsymbol{x}(t)\|_{\boldsymbol{R}}^2 < \epsilon^2 \tag{5.61}$$

provided

$$\|\boldsymbol{x}_0\|_{\boldsymbol{R}}^2 < \delta^2 = \delta^2(\epsilon) \tag{5.62}$$

holds. Considering the variation in time of $\|\boldsymbol{x}(t)\|_{\boldsymbol{R}}^2$ along a trajectory, one gets

$$\begin{aligned} \frac{\mathrm{d}}{\mathrm{d}t}\|\boldsymbol{x}(t)\|_{\boldsymbol{R}}^2 &= \dot{\boldsymbol{x}}^T(t)\boldsymbol{R}\boldsymbol{x}(t) + \boldsymbol{x}^T(t)\boldsymbol{R}\dot{\boldsymbol{x}}(t) \\ &= \boldsymbol{x}^T(t)(\boldsymbol{A}^T\boldsymbol{R} + \boldsymbol{R}\boldsymbol{A})\boldsymbol{x}(t). \end{aligned} \tag{5.63}$$

If it is possible to find $\boldsymbol{R}$ so that the time derivative (5.63) is nonpositive for all times $t \geqslant 0$,

$$\frac{\mathrm{d}}{\mathrm{d}t}\|\boldsymbol{x}(t)\|_{\boldsymbol{R}}^{2} \leqslant 0, \tag{5.64}$$

then, taking $\delta = \epsilon$, one sees that (5.62) implies (5.61), which means that the system (5.6) is stable. In view of the quadratic form in (5.63), condition (5.64) can be satisfied by

$$\frac{\mathrm{d}}{\mathrm{d}t}\|\boldsymbol{x}(t)\|_{\boldsymbol{R}}^{2} = -\boldsymbol{x}^{T}(t)\boldsymbol{S}\boldsymbol{x}(t), \quad \boldsymbol{S} = \boldsymbol{S}^{T} \geqslant \boldsymbol{0}. \tag{5.65}$$

Thus (5.63) and (5.65) imply the relation

$$\boldsymbol{x}^{T}(t)[\boldsymbol{A}^{T}\boldsymbol{R} + \boldsymbol{R}\boldsymbol{A}]\boldsymbol{x}(t) = -\boldsymbol{x}^{T}(t)\boldsymbol{S}\boldsymbol{x}(t). \tag{5.66}$$

Since (5.66) should be valid for all trajectories $\boldsymbol{x}(t)$, it follows that (5.66) must also be satisfied for the matrices and thus, precisely Ljapunov's equation (5.59) is found. Hence, if there exists for a symmetric positive semidefinite matrix $\boldsymbol{S} = \boldsymbol{S}^{T} \geqslant \boldsymbol{0}$ a symmetric positive definite matrix $\boldsymbol{R} = \boldsymbol{R}^{T} > \boldsymbol{0}$ solving (5.59), then the stability inequalities (5.61) and (5.62) hold for $\delta = \epsilon$, i.e., the dynamic system (5.6) is stable.

Moreover, if the time derivative (5.65) does not vanish identically along any trajectory,

$$\frac{\mathrm{d}}{\mathrm{d}t}\|\boldsymbol{x}(t)\|_{\boldsymbol{R}}^{2} \not\equiv 0, \qquad t \geqslant 0, \tag{5.67}$$

then $\|\boldsymbol{x}(t)\|_{\boldsymbol{R}}^{2} = \boldsymbol{x}^{T}(t)\boldsymbol{R}\boldsymbol{x}(t)$ is decreasing as t grows. On the other hand, by (5.1a, b), it is

$$\|\boldsymbol{x}(t)\|_{\boldsymbol{R}}^{2} \geqslant 0. \tag{5.68}$$

Thus $\|\boldsymbol{x}(t)\|_{\boldsymbol{R}}^{2}$ decreases to its minimum value as $t \to \infty$,

$$\lim_{t\to\infty} \|\boldsymbol{x}(t)\|_{\boldsymbol{R}}^{2} = 0. \tag{5.69}$$

But then (5.1a) implies

$$\lim_{t\to\infty} \boldsymbol{x}(t) = \boldsymbol{0},$$

i.e., the additional condition (5.9) for asymptotic stability is satisfied.

Asymptotic stability differs from stability by the additional requirement (5.67). This can be satisfied e.g. by using a symmetric positive definite matrix $S = S^T > 0$. However for many mechanical vibration problems the practical stability proof is considerably more difficult by this approach (Section 5.2.4). Thus it is frequently more convenient to retain the requirement (5.65) with $S = S^T \geqslant 0$ and to check that (5.67) holds. This can be done by applying the observability condition

$$\operatorname{rank}\left[S \mid A^T S \mid A^{T^2} S \mid \ldots \mid A^{T^{n-1}} S\right] = n. \tag{5.70}$$

(There will be some hints in Chapter 12 about the problem of observability and about further possibilities of verifying (5.70)). This requirement can be explained by the following considerations:

Suppose there exists at least one trajectory $x(t) \not\equiv 0$ for which (5.67) does not hold, i.e.,

$$x^T(t) S x(t) \equiv 0. \tag{5.71}$$

Since $S = S^T \geqslant 0$, one also has

$$x^T(t) S \equiv 0. \tag{5.72}$$

Repeated differentiation of (5.72) with respect to time along this trajectory leads to

$$x^T(t) A^{T^k} S \equiv 0, \qquad k = 0(1)\infty. \tag{5.73}$$

Because of the theorem (4.130) of Cayley and Hamilton only powers $k = 0(1)n-1$ are of interest in (5.73). Thus another way to state (5.73) is given by

$$x^T(t)\left[S \mid A^T S \mid A^{T^2} S \mid \ldots \mid A^{T^{n-1}} S\right] \equiv 0. \tag{5.74}$$

Hence if there exists a trajectory with the property (5.71), then (5.74) implies that the observability condition (5.70) cannot hold.

Having made plausible the equivalence of the conditions (5.67) and (5.70), one can collect the results of this section in two theorems about stability and asymptotical stability. An instability theorem will also be included, but its proof will be omitted.

Asymptotical stability:

The linear time-invariant system (5.6) is asymptotically stable iff there exists for at least one (and then for each) symmetric positive semidefinite matrix $S = S^T \geqslant 0$, satisfying the observability condition (5.70), a unique symmetric positive definite matrix $R = R^T > 0$ solving the Ljapunov equation (5.59).

Marginal stability:

The linear time-invariant system (5.6) is marginally stable iff there exists for at least one positive semidefinite matrix $\boldsymbol{S}=\boldsymbol{S}^T \geqslant \mathbf{0}$, which does not satisfy (5.70), a (not necessarily unique) symmetric positive definite matrix $\boldsymbol{R}=\boldsymbol{R}^T > \mathbf{0}$ solving Ljapunov's equation (5.59).

Instability:

The linear time-invariant system (5.6) is marginally stable or unstable when for at least one symmetric positive semidefinite matrix $\boldsymbol{S}=\boldsymbol{S}^T \geqslant \mathbf{0}$ there exists a (not necessarily unique) symmetric matrix $\boldsymbol{R}=\boldsymbol{R}^T$, which satisfies the Ljapunov equation (5.59), such that for at least one observable state $\boldsymbol{x} \neq \mathbf{0}$,

$$\boldsymbol{x}=\left[\boldsymbol{S} \,\vdots\, \boldsymbol{A}^T\boldsymbol{S} \,\vdots\, \dots \,\vdots\, \boldsymbol{A}^{T^{n-1}}\boldsymbol{S}\right]\boldsymbol{\mathfrak{X}} \tag{5.75}$$

with an arbitrary $n^2 \times 1$ vector $\boldsymbol{\mathfrak{X}}$ the inequality

$$\boldsymbol{x}^T\boldsymbol{R}\boldsymbol{x} < 0 \tag{5.76}$$

holds.

Example 5.9. Force-free particle

Ljapunov's equation (5.59) will be used to investigate the stability behaviour of a force-free particle with the scalar differential equation

$$\ddot{y}(t)=0. \tag{5.77}$$

The state space representation corresponding to (5.77) is

$$\boldsymbol{x}=\begin{bmatrix} y \\ \dot{y} \end{bmatrix}, \quad \dot{\boldsymbol{x}}(t)=\begin{bmatrix} 0 & 1 \\ 0 & 0 \end{bmatrix}\boldsymbol{x}(t)=\boldsymbol{A}\boldsymbol{x}(t). \tag{5.78}$$

For the matrices

$$\boldsymbol{R}=\begin{bmatrix} R_{11} & R_{12} \\ R_{12} & R_{22} \end{bmatrix}, \quad \boldsymbol{S}=\begin{bmatrix} S_{11} & S_{12} \\ S_{12} & S_{22} \end{bmatrix} \tag{5.79}$$

one has, by (5.59),

$$0=-S_{11}, \qquad R_{11}=-S_{12}, \qquad 2R_{12}=-S_{22}. \tag{5.80}$$

For (5.59) to have a solution, $S_{11}=0$ is necessary. Then the matrix $\boldsymbol{S}$ cannot be chosen positive definite, in the best case positive semidefinite, with $S_{12}=0$ and $S_{22}>0$. Hence (5.80) implies

$$R_{11}=0, \qquad R_{12}=-\tfrac{1}{2}S_{22}<0, \qquad R_{22} \text{ arbitrary}. \tag{5.81}$$

Since the matrix R with the elements (5.81) certainly cannot be positive definite, (5.77) is unstable. The relations (5.75) and (5.76) are satisfied e.g. when $\mathfrak{X} = [0\ 1\ 0\ 0]^T$, $\boldsymbol{x} = [0\ S_{22}]^T$ and $\boldsymbol{x}^T\boldsymbol{R}\boldsymbol{x} = S_{22}^2 R_{22} < 0$ for $R_{22} < 0$ holds. ◆

This simple example shows that Ljapunov's equation does not always have a solution. The properties of (5.59), especially the existence and uniqueness assertions as well as the analytical and numerical procedures will be collected in Sections 13.2 and 14.2. Ljapunov's equation (5.59) will play also an essential role in the covariance analysis of random vibrations (Section 9.4).

Example 5.10. Double pendulum

The state equations (4.135) of a double pendulum (Fig. 3.3) are

$$\dot{\boldsymbol{x}}(t) = \boldsymbol{A}\boldsymbol{x}(t) = \left[\begin{array}{c|c} 0 & \boldsymbol{E} \\ \hline -\boldsymbol{K} & \boldsymbol{0} \end{array}\right]\boldsymbol{x}(t), \quad \boldsymbol{K} = \frac{g}{L}\begin{bmatrix} 3 & -1 \\ -1 & 1 \end{bmatrix}. \tag{5.82}$$

Ljapunov's equation (5.59) is satisfied with $\boldsymbol{S} = \boldsymbol{0}$ and

$$\boldsymbol{R} = \left[\begin{array}{c|c} \boldsymbol{K} & \boldsymbol{0} \\ \hline \boldsymbol{0} & \boldsymbol{E} \end{array}\right]. \tag{5.83}$$

By the theorem on marginal stability, (5.82) is certainly marginally stable when $\boldsymbol{R} > \boldsymbol{0}$ and thus $\boldsymbol{K} > \boldsymbol{0}$ as is the case here. Contrary to the stability theorems in Section 5.2.2 (Example 5.6), Ljapunov's equation asserts here that the motions of the double pendulum are stable. ◆

Example 5.11. Centrifuge

Consider the centrifuge which was dealt with in Problems 3.1, 4.1, and in Example 5.8. The state space representation of its equations of motion is

$$\dot{\boldsymbol{x}}(t) = \left[\begin{array}{c|c} \boldsymbol{0} & \boldsymbol{E} \\ \hline -\kappa\boldsymbol{E} & -\vartheta\boldsymbol{E} - g\Omega\boldsymbol{S} \end{array}\right]\boldsymbol{x}(t). \tag{5.84}$$

Ljapunov's equation (5.59) holds for

$$\boldsymbol{R} = \left[\begin{array}{c|c} \kappa\boldsymbol{E} & \boldsymbol{0} \\ \hline \boldsymbol{0} & \boldsymbol{E} \end{array}\right], \qquad \boldsymbol{S}' = \left[\begin{array}{c|c} \boldsymbol{0} & \boldsymbol{0} \\ \hline \boldsymbol{0} & 2\vartheta\boldsymbol{E} \end{array}\right]. \tag{5.85}$$

If $\vartheta > 0$, $\kappa > 0$, one has $\boldsymbol{R} > \boldsymbol{0}$ and $\boldsymbol{S}' \geqslant \boldsymbol{0}$, so at least the system is marginally stable. Verification of the observability condition yields

$$\operatorname{rank}\left[\boldsymbol{S}' \,\vdots\, \boldsymbol{A}^T\boldsymbol{S}' \,\vdots\, \boldsymbol{A}^{T^2}\boldsymbol{S}' \,\vdots\, \boldsymbol{A}^{T^3}\boldsymbol{S}'\right]$$

$$= \operatorname{rank}\left\{2\vartheta\left[\begin{array}{c|c|c} \boldsymbol{0} & -\kappa\boldsymbol{E} & \cdots \\ \hline \boldsymbol{E} & -\vartheta\boldsymbol{E} - g\Omega\boldsymbol{S} & \cdots \end{array}\right]\right\} = 4 \tag{5.86}$$

for $\vartheta \neq 0$, $\kappa \neq 0$. Therefore the system is even asymptotically stable. For the boundary case $\vartheta = 0$, one finds from (5.85) with $\kappa > 0$ that the system is marginally stable. However the boundary case $\kappa = 0$, $\vartheta > 0$ cannot be checked with the matrices $\boldsymbol{R}$, $\boldsymbol{S}'$ (5.85). Instead, the following matrices are suitable

$$\boldsymbol{R} = \left[\begin{array}{c|c} (\vartheta^2 + g^2\Omega^2)\boldsymbol{E} & \vartheta\boldsymbol{E} - g\Omega\boldsymbol{S} \\ \hline \vartheta\boldsymbol{E} + g\Omega\boldsymbol{S} & 2\boldsymbol{E} \end{array}\right] > \boldsymbol{0}, \quad \boldsymbol{S}' = \left[\begin{array}{c|c} \boldsymbol{0} & \boldsymbol{0} \\ \hline \boldsymbol{0} & 2\vartheta\boldsymbol{E} \end{array}\right] \geqslant \boldsymbol{0}. \tag{5.87}$$

Since the observability condition (5.70) does not hold for $\kappa = 0$, $\vartheta \neq 0$ (rank$[\ldots] = 2$), the case $\kappa = 0$, $\vartheta > 0$ yields marginal stability. If simultaneously $\vartheta = 0$ and $\kappa = 0$, then (5.87) also implies marginal stability, provided $g\Omega \neq 0$. Also if $g\Omega = 0$, then the problem is related to Example 5.9. In this case the instability of the system is proved with

$$\boldsymbol{R} = \left[\begin{array}{c|c} \boldsymbol{0} & -\boldsymbol{E} \\ \hline -\boldsymbol{E} & -\boldsymbol{E} \end{array}\right], \qquad \boldsymbol{S}' = \left[\begin{array}{c|c} \boldsymbol{0} & \boldsymbol{0} \\ \hline \boldsymbol{0} & 2\boldsymbol{E} \end{array}\right] \geqslant \boldsymbol{0} \tag{5.88}$$

and $\kappa = 0$, $\vartheta = 0$, $g\Omega = 0$.

Note that in Example 5.8 it was possible to discuss the boundary cases $\vartheta = 0$ or $\kappa = 0$ only by considering the corresponding eigenvalues. In comparison, the stability investigation by means of the Ljapunov equation (5.59) yields a result in each case. Here the difficulty lies in the necessity of using different matrices $\boldsymbol{R}$ and $\boldsymbol{S}'$ for different values of the parameters. ◆

Example 5.12. Magnetically levitated vehicle and vibrating system with elastic damper

Examples 5.5 and 5.7 were devoted to the stability investigation of a magnetically levitated vehicle and a vibrating system with elastic damper by means of the Routh-Hurwitz stability criteria. In both cases the system matrix has the form

$$\boldsymbol{A}_F = \begin{bmatrix} 0 & 1 & 0 \\ 0 & 0 & 1 \\ -a_3 & -a_2 & -a_1 \end{bmatrix} \tag{5.89}$$

(cf. (4.153) and (4.157)). Stability can be verified by means of the Ljapunov equation (5.59) with the matrices

$$\boldsymbol{R} = \frac{1}{2}\begin{bmatrix} a_2 a_3 & 0 & a_3 \\ 0 & a_1 a_2 - a_3 & 0 \\ a_3 & 0 & a_1 \end{bmatrix}, \quad \boldsymbol{S} = \begin{bmatrix} a_3^2 & 0 & a_3 a_1 \\ 0 & 0 & 0 \\ a_3 a_1 & 0 & a_1^2 \end{bmatrix}. \tag{5.90}$$

The matrix $\boldsymbol{S}$ is positive semidefinite since

$$\boldsymbol{S}=\begin{bmatrix} a_3 \\ 0 \\ a_1 \end{bmatrix}[a_3 \quad 0 \quad a_1] \tag{5.91}$$

and thus it satisfies the assumptions in the theorems about stability and asymptotical stability. The final decision is given by the observability criterion (5.70):

$$\left.\begin{aligned} &\operatorname{rank}\left[\boldsymbol{S} \vdots \boldsymbol{A}_F^T\boldsymbol{S} \vdots \boldsymbol{A}_F^{T^2}\boldsymbol{S}\right] = \operatorname{rank}\begin{bmatrix} a_3 & -a_1a_3 & a_1^2a_3 \\ 0 & a_3-a_2a_1 & -a_1(a_3-a_1a_2) \\ a_1 & -a_1^2 & a_3-a_2a_1+a_1^3 \end{bmatrix} = 3 \\ &\text{for } a_3(a_3-a_2a_1) \not\equiv 0. \end{aligned}\right\} \tag{5.92}$$

According to the theorem about asymptotic stability, (5.89) is then precisely stable under the assumption (5.92) when the matrix $\boldsymbol{R}$ (5.90) is positive definite:

$$a_1>0, \quad a_1(a_1a_2-a_3)>0, \quad a_3(a_1a_2-a_3)^2>0. \tag{5.93}$$

Since (5.93) implies (5.92), the final condition for asymptotic stability is given by (5.93):

$$a_1>0, \quad a_1a_2-a_3>0, \quad a_3>0. \tag{5.94}$$

These conditions coincide precisely with the Liénard-Chipart requirements (5.42) for a third order system. This confirms the stability conditions of the Examples 5.5 and 5.7. ◆

5.2.4. Stability criteria for mechanical systems

The stability criteria presented in Section 5.2.3 can be applied to linear time-invariant ordinary mechanical vibrating systems

$$\boldsymbol{M}\ddot{\boldsymbol{y}}(t)+(\boldsymbol{D}+\boldsymbol{G})\dot{\boldsymbol{y}}(t)+(\boldsymbol{K}+\boldsymbol{N})\boldsymbol{y}(t)=\boldsymbol{0}. \tag{5.95}$$

To do this the state vector $\boldsymbol{x}(t)=[\boldsymbol{y}^T(t) \vdots \dot{\boldsymbol{y}}^T(t)]^T$ is introduced and one considers the state space representation (3.37, 3.38) of (5.95):

$$\dot{\boldsymbol{x}}(t)=\boldsymbol{A}\boldsymbol{x}(t), \qquad \boldsymbol{A}=\left[\begin{array}{c|c} \boldsymbol{0} & \boldsymbol{E} \\ \hline -\boldsymbol{M}^{-1}(\boldsymbol{K}+\boldsymbol{N}) & -\boldsymbol{M}^{-1}(\boldsymbol{D}+\boldsymbol{G}) \end{array}\right]. \tag{5.96}$$

Comparing the various possibilities of stability analysis for (5.95), i.e. with the help of eigenvalues and characteristic coefficients or Ljapunov's equation, conclusion is that the latter is much advantageous for many mechanical systems. When applying the eigenvalue criteria and the Routh-Hurwitz criteria, the special structure of the system matrix (5.96) isn't used, i.e. no consideration is given to the physical meaning of the individual matrices $\boldsymbol{M}$, $\boldsymbol{D}$, $\boldsymbol{G}$, $\boldsymbol{K}$ and $\boldsymbol{N}$ in the equation of motion (5.95) (Section 3.1). In contrast, if the stability problem of the system (5.95) is solved by the Ljapunov's matrix equation (5.59), one can give a physical interpretation of the stability conditions in terms of the individual matrices. The influence of the various categories of forces upon the stability behavior then becomes recognizable.

If nonconservative position forces don't appear in (5.95), i.e. $\boldsymbol{N}=\boldsymbol{0}$, then it is usually possible to characterize the stability behavior of (5.95) by the Hamilton function

$$H = \tfrac{1}{2}(\dot{\boldsymbol{y}}^T\boldsymbol{M}\dot{\boldsymbol{y}} + \boldsymbol{y}^T\boldsymbol{K}\boldsymbol{y})$$

$$= \tfrac{1}{2}\boldsymbol{x}^T\left[\begin{array}{c|c} \boldsymbol{K} & \boldsymbol{0} \\ \hline \boldsymbol{0} & \boldsymbol{M} \end{array}\right]\boldsymbol{x} = \boldsymbol{x}^T\boldsymbol{R}\boldsymbol{x}. \tag{5.97}$$

The time variation of H along the trajectories yields

$$\dot{H} = -\dot{\boldsymbol{y}}^T\boldsymbol{D}\dot{\boldsymbol{y}} = -\boldsymbol{x}^T\left[\begin{array}{c|c} \boldsymbol{0} & \boldsymbol{0} \\ \hline \boldsymbol{0} & \boldsymbol{D} \end{array}\right]\boldsymbol{x} = -\boldsymbol{x}^T\boldsymbol{S}\boldsymbol{x}. \tag{5.98}$$

The matrices $\boldsymbol{R}$ and $\boldsymbol{S}$ thus defined satisfy Ljapunov's equation (5.59) with (5.96) for $\boldsymbol{N}=\boldsymbol{0}$. One can now apply without any difficulty the stability theorems of Section 5.2.3. The distinction between stable and asymptotically stable systems (5.95) with $\boldsymbol{N}=\boldsymbol{0}$ is performed by investigating the behaviour of Rayleigh's *dissipation function*

$$R = \tfrac{1}{2}\dot{\boldsymbol{y}}^T\boldsymbol{D}\dot{\boldsymbol{y}} \tag{5.99}$$

according to (5.67) with (5.97, 5.98). If $\boldsymbol{D}=\boldsymbol{D}^T>\boldsymbol{0}$, then the Rayleigh function is positive for $\dot{\boldsymbol{y}}\neq\boldsymbol{0}$ and the system (5.95) is called *completely damped*; however if $\boldsymbol{D}=\boldsymbol{D}^T\geqslant\boldsymbol{0}$, i.e., the damping applies only some position coordinates of the mechanical system, then (5.95) is called *pervasively damped* if the Rayleigh function does not vanish identically for every trajectory $\boldsymbol{y}(t)\not\equiv\boldsymbol{0}$. Thus, according to (5.98) a mechanical system (5.95) with $\boldsymbol{N}=\boldsymbol{0}$ is pervasively damped iff the observability condition (5.70) holds for

$$\boldsymbol{A} = \left[\begin{array}{c|c} \boldsymbol{0} & \boldsymbol{E} \\ \hline -\boldsymbol{M}^{-1}\boldsymbol{K} & -\boldsymbol{M}^{-1}(\boldsymbol{D}+\boldsymbol{G}) \end{array}\right], \quad \boldsymbol{S} = \left[\begin{array}{c|c} \boldsymbol{0} & \boldsymbol{0} \\ \hline \boldsymbol{0} & \boldsymbol{D} \end{array}\right]. \tag{5.100}$$

At this point one can clearly see the advantage in formulating the criterion for asymptotical stability relative to a positive semidefinite matrix $\boldsymbol{S}$, with the additional requirement (5.70) instead of requiring $\boldsymbol{S} > \boldsymbol{0}$. Indeed, with this approach one can verify stability of mechanical systems by means of the Hamilton function (5.97) and the dissipation function (5.99).

Following these considerations, one can set up various criteria for the stability of mechanical vibrating systems. Here it is taken into account that for ordinary mechanical systems the inertia matrix is always positive definite:

$$\boldsymbol{M} = \boldsymbol{M}^T > \boldsymbol{0}. \tag{5.101}$$

Conservative systems

Nongyroscopic conservative systems

$$\boldsymbol{M}\ddot{\boldsymbol{y}}(t) + \boldsymbol{K}\boldsymbol{y}(t) = \boldsymbol{0}, \tag{5.102}$$

are marginally stable if and only if the stiffness matrix is positive definite,

$$\boldsymbol{K} = \boldsymbol{K}^T > \boldsymbol{0}. \tag{5.103}$$

If $\boldsymbol{K} \ngtr \boldsymbol{0}$, then (5.102) is unstable. Gyroscopic conservative systems

$$\boldsymbol{M}\ddot{\boldsymbol{y}}(t) + \boldsymbol{G}\dot{\boldsymbol{y}}(t) + \boldsymbol{K}\boldsymbol{y}(t) = \boldsymbol{0} \tag{5.104}$$

which satisfy (5.103), are always marginally stable, independently of $\boldsymbol{G} = -\boldsymbol{G}^T$. Thus, statically stable systems (5.104) remain marginally stable, also under the influence of gyroscopic forces. On the other hand, statically unstable systems (5.104), i.e. satisfying $\boldsymbol{K} < \boldsymbol{0}$, can be gyroscopically stabilized by the application of suitable gyroscopic forces $(-\boldsymbol{G}\dot{\boldsymbol{y}})$ iff the number of degrees of freedom f is even and the gyroscopic effect is sufficiently large

$$\left.\begin{array}{l} \boldsymbol{K} < \boldsymbol{0}, \quad \det \boldsymbol{K} > 0, \\ \det \boldsymbol{G} > 0 \text{ (sufficiently large)}. \end{array}\right\} \tag{5.105}$$

Damped systems

Consider a pervasively damped system

$$\boldsymbol{M}\ddot{\boldsymbol{y}}(t) + (\boldsymbol{D} + \boldsymbol{G})\dot{\boldsymbol{y}}(t) + \boldsymbol{K}\boldsymbol{y}(t) = \boldsymbol{0} \tag{5.106}$$

with a regular inertia matrix (5.101) and a positive semidefinite damping matrix $\boldsymbol{D} = \boldsymbol{D}^T \geqslant \boldsymbol{0}$ which satisfies the observability condition (5.70) with the matrices (5.100). Independently of the gyroscopic matrix $\boldsymbol{G} = -\boldsymbol{G}^T$, such a system is asymptotically stable if the stiffness matrix $\boldsymbol{K}$ is positive definite,

$$\boldsymbol{K} = \boldsymbol{K}^T > \boldsymbol{0}. \tag{5.107}$$

A system (5.106) which is not pervasively damped is certainly marginally stable when

$$\boldsymbol{D} = \boldsymbol{D}^T \geqslant \boldsymbol{0}, \quad \boldsymbol{K} = \boldsymbol{K}^T \geqslant \boldsymbol{0}, \quad \boldsymbol{K} + (\boldsymbol{D} - \boldsymbol{G})(\boldsymbol{D} + \boldsymbol{G}) > \boldsymbol{0} \tag{5.108}$$

holds.

It may be quite tedious to verify by means of the matrices (5.100) the presence or absence of pervasive damping. However in this case it is possible to simplify the observability condition (5.70) and to replace it by a controllability condition. The system (5.95) with $\boldsymbol{N} = \boldsymbol{0}$ is for $\boldsymbol{D} = \boldsymbol{D}^T \geqslant \boldsymbol{0}$ pervasively damped iff

$$\left.\begin{array}{l} \det \boldsymbol{K} \neq 0, \\ \operatorname{rank}\left[\boldsymbol{B}_0 \vdots \boldsymbol{A}_0 \boldsymbol{B}_0 \vdots \boldsymbol{A}_0^2 \boldsymbol{B}_0 \vdots \ldots \vdots \boldsymbol{A}_0^{2f-1} \boldsymbol{B}_0\right] = 2f \end{array}\right\} \tag{5.109}$$

where the matrices $\boldsymbol{A}_0$ and $\boldsymbol{B}_0$ are defined by

$$\boldsymbol{A}_0 = \left[\begin{array}{c:c} \boldsymbol{0} & \boldsymbol{E} \\ \hdashline -\boldsymbol{M}^{-1}\boldsymbol{K} & -\boldsymbol{M}^{-1}\boldsymbol{G} \end{array}\right], \qquad \boldsymbol{B}_0 = \left[\begin{array}{c} \boldsymbol{0} \\ \hdashline \boldsymbol{M}^{-1}\boldsymbol{D} \end{array}\right]. \tag{5.110}$$

If $\boldsymbol{G} = \boldsymbol{0}$, (5.109) can be written more simply as

$$\left.\begin{array}{l} \det \boldsymbol{K} \neq 0 \\ \operatorname{rank}\left[\boldsymbol{M}^{-1}\boldsymbol{D} \vdots (\boldsymbol{M}^{-1}\boldsymbol{K})\boldsymbol{M}^{-1}\boldsymbol{D} \vdots \ldots \vdots (\boldsymbol{M}^{-1}\boldsymbol{K})^{f-1}\boldsymbol{M}^{-1}\boldsymbol{D}\right] = f. \end{array}\right\} \tag{5.111}$$

In the case when the stiffness matrix is regular, the condition (5.109) and respectively (5.111) always holds for completely damped systems, i.e. $\boldsymbol{D} = \boldsymbol{D}^T > 0$.

For complete damping the theorem about asymptotic stability for the system (5.106) is known as the *theorem of Thomson and Tait* (1879).

The theorem about asymptotic stability for pervasively damped systems is of great importance for the engineering practice. Since in practice, all mechanical systems are pervasively damped due to friction or material damping, a stable behaviour can take place in the case $\boldsymbol{N} = \boldsymbol{0}$ only if a stable stiffness (5.107) is provided. Thus a gyroscopic stabilization (5.105) of a conservative system can be applied only to vibrating systems whose operation is restricted to a small finite time interval, so that the small, in reality always present damping influences will have no effect.

Example 5.13. Double pendulum

The stability of the double pendulum was shown in Examples 5.6 and 5.10 by means of the characteristic coefficients and with the help of Ljapunov's equation. If the stability criterion (5.103) is applied, one finds immediately without further calculation from

$$\boldsymbol{K} = \frac{g}{L}\begin{bmatrix} 3 & -1 \\ -1 & 1 \end{bmatrix} > \mathbf{0} \tag{5.112}$$

that the motion of a suspended double pendulum is stable.

Example 5.14. Centrifuge

The equation of motion (4.149) of a centrifuge (Problem 4.1) yields the following values for the matrices $\boldsymbol{M}$, $\boldsymbol{D}$, $\boldsymbol{G}$, $\boldsymbol{K}$ in (5.106):

$$\boldsymbol{M} = \boldsymbol{E}, \quad \boldsymbol{D} = \vartheta \boldsymbol{E}, \quad \boldsymbol{G} = g\Omega \boldsymbol{S}, \quad \boldsymbol{K} = \kappa \boldsymbol{E}. \tag{5.113}$$

If $\vartheta > 0$ and $\kappa \neq 0$, then due to $\boldsymbol{D} > \mathbf{0}$ and det $\boldsymbol{K} \neq 0$ the motion of the centrifuge is completely damped. Thus, by (5.107), the system (4.149) is asymptotically stable precisely when $\vartheta > 0$, $\kappa > 0$.

For the boundary case $\kappa = 0$, $\vartheta > 0$, it follows from (5.108) that the system is marginally stable. For $\vartheta = 0$ one has a gyroscopically conservative system (5.104) which for $\kappa > 0$ is certainly marginally stable; for $\kappa = 0$ one obtains from (5.108) with $g\Omega \neq 0$ marginal stability, due to $-\boldsymbol{G}\boldsymbol{G} = g^2\Omega^2\boldsymbol{E} > 0$. Also for $\kappa < 0$ one can theoretically attain for $\vartheta = 0$ a marginally stable motion as a result of gyroscopic stabilization (5.105). The eigenvalues (5.57) show how large $g\Omega$ must be for this purpose,

$$g^2\Omega^2 > -4\kappa \quad (\kappa < 0). \tag{5.114}$$

However the above results imply that this gyroscopic stabilization fails when the stiffness is statically unstable ($\kappa < 0$) and damping ($\vartheta > 0$) is present. For $\vartheta > 0$, $\kappa < 0$ the centrifuge is unstable, by (5.107).

5.3. Boundedness

To give qualitative estimates for the behavior of externally excited vibrating systems (5.12) it was necessary to introduce, in addition to the concept of stability for homogeneous systems (5.6), the definitions (5.13 to 5.16) of general boundedness and boundedness relative to a given excitation function. However the equations in these definitions are not very suitable for the practical verification of these properties. Thus, by analogy with the procedure for the stability problem, one needs to have simple criteria permitting assertions about the boundedness of the trajectories of (5.12), without calculating these trajectories.

This problem can be solved very satisfactorily in the case of linear systems, since there exists a close connection between the boundedness of inhomogeneous systems (5.12) and the stability of homogeneous systems (5.6).

For generally bounded systems (5.12) the solutions $\boldsymbol{x}(t)$ of (5.12) are also bounded for the identically vanishing excitation function $b(t) \equiv \mathbf{0}$. Thus, by (5.10, 5.11) the homogeneous system (5.6) is then stable. On the other hand there exist homogeneous stable systems (5.6) which can be excited by a constant or periodic function in such a manner that the trajectories of (5.12) are unbounded (see Chapter 7). Accordingly, the homogeneous system (5.6) must be asymptotically stable if the inhomogeneous system (5.12) is generally bounded.

Conversely, if the homogeneous system (5.6) is asymptotically stable, then (4.90) and (5.24) imply the estimation

$$\|\boldsymbol{\Phi}(t)\| < c_1\,\mathrm{e}^{-c_2 t}, \quad c_1 > 0, \quad c_2 > 0, \tag{5.115}$$

and thus, in view of the general solution (4.42) and the assumptions (5.13, 5.14), one has

$$\begin{aligned} \|\boldsymbol{x}(t)\| &\leqslant \|\boldsymbol{\Phi}(t)\| \cdot \|\boldsymbol{x}_0\| + \int_0^t \|\boldsymbol{\Phi}(t-\tau)\| \cdot \|\boldsymbol{b}(\tau)\| \mathrm{d}\tau \\ &\leqslant c_1\delta + c_1\gamma \int_0^t \mathrm{e}^{-c_2(t-\tau)} \mathrm{d}\tau \\ &\leqslant c_1\delta + \frac{c_1\gamma^0}{c_2} = \epsilon(\delta, \gamma). \end{aligned} \tag{5.116}$$

Hence the solutions of an externally excited system (5.12) are bounded. Therefore, one gets the following important connection between

Boundedness and asymptotical stability (I):
The inhomogeneous system (5.12) is generally bounded iff the homogeneous system (5.6) is asymptotically stable.

Of course if (5.6) is asymptotically stable, then (5.12) is also bounded relative to an excitation function $\boldsymbol{b}(t)$. But the converse here is not true: a system (5.12) can be bounded relative to a special function $\boldsymbol{b}(t)$ while also being marginally stable.

Especially important for mechanical systems is the situation when the

n coordinates $b_i(t)$ of the excitation function $\boldsymbol{b}(t)$ are not independent, but rather depend on r $(r < n)$ independent time functions $u_i(t)$:

$$\boldsymbol{b}(t) = \boldsymbol{B}\boldsymbol{u}(t). \tag{5.117}$$

Here $\boldsymbol{u}(t)$ is the r-dimensional excitation vector and the $n \times r$ matrix $\boldsymbol{B}$ shows how the functions $u_i(t)$, $i = 1(1)r$ enter the dynamic system. Thus for ordinary mechanical systems with

$$\boldsymbol{B} = \begin{bmatrix} \boldsymbol{0} \\ \hdashline \boldsymbol{M}^{-1} \end{bmatrix}, \qquad \boldsymbol{u}(t) = \boldsymbol{h}(t) \tag{5.118}$$

the excitation function $\boldsymbol{b}(t)$ is given by (3.39):

$$\boldsymbol{b}(t) = \begin{bmatrix} \boldsymbol{0} \\ \hdashline \boldsymbol{M}^{-1} \end{bmatrix} \boldsymbol{h}(t).$$

For the special excitation functions (5.117) the above results hold in a slightly altered form.

Boundedness and asymptotical stability (II):

If the homogeneous system (5.6) is asymptotically stable, then the inhomogeneous system (5.12) with the excitation function (5.117) is generally bounded. Conversely, if (5.12) with (5.117) is generally bounded, then (5.6) is at least marginally stable; and (5.6) is asymptotically stable even if the controllability condition

$$\operatorname{rank}\left[\boldsymbol{B} \,\vdots\, \boldsymbol{A}\boldsymbol{B} \,\vdots\, \boldsymbol{A}^2\boldsymbol{B} \,\vdots\, \dots \,\vdots\, \boldsymbol{A}^{n-1}\boldsymbol{B}\right] = n \tag{5.119}$$

holds.

It is not possible to make equally far ranging statements about boundedness relative to given excitation functions as it is about general boundedness. In spite of this, for certain classes of excitating functions it is still possible to find connections with the stability of the homogeneous system.

Boundedness and stability:

As it has been observed above, the asymptotical stability of (5.6) always implies the boundedness of (5.12) with respect to every excitation function $\boldsymbol{b}(t)$, but the boundedness of (5.12) with respect to a given excitation function merely allows one to conclude that (5.6) is marginally stable. Conversely, the marginal stability of (5.6) does not imply the boundedness of (5.12); this implication holds only for certain excitation functions. Thus a stable system is certainly bounded relative to periodical excitation functions $\boldsymbol{b}(t)=\boldsymbol{b}(t+T)$, provided the frequencies of the purely imaginary eigenvalues of $\boldsymbol{A}$ do not coincide with the frequencies of the nonvanishing Fourier terms of $\boldsymbol{b}(t)$. Boundedness can be also concluded from stability relative to decreasing functions $\boldsymbol{b}(t)$ satisfying $\int_0^\infty \|\boldsymbol{b}(\tau)\| \mathrm{d}\tau < \infty$.

To conclude, consider the following simple statement about

Unboundedness and instability:

The inhomogeneous system (5.12) is certainly unbounded if the homogeneous system (5.6) is unstable.

Summarizing, it is found: If the excitation function is largely unknown, then bounded systems can be reached only by asymptotic stability. But if the excitation function is known, then frequently stable systems are sufficient. Independently of the excitation, unstable systems are excluded from technical applications.

Problem 5.1. Eigenvalues

What is the stability behavior of the linear systems with the following eigenvalues?

					asympt. stable	marg. stable	unstable	
a)	$\lambda_{1,2}=0$,	$\lambda_{3,4}=-5\pm100\,\mathrm{i}$,		$d_{1,2}=1$	□	□	□	
b)	$\lambda_{1,2}=1$,	$\lambda_{3,4}=-100\pm20\,\mathrm{i}$,		$d_{1,2}=2$	□	□	□	
c)	$\lambda_1=-1$,	$\lambda_2=1$,	$\lambda_{3,4}=\pm20\,\mathrm{i}$		□	□	□	
d)	$\lambda_1=-1$,	$\lambda_{2,3}=1\pm100\,\mathrm{i}$,	$\lambda_4=-10$		□	□	□	
e)	$\lambda_{1,2,3,4}=-1$,			$d_{1,2,3,4}=1$	□	□	□	
f)	$\lambda_{1,2,3,4}=0$,			$d_{1,2,3,4}=4$	□	□	□	
g)	$\lambda_1=-1$,	$\lambda_2=-2$,	$\lambda_3=-5$,	$\lambda_4=-10$	□	□	□	
h)	$\lambda_{1,2}=\pm\mathrm{i}$,	$\lambda_{3,4}=\pm10\,\mathrm{i}$			□	□	□	
i)	$\lambda_1=-0,1$,	$\lambda_{2,3}=-1\pm10\,\mathrm{i}$,	$\lambda_4=-10$		□	□	□	
j)	$\lambda_{1,2}=-1$,	$\lambda_3=1$,	$\lambda_4=-10$,	$d_{1,2}=1$	□	□	□	◆

Problem 5.2. Vertical vibrations of an automobile

The stability behavior of the system (4.99),

$$\begin{bmatrix} m_1 & 0 \\ 0 & m_2 \end{bmatrix} \ddot{y}(t) + \begin{bmatrix} d_1 & -d_1 \\ -d_1 & d_1 \end{bmatrix} \dot{y}(t) + \begin{bmatrix} k_1 & -k_1 \\ -k_1 & k_1+k_2 \end{bmatrix} y(t) = \mathbf{0},$$

is considered. Using the abbreviations (4.101), the corresponding characteristic polynomial, according to (4.100), reads as

$$p(\lambda) \equiv \lambda^4 + (\vartheta_1+\vartheta_2)\lambda^3 + (\kappa_{11}+\kappa_{22}+\kappa_{12})\lambda^2 + \vartheta_1\kappa_{22}\lambda + \kappa_{11}\kappa_{22}.$$

a) Which are the stability conditions of Liénard and Chipart?

□ $\vartheta_1 + \vartheta_2 > 0,\ \kappa_{11} + \kappa_{22} + \kappa_{12} > 0,\ \vartheta_1 \kappa_{22} > 0,\ \kappa_{11}\kappa_{22} > 0;$

□ $\vartheta_1 + \vartheta_2 > 0,\ \kappa_{11} + \kappa_{22} + \kappa_{12} > 0,\ \vartheta_1 \vartheta_2 \kappa_{22}^2 > 0,\ \kappa_{11}\kappa_{22} > 0;$

□ $d_1\left(\frac{1}{m_1} + \frac{1}{m_2}\right) > 0,\ k_1\left(\frac{1}{m_1} + \frac{1}{m_2}\right) + \frac{k_2}{m_2} > 0, \frac{d_1 k_1}{m_1 m_2} > 0, \frac{k_1 k_2}{m_1 m_2} > 0;$

□ $d_1\left(\frac{1}{m_1} + \frac{1}{m_2}\right) > 0,\ k_1\left(\frac{1}{m_1} + \frac{1}{m_2}\right) + \frac{k_2}{m_2} > 0, \frac{d_1^2 k_2^2}{m_1 m_2^3} > 0, \frac{k_1 k_2}{m_1 m_2} > 0.$

b) Which conditions characterize asymptotic stability?

□ $d_1 > 0, \quad k_1 = 0, \quad k_2 > 0;$

□ $d_1 \geqslant 0, \quad k_1 > 0, \quad k_2 = 0;$

□ $d_1 > 0, \quad k_1 > 0, \quad k_2 > 0;$

□ $d_1 \geqslant 0, \quad k_1 \geqslant 0, \quad k_2 \geqslant 0.$

c) Is the damping complete, pervasive, or not pervasive?

□ complete, since $\boldsymbol{D} > \boldsymbol{0}$;

□ not pervasive, since $\boldsymbol{D} \geqslant \boldsymbol{0}$;

□ pervasive, since $\boldsymbol{D} \geqslant \boldsymbol{0}$ and

$$\operatorname{rank}\left[\boldsymbol{M}^{-1}\boldsymbol{D} \vdots \boldsymbol{M}^{-1}\boldsymbol{K}\boldsymbol{M}^{-1}\boldsymbol{D}\right] = 1 \quad \text{for} \quad k_2 = 0;$$

□ pervasive, since $\boldsymbol{D} \geqslant \boldsymbol{0}$ and

$$\operatorname{rank}\left[\boldsymbol{M}^{-1}\boldsymbol{D} \vdots \boldsymbol{M}^{-1}\boldsymbol{K}\boldsymbol{M}^{-1}\boldsymbol{D}\right] = 2 \quad \text{for} \quad d_1 k_2 \neq 0.$$

d) Is the system (4.99) stable, according to the stability theorems for mechanical systems?

$d_1 > 0$ $k_1 > 0$ $k_2 > 0$	$d_1 = 0$ $k_1 > 0$ $k_2 > 0$	$d_1 > 0$ $k_1 = 0$ $k_2 > 0$	$d_1 > 0$ $k_1 > 0$ $k_2 = 0$	
□	□	□	□	marginally stable because of (5.103)
□	□	□	□	marginally stable because of (5.108)
□	□	□	□	asymptotically stable because of (5.107), (5.111)
□	□	□	□	cannot be determined with the stability theorems of section 5.2.4.

◆

Problem 5.3. Magnetically levitated vehicle

Which conditions must be satisfied by the parameters a and b of the system (4.119)

$$\dot{\hat{x}}(t)=\begin{bmatrix} 0 & 1 & 0 \\ 0 & 0 & 1 \\ \dfrac{c+ad}{mT} & \dfrac{c}{m}+\dfrac{d}{m}\dfrac{b-k}{T} & -\dfrac{1}{T} \end{bmatrix}\hat{x}(t),$$

so that the motion of the magnetically levitated vehicle is stable?

□ $a<-\dfrac{c}{d}$, $\quad b>k-\dfrac{cT}{d}$;

□ $a<-\dfrac{c}{d}$, $\quad b>-aT+k-\dfrac{cT}{d}$;

□ $a<-\dfrac{c}{d}$, $\quad b<-aT+k+\dfrac{cT}{d}$;

□ $a<-\dfrac{c}{d}$, $\quad b<aT+k$. ◆

Problem 5.4. Centrifuge

The centrifuge with radially symmetric damping discussed in Problems 3.1 and 4.1, is replaced by a centrifuge with a technically simpler, axially operating damping. With the previous relations (4.149, 4.150) the following equations of motion are obtained:

$$\ddot{y}(t)+\left(\begin{bmatrix} \vartheta & 0 \\ 0 & 0 \end{bmatrix}+g\Omega S\right)\dot{y}(t)+\kappa y(t)=0. \tag{5.120}$$

a) What is, according to (5.109), the controllability matrix $[B_0 \vdots A_0 B_0 \vdots A_0^2 B_0 \vdots A_0^3 B_0]$ and what is its rank?

$$\operatorname{rank}[B_0 \vdots A_0 B_0 \vdots A_0^2 B_0 \vdots A_0^3 B_0]=\operatorname{rank}\begin{bmatrix} \quad \end{bmatrix}.$$

□ rank = 4 for $\kappa\neq 0,\ g\Omega\neq 0,\ \vartheta\neq 0$;

□ rank = 2 for $\kappa\neq 0,\ g\Omega\neq 0,\ \vartheta\neq 0$;

□ rank = 3 for $\kappa=0,\ g\Omega\neq 0,\ \vartheta\neq 0$;

□ rank = 2 for $\kappa\neq 0,\ g\Omega=0,\ \vartheta\neq 0$.

b) For which parameters ϑ, $g\Omega$, κ is the system pervasively damped?

☐ $\kappa \neq 0$, $g\Omega$ any, $\vartheta > 0$;
☐ κ any, $g\Omega \neq 0$, $\vartheta > 0$;
☐ κ any, $g\Omega$ any, $\vartheta > 0$;
☐ $\kappa \neq 0$, $g\Omega \neq 0$, $\vartheta > 0$.

c) For which values of the parameters is the system, according to the stability theorems for mechanical systems, asymptotically stable or marginally stable?

asympt. stable	margin. stable			
☐	☐	$\kappa > 0$,	$g\Omega = 0$,	$\vartheta > 0$;
☐	☐	$\kappa = 0$,	$g\Omega = 0$,	$\vartheta > 0$;
☐	☐	$\kappa > 0$,	$g\Omega \neq 0$,	$\vartheta > 0$;
☐	☐	$\kappa = 0$,	$g\Omega \neq 0$,	$\vartheta = 0$. ◆

6. Free vibrations

Free vibrations, also called natural vibrations, characterize the motions of homogeneous vibrating systems not subject to external disturbances. Given the initial conditions, the natural vibrations can be determined from the general solution given in Chapter 4. Qualitative statements about the behavior of natural vibrations are obtained from the stability theorems of Chapter 5. The present chapter is devoted to the investigation of the *natural modes* which allow a quantitative view of natural vibrations by a more detailed investigation of the eigenvectors originated in Section 4.3. Extremely simple results are obtained for the nongyroscopic conservative vibrating systems. Here it also becomes visible that the amplitudes of natural modes decrease with increasing natural frequencies. Moreover, the natural modes for weakly damped and ordinary mechanical systems will be presented.

Apart from the natural modes, *cost functionals* can be conveniently applied to the analysis of asymptotically stable vibrating system. The cost functionals can be defined either by eigenvalues or by integrals along the trajectories. With the help of the cost functionals it is also possible to optimize the free parameters of the system such as damping and spring constants. Various cost functionals will be introduced and their calculation and the corresponding parameter optimizations will be discussed.

6.1. Natural modes

Among the descriptions of the solutions of homogeneous vibrating systems given in Chapter 4 were those by eigenvalues and eigenvectors. For distinct eigenvalues the solution (4.69) is given by a linear combination of the eigenvectors weighted by exponential functions of the eigenvalues. These *eigenvectors* characterize the *natural modes* of the system. The description of the natural modes is especially clear for mechanical systems, and it permits a good insight into the process of natural vibrations.

6.1.1. Nongyroscopic conservative systems

A nongyroscopic conservative vibrating system is given by the $f \times f$ inertia matrix $\boldsymbol{M}$ and the $f \times f$ matrix $\boldsymbol{K}$ of the conservative position forces. The equation of motion is

$$\boldsymbol{M}\ddot{\boldsymbol{y}}(t)+\boldsymbol{K}\boldsymbol{y}(t)=\boldsymbol{0}, \tag{6.1}$$

and the state equation is determined, according to (3.37, 3.38), by the $n \times n$ system matrix $\boldsymbol{A}$ $(n=2f)$

$$\dot{\boldsymbol{x}}(t)=\boldsymbol{A}\boldsymbol{x}(t) \quad \text{with} \quad \boldsymbol{A}=\left[\begin{array}{c|c} \boldsymbol{0} & \boldsymbol{E} \\ \hline -\boldsymbol{M}^{-1}\boldsymbol{K} & \boldsymbol{0} \end{array}\right]. \tag{6.2}$$

By (4.98), the characteristic equation for (6.1) or (6.2) is of the form

$$p(\lambda)=\frac{1}{\det \boldsymbol{M}}\det(\boldsymbol{M}\lambda^2+\boldsymbol{K})=0, \tag{6.3}$$

i.e., one gets a polynomial of degree f in λ^2. Under the assumption that the stiffness matrix $\boldsymbol{K}$ is positive definite, the system (6.1) is marginally stable, by the stability condition (5.103). The eigenvalues then have the form

$$\lambda_k=\pm\mathrm{i}\omega_j, \qquad k=1(1)n, \qquad j=1(1)f, \tag{6.4}$$

where ω_j are the natural frequencies of the system. By (5.25), the corresponding Jordan matrix (4.86) is for marginally stable systems a diagonal matrix (4.72) in which the eigenvalues λ_k will be arranged as follows:

$$\boldsymbol{\Lambda}=\left[\begin{array}{c|c} \mathrm{i}\boldsymbol{\Omega} & \boldsymbol{0} \\ \hline \boldsymbol{0} & -\mathrm{i}\boldsymbol{\Omega} \end{array}\right], \quad \boldsymbol{\Omega}=\operatorname{diag}\{\omega_j\},\ j=1(1)f. \tag{6.5}$$

As a further step towards the solution, the eigenvectors and hence the modal matrix (4.71) have to be calculated. According to (4.96, 4.97), one has

$$\left.\begin{aligned} \tilde{\boldsymbol{x}}_j&=\left[\begin{array}{c} \tilde{\boldsymbol{y}}_j \\ \hline \mathrm{i}\omega_j\tilde{\boldsymbol{y}}_j \end{array}\right], & j&=1(1)f, \\ \tilde{\boldsymbol{x}}_{f+j}&=\left[\begin{array}{c} \tilde{\boldsymbol{y}}_{f+j} \\ \hline -\mathrm{i}\omega_j\tilde{\boldsymbol{y}}_{f+j} \end{array}\right], & j&=1(1)f, \end{aligned}\right\} \tag{6.6}$$

where the vectors $\tilde{y}_j$ and $\tilde{y}_{f+j}$, $j=1(1)f$ are to be determined from

$$\left.\begin{aligned}\left(-\boldsymbol{M}\omega_j^2+\boldsymbol{K}\right)\tilde{\boldsymbol{y}}_j=\mathbf{0}, \qquad & j=1(1)f,\\ \left(-\boldsymbol{M}\omega_j^2+\boldsymbol{K}\right)\tilde{\boldsymbol{y}}_{f+j}=\mathbf{0}, \quad & j=1(1)f.\end{aligned}\right\} \tag{6.7}$$

Since the eigenvectors belonging to complex conjugate eigenvectors are also complex conjugate, it follows that

$$\left.\begin{aligned}\tilde{\boldsymbol{x}}_{f+j}=\bar{\tilde{\boldsymbol{x}}}_j, \quad & j=1(1)f,\\ \tilde{\boldsymbol{y}}_{f+j}=\bar{\tilde{\boldsymbol{y}}}_j, \quad & j=1(1)f.\end{aligned}\right\} \tag{6.8}$$

As the system of equations (6.7) with the real matrix $(-\boldsymbol{M}\omega_j^2+\boldsymbol{K})$ must be satisfied by both $\tilde{\boldsymbol{y}}_j$ and $\bar{\tilde{\boldsymbol{y}}}_j$, the solution $\tilde{\boldsymbol{y}}_j$ can be chosen real:

$$\tilde{\boldsymbol{y}}_{f+j}=\tilde{\boldsymbol{y}}_j, \qquad j=1(1)f. \tag{6.9}$$

This shows that the original eigenvector problem for the $\tilde{\boldsymbol{x}}_i$, $i=1(1)2f$, reduces to a real eigenvector problem of order f for the vectors $\tilde{\boldsymbol{y}}_j$, $j=1(1)f$. If one builds a real $f\times f$ matrix from the vectors $\tilde{\boldsymbol{y}}_j$,

$$\boldsymbol{Y}=\left[\,\tilde{\boldsymbol{y}}_1 \,\vdots\, \tilde{\boldsymbol{y}}_2 \,\vdots\, \dots \,\vdots\, \tilde{\boldsymbol{y}}_f\right], \tag{6.10}$$

then the $n\times n$-modal matrix (4.71) can be expressed as follows by submatrices:

$$\boldsymbol{X}=\begin{bmatrix}\boldsymbol{Y} & \boldsymbol{Y}\\ \mathrm{i}\boldsymbol{Y}\boldsymbol{\Omega} & -\mathrm{i}\boldsymbol{Y}\boldsymbol{\Omega}\end{bmatrix}=\begin{bmatrix}\boldsymbol{Y} & \mathbf{0}\\ \mathbf{0} & \boldsymbol{Y}\end{bmatrix}\begin{bmatrix}\boldsymbol{E} & \boldsymbol{E}\\ \mathrm{i}\boldsymbol{\Omega} & -\mathrm{i}\boldsymbol{\Omega}\end{bmatrix}. \tag{6.11}$$

Using the similarity transformation (4.81), one can visualize an interpretation of the matrix $\boldsymbol{Y}$. First, (4.81) together with (6.2), (6.5) and (6.11) imply the relation

$$\begin{bmatrix}\boldsymbol{E} & \boldsymbol{E}\\ \mathrm{i}\boldsymbol{\Omega} & -\mathrm{i}\boldsymbol{\Omega}\end{bmatrix}\begin{bmatrix}\mathrm{i}\boldsymbol{\Omega} & \mathbf{0}\\ \mathbf{0} & -\mathrm{i}\boldsymbol{\Omega}\end{bmatrix}\begin{bmatrix}\tfrac{1}{2}\boldsymbol{E} & -\tfrac{1}{2}\mathrm{i}\boldsymbol{\Omega}^{-1}\\ \tfrac{1}{2}\boldsymbol{E} & \tfrac{1}{2}\mathrm{i}\boldsymbol{\Omega}^{-1}\end{bmatrix}$$

$$=\begin{bmatrix}\boldsymbol{Y}^{-1} & \mathbf{0}\\ \mathbf{0} & \boldsymbol{Y}^{-1}\end{bmatrix}\begin{bmatrix}\mathbf{0} & \boldsymbol{E}\\ -\boldsymbol{M}^{-1}\boldsymbol{K} & \mathbf{0}\end{bmatrix}\begin{bmatrix}\boldsymbol{Y} & \mathbf{0}\\ \mathbf{0} & \boldsymbol{Y}\end{bmatrix} \tag{6.12}$$

and consequently

$$\left[\begin{array}{c|c} \mathbf{0} & \boldsymbol{E} \\ \hline -\boldsymbol{\Omega}^2 & \mathbf{0} \end{array}\right] = \left[\begin{array}{c|c} \mathbf{0} & \boldsymbol{E} \\ \hline -\boldsymbol{Y}^{-1}\boldsymbol{M}^{-1}\boldsymbol{K}\boldsymbol{Y} & \mathbf{0} \end{array}\right]. \tag{6.13}$$

From the essential submatrix equation in (6.13) it follows

$$\boldsymbol{M}^{-1}\boldsymbol{K} = \boldsymbol{Y}\boldsymbol{\Omega}^2\boldsymbol{Y}^{-1} \tag{6.14}$$

and if this equation is multiplied from the left by $\boldsymbol{Y}^{\mathrm{T}}\boldsymbol{M}$ and from the right by $\boldsymbol{Y}$, one finally gets

$$\boldsymbol{Y}^{\mathrm{T}}\boldsymbol{K}\boldsymbol{Y} = \boldsymbol{Y}^{\mathrm{T}}\boldsymbol{M}\boldsymbol{Y}\boldsymbol{\Omega}^2. \tag{6.15}$$

Since $\boldsymbol{Y}^{\mathrm{T}}\boldsymbol{K}\boldsymbol{Y}$ and $\boldsymbol{Y}^{\mathrm{T}}\boldsymbol{M}\boldsymbol{Y}$ are symmetric matrices and $\boldsymbol{\Omega} = \mathrm{diag}\{\omega_j\}$, (6.15) implies the relations

$$\boldsymbol{Y}^{\mathrm{T}}\boldsymbol{M}\boldsymbol{Y} = \mathrm{diag}\{m_j\}, \qquad j = 1(1)f, \tag{6.16}$$

and

$$\boldsymbol{Y}^{\mathrm{T}}\boldsymbol{K}\boldsymbol{Y} = \mathrm{diag}\{k_j\}, \qquad j = 1(1)f. \tag{6.17}$$

Introducing further the normalized modal matrix

$$\hat{\boldsymbol{Y}} = \boldsymbol{Y}\,\mathrm{diag}\left\{\frac{1}{\sqrt{m_j}}\right\} \tag{6.18}$$

one obtains

$$\hat{\boldsymbol{Y}}^{\mathrm{T}}\boldsymbol{M}\hat{\boldsymbol{Y}} = \boldsymbol{E}, \qquad \hat{\boldsymbol{Y}}^{\mathrm{T}}\boldsymbol{K}\hat{\boldsymbol{Y}} = \boldsymbol{\Omega}^2. \tag{6.19}$$

This means that the matrix $\hat{\boldsymbol{Y}}$ is orthogonal relative to the positive definite inertia matrix $\boldsymbol{M}$. (A matrix $\boldsymbol{A}$ is called orthogonal if it satisfies $\boldsymbol{A}\boldsymbol{A}^{\mathrm{T}} = \boldsymbol{E}$, i.e. $\boldsymbol{A}^{\mathrm{T}} = \boldsymbol{A}^{-1}$). Further, one observes that under the *congruence transformation*

$$\boldsymbol{y}(t) = \hat{\boldsymbol{Y}}\hat{\boldsymbol{y}}(t) \tag{6.20}$$

the system (6.1) is carried over into the uncoupled system

$$\ddot{\hat{\boldsymbol{y}}}(t) + \boldsymbol{\Omega}^2\hat{\boldsymbol{y}}(t) = \mathbf{0}. \tag{6.21}$$

The system (6.21) is called the *normal form* of the vibrating system (6.1) and the coordinates $\hat{y}_j$, $j = 1(1)f$ are termed *normal* or *principal coordinates.*

Due to (5.11) and (4.74), the general solution of the vibrating system (6.1) can be written as

$$\begin{bmatrix} y(t) \\ \hline \dot{y}(t) \end{bmatrix} = \left[\begin{array}{c|c} Y & Y \\ \hline \mathrm{i}Y\Omega & -\mathrm{i}Y\Omega \end{array}\right] \left[\begin{array}{c|c} \mathrm{e}^{\mathrm{i}\Omega t} & 0 \\ \hline 0 & \mathrm{e}^{-\mathrm{i}\Omega t} \end{array}\right] \begin{bmatrix} c_1 \\ \hline c_2 \end{bmatrix} \tag{6.22}$$

or, by (4.75),

$$y(t) = Y\tfrac{1}{2}(\mathrm{e}^{\mathrm{i}\Omega t} + \mathrm{e}^{-\mathrm{i}\Omega t})Y^{-1}y_0 + Y\Omega^{-1}\tfrac{1}{2}\mathrm{i}(-\mathrm{e}^{\mathrm{i}\Omega t} + \mathrm{e}^{-\mathrm{i}\Omega t})Y^{-1}\dot{y}_0. \tag{6.23}$$

Since one has

$$\tfrac{1}{2}(\mathrm{e}^{\mathrm{i}\Omega t} + \mathrm{e}^{-\mathrm{i}\Omega t}) = \mathrm{diag}\{\tfrac{1}{2}(\mathrm{e}^{\mathrm{i}\omega_j t} + \mathrm{e}^{-\mathrm{i}\omega_j t})\}$$
$$= \mathrm{diag}\{\cos \omega_j t\} = \cos \Omega t, \tag{6.24a}$$

$$\tfrac{1}{2}\mathrm{i}(-\mathrm{e}^{\mathrm{i}\Omega t} + \mathrm{e}^{-\mathrm{i}\Omega t}) = \mathrm{diag}\{\tfrac{1}{2}\mathrm{i}(-\mathrm{e}^{\mathrm{i}\omega_j t} + \mathrm{e}^{-\mathrm{i}\omega_j t})\}$$
$$= \mathrm{diag}\{\sin \omega_j t\} = \sin \Omega t, \tag{6.24b}$$

the solution (6.23) takes the final form

$$y(t) = Y\cos \Omega t Y^{-1}y_0 + Y\Omega^{-1} \sin \Omega t Y^{-1}\dot{y}_0. \tag{6.25}$$

The solution (6.25) is a matrix generalization of the solution (4.47) of the spring-mass vibrating system (4.46). By (6.25), the vibrations of (6.1) arise by superposition of harmonic functions with frequencies ω_j, $j = 1(1)f$.

Should excitation of the j-th natural mode take place, i.e., of the partial vibration with frequency ω_j, then (6.22) has to be taken with

$$c_1 = \alpha e_j, \qquad c_2 = \beta e_j \tag{6.26}$$

where e_j is the j-th unit vector. By (4.75), the initial conditions for the j-th natural mode are

$$\left.\begin{aligned} y_0 &= (\alpha + \beta)Ye_j = (\alpha + \beta)\tilde{y}_j, \\ \dot{y}_0 &= i(\alpha - \beta)Y\Omega e_j = i(\alpha - \beta)\omega_j\tilde{y}_j. \end{aligned}\right\} \tag{6.27}$$

Employing the real numbers

$$\gamma_0 = \alpha + \beta, \qquad \frac{\dot{\gamma}_0}{\omega_j} = i(\alpha - \beta) \tag{6.28}$$

one can write the corresponding vibration as

$$y_j(t) = \left(\gamma_0 \cos \omega_j t + \frac{\dot{\gamma}_0}{\omega_j} \sin \omega_j t \right) \tilde{y}_j. \tag{6.29}$$

The j-th *natural mode* is a vibration with natural frequency ω_j in the direction of the j-th eigenvector $\tilde{y}_j$. If this natural mode is excited only by an initial displacement of the system (6.1) in its position coordinate $y_0 = \tilde{y}_j$ with vanishing initial velocities, $\dot{y}_0 = \mathbf{0}$, then one has $\alpha = \beta = \frac{1}{2}$ and (6.29) yields

$$y_j(t) = \cos \omega_j t \, \tilde{y}_j = \cos \omega_j t \, \boldsymbol{a}_j. \tag{6.30}$$

Thus the j-th eigenvector $\tilde{y}_j$ represents a simple initial condition for the j-th natural mode and it also characterizes the amplitudes $(a_j)_k$, $k = 1(1)f$, of the individual coordinates of this natural mode. Hence the j-th eigenvector is often called the j-th natural mode.

For nongyroscopic conservative systems one can estimate the amplitudes of the individual natural modes using the energy theorem. Indeed, by (3.11) and (5.97), the energy conservation theorem can be written as

$$T + U = \tfrac{1}{2}\left(\dot{y}^T(t) \boldsymbol{M} \dot{y}(t) + y^T(t) \boldsymbol{K} y(t) \right) = H = H_0 = \text{const}, \tag{6.31}$$

which in this case leads to a constant Hamilton function. Employing the congruence transformation (6.20) and the relations (6.19), one obtains from (6.31)

$$\begin{aligned} T + U &= \tfrac{1}{2}\left(\dot{\hat{y}}^T(t) \dot{\hat{y}}(t) + \hat{y}^T(t) \boldsymbol{\Omega}^2 \hat{y}(t) \right) \\ &= \tfrac{1}{2} \sum_{j=1}^{f} \left(\dot{\hat{y}}_j^2(t) + \omega_j^2 \hat{y}_j^2(t) \right) = H_0. \end{aligned} \tag{6.32}$$

This implies the following estimates for the maximum possible amplitudes:

$$\max_t |\hat{y}_j(t)| \leqslant \sqrt{\frac{2H_0}{\omega_j^2}} = \frac{1}{\omega_j}\sqrt{2H_0}, \; j = 1(1)f. \tag{6.33}$$

For a given initial energy the amplitudes of the natural modes with higher frequencies $\omega_j \geqslant \omega_1 > 0,\ j = 2(1)f$, are smaller by at least the ratio ω_1/ω_j than the amplitude of the basic natural mode with frequency ω_1. This fact is of great practical importance because it allows one to disregard in many engineering situations the natural modes with higher frequencies. In a variety of vibrating systems with many degrees of freedom it is sufficient to consider the first three to five natural modes.

Example 6.1. Double pendulum

Consider the double pendulum (Fig. 3.3) with the equation of motion

$$\ddot{y}(t) + \boldsymbol{K}y(t) = \boldsymbol{0},\ \boldsymbol{K} = \frac{g}{L}\begin{bmatrix} 3 & -1 \\ -1 & 1 \end{bmatrix}. \tag{6.34}$$

By (4.137) the two natural frequencies are

$$\omega_1 = \sqrt{\frac{g}{L}(2 - \sqrt{2})}, \qquad \omega_2 = \sqrt{\frac{g}{L}(2 + \sqrt{2})}. \tag{6.35}$$

The eigenvalue problem (6.7) yields for the first natural frequency

$$\frac{g}{L}\left\{\begin{bmatrix} -2 + \sqrt{2} & 0 \\ 0 & -2 + \sqrt{2} \end{bmatrix} + \begin{bmatrix} 3 & -1 \\ -1 & 1 \end{bmatrix}\right\}\begin{bmatrix} \tilde{y}_{11} \\ \tilde{y}_{21} \end{bmatrix} = \boldsymbol{0}, \tag{6.36}$$

and thus the first eigenvector is

$$\tilde{y}_1 = \begin{bmatrix} 1 \\ 1 + \sqrt{2} \end{bmatrix}. \tag{6.37}$$

In the same manner one obtains for the second natural frequency, the second eigenvector

$$\tilde{y}_2 = \begin{bmatrix} 1 \\ 1 - \sqrt{2} \end{bmatrix}. \tag{6.38}$$

The 2×2 modal matrix (6.10) is

$$\boldsymbol{Y} = \begin{bmatrix} 1 & 1 \\ 1 + \sqrt{2} & 1 - \sqrt{2} \end{bmatrix}. \tag{6.39}$$

It can now be shown that the eigenvectors are orthogonal with respect to the inertia matrix. For $\boldsymbol{M} = \boldsymbol{E}$ it follows that

$$\boldsymbol{Y}^T\boldsymbol{Y} = \begin{bmatrix} 1 & 1 + \sqrt{2} \\ 1 & 1 - \sqrt{2} \end{bmatrix}\begin{bmatrix} 1 & 1 \\ 1 + \sqrt{2} & 1 - \sqrt{2} \end{bmatrix} = \begin{bmatrix} 4 + 2\sqrt{2} & 0 \\ 0 & 4 - 2\sqrt{2} \end{bmatrix}. \tag{6.40}$$

or by (6.16)

$$m_1 = 4 + 2\sqrt{2}\,,\; m_2 = 4 - 2\sqrt{2}\,. \tag{6.41}$$

If the modal matrix (6.39) is normalized with (6.41),

$$\hat{Y} = \begin{bmatrix} \dfrac{1}{\sqrt{4+2\sqrt{2}}} & \dfrac{1}{\sqrt{4-2\sqrt{2}}} \\ \dfrac{1+\sqrt{2}}{\sqrt{4+2\sqrt{2}}} & \dfrac{1-\sqrt{2}}{\sqrt{4-2\sqrt{2}}} \end{bmatrix} \tag{6.42}$$

then it becomes orthogonal with respect to $\boldsymbol{M} = \boldsymbol{E}$, $\hat{\boldsymbol{Y}}^T\hat{\boldsymbol{Y}} = \boldsymbol{E}$, and moreover it transforms the matrix $\boldsymbol{K}$ into the diagonal matrix of the squares of natural frequencies,

$$\hat{\boldsymbol{Y}}^T\boldsymbol{K}\hat{\boldsymbol{Y}} = \frac{g}{L}\begin{bmatrix} 2-\sqrt{2} & 0 \\ 0 & 2+\sqrt{2} \end{bmatrix} = \boldsymbol{\Omega}^2. \tag{6.43}$$

The solution (6.30) for the first natural mode is

$$y_1(t) = \cos\sqrt{\frac{g}{L}(2-\sqrt{2})}\; t \begin{bmatrix} 1 \\ 1+\sqrt{2} \end{bmatrix}, \tag{6.44}$$

and for the second natural mode one gets

$$y_2(t) = \cos\sqrt{\frac{g}{L}(2+\sqrt{2})}\; t \begin{bmatrix} 1 \\ 1-\sqrt{2} \end{bmatrix} \tag{6.45}$$

For a given total energy, the amplitudes of the second natural mode are smaller than the amplitudes of the first natural mode by the factor

$$\frac{\omega_1}{\omega_2} = \sqrt{\frac{2-\sqrt{2}}{2+\sqrt{2}}} = \sqrt{2} - 1 = 0.414. \tag{6.46}$$

If for constant total energy both natural modes are plotted in a $(\hat{y}_j, \dot{\hat{y}}_j)$-diagram, one obtains Fig. 6.1. The time dependence of $\hat{y}_1(t)$ and $\hat{y}_2(t)$ is shown in Fig. 6.2. Here both natural modes for ω_1 and ω_2 are shown separately, while in Fig. 4.4 one can see the superposition of these natural modes, according to (6.20). ◆

6.1.2. Lightly damped mechanical systems

By (3.10), the equation of motion of a homogeneous mechanical system is

$$\boldsymbol{M}\ddot{y}(t)+(\boldsymbol{D}+\boldsymbol{G})\dot{y}(t)+(\boldsymbol{K}+\boldsymbol{N})y(t)=\boldsymbol{0}, \tag{6.47}$$

and the corresponding state equation is given by

$$\dot{x}(t)=\boldsymbol{A}x(t), \qquad \boldsymbol{A}=\left[\begin{array}{c|c} \boldsymbol{0} & \boldsymbol{E} \\ \hline -\boldsymbol{M}^{-1}(\boldsymbol{K}+\boldsymbol{N}) & -\boldsymbol{M}^{-1}(\boldsymbol{D}+\boldsymbol{G}) \end{array}\right]. \tag{6.48}$$

The corresponding characteristic equation (4.98) is

$$\det\{\boldsymbol{M}\lambda^2+(\boldsymbol{D}+\boldsymbol{G})\lambda+(\boldsymbol{K}+\boldsymbol{N})\}=0. \tag{6.49}$$

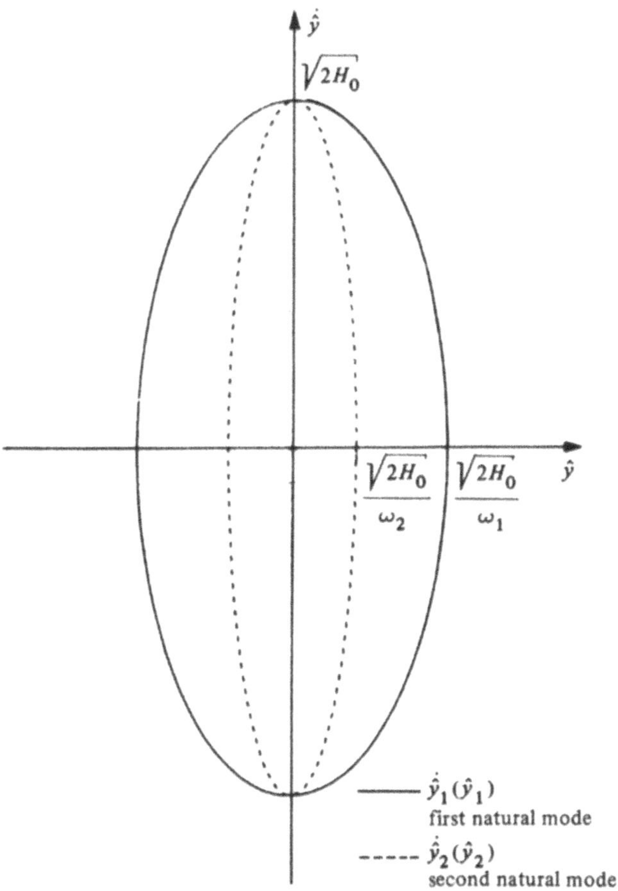

Fig. 6.1. State plane representation of the natural modes of a double pendulum at constant total energy H_0

The mechanical system (6.47) is called *lightly damped* if all eigenvalues are complex and their real parts δ_j are negative,

$$\lambda_i = \delta_j \pm \mathrm{i}\omega_j, \quad \delta_j < 0, \quad \omega_j \neq 0, \quad i = 1(1)n, \quad j = 1(1)f, \tag{6.50}$$

where for multiple eigenvalues the multiplicity equals the nullity of the characteristic matrix ($d_i = v_i$, according to Section 4.3). The Jordan matrix of the system is then a diagonal matrix (4.72) of the form

$$\boldsymbol{\Lambda} = \left[\begin{array}{c|c} \boldsymbol{\Delta} + \mathrm{i}\boldsymbol{\Omega} & \mathbf{0} \\ \hline \mathbf{0} & \boldsymbol{\Delta} - \mathrm{i}\boldsymbol{\Omega} \end{array}\right], \tag{6.51}$$

$$\boldsymbol{\Delta} = \operatorname{diag}\{\delta_j\} < \mathbf{0}, \quad \boldsymbol{\Omega} = \operatorname{diag}\{\omega_j\}. \tag{6.52}$$

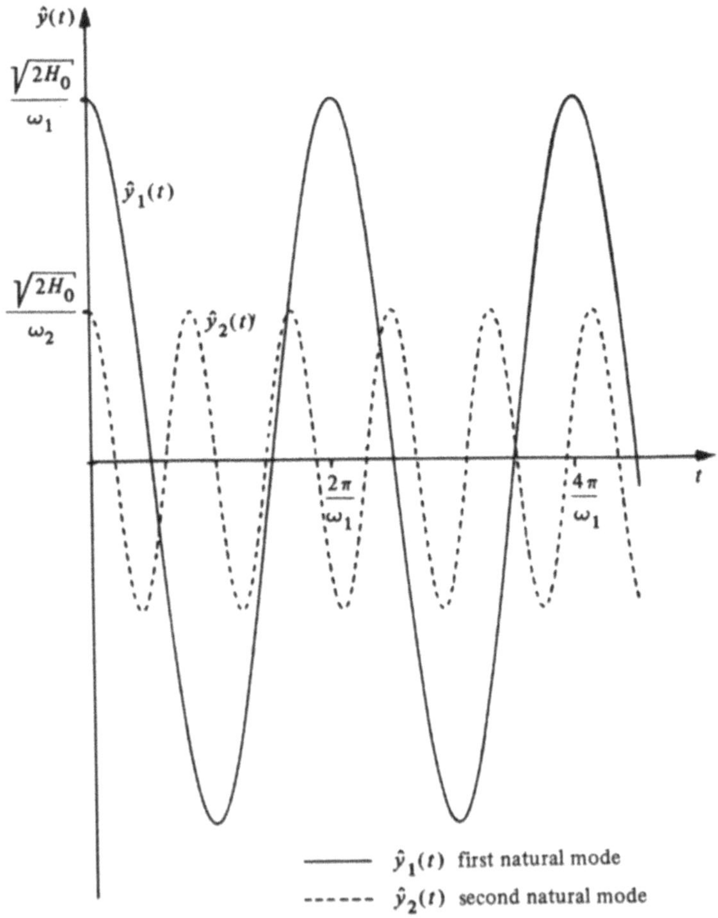

Fig. 6.2. Amplitude-time diagram of the natural modes of the double pendulum at constant total energy H_0

By (4.96, 4.97), the eigenvalue problem (4.53) implies

$$\left[\boldsymbol{M}(\delta_j+\mathrm{i}\omega_j)^2+(\boldsymbol{D}+\boldsymbol{G})(\delta_i+\mathrm{i}\omega_j)+(\boldsymbol{K}+\boldsymbol{N})\right]\tilde{\boldsymbol{y}}_j=\boldsymbol{0}, \tag{6.53}$$

$$\tilde{\boldsymbol{x}}_j=\left[\begin{array}{c}\tilde{\boldsymbol{y}}_j\\ \hdashline (\delta_j+\mathrm{i}\omega_j)\,\tilde{\boldsymbol{y}}_j\end{array}\right], \qquad j=1(1)f, \tag{6.54}$$

$$\left[\boldsymbol{M}(\delta_j-\mathrm{i}\omega_j)^2+(\boldsymbol{D}+\boldsymbol{G})(\delta_j-\mathrm{i}\omega_j)+(\boldsymbol{K}+\boldsymbol{N})\right]\tilde{\boldsymbol{y}}_{f+j}=\boldsymbol{0}, \tag{6.55}$$

$$\tilde{\boldsymbol{x}}_{f+j}=\left[\begin{array}{c}\tilde{\boldsymbol{y}}_{f+j}\\ \hdashline (\delta_j-\mathrm{i}\omega_j)\,\tilde{\boldsymbol{y}}_{f+j}\end{array}\right], \qquad j=1(1)f. \tag{6.56}$$

In contrast to the situation in Section 6.1.1, the eigenvalue problem (4.53) now splits into two mutually conjugate complex eigenvalue problems (6.53) and (6.55) of order $f=n/2$. If one now introduces the complex $f\times f$ matrix

$$\boldsymbol{Y}=\left[\,\tilde{\boldsymbol{y}}_1 \,\vdots\, \tilde{\boldsymbol{y}}_2 \,\vdots\, \cdots \,\vdots\, \tilde{\boldsymbol{y}}_f\right] \tag{6.57}$$

one can write the $n\times n$ modal matrix (4.71) as

$$\boldsymbol{X}=\left[\begin{array}{c:c}\boldsymbol{Y}\;\cdot & \overline{\boldsymbol{Y}}\\ \hdashline \boldsymbol{Y}(\boldsymbol{\Delta}+\mathrm{i}\boldsymbol{\Omega}) & \overline{\boldsymbol{Y}}(\boldsymbol{\Delta}-\mathrm{i}\boldsymbol{\Omega})\end{array}\right], \tag{6.58}$$

where $\overline{\boldsymbol{Y}}$ denotes the complex conjugate of the matrix $\boldsymbol{Y}$. But since in general $\boldsymbol{Y}\neq\overline{\boldsymbol{Y}}$, the further properties shown in Section 6.1.1 are now lost. This becomes clearly visible if one applies to the system (6.48) the real similarity transformation $\boldsymbol{x}(t)=\hat{\boldsymbol{X}}\hat{\boldsymbol{x}}(t)$ with

$$\begin{aligned}\hat{\boldsymbol{X}}&=\boldsymbol{X}\left[\begin{array}{c:c}\frac{1}{2}\boldsymbol{E} & -\frac{1}{2}\mathrm{i}\boldsymbol{\Omega}^{-1}\\ \hdashline \frac{1}{2}\boldsymbol{E} & \frac{1}{2}\mathrm{i}\boldsymbol{\Omega}^{-1}\end{array}\right]\\ &=\left[\begin{array}{c:c}\frac{1}{2}(\boldsymbol{Y}+\overline{\boldsymbol{Y}}) & -\frac{1}{2}\mathrm{i}(\boldsymbol{Y}-\overline{\boldsymbol{Y}})\boldsymbol{\Omega}^{-1}\\ \hdashline \frac{1}{2}(\boldsymbol{Y}+\overline{\boldsymbol{Y}})\boldsymbol{\Delta}+\frac{1}{2}\mathrm{i}(\boldsymbol{Y}-\overline{\boldsymbol{Y}})\boldsymbol{\Omega} & \frac{1}{2}(\boldsymbol{Y}+\overline{\boldsymbol{Y}})-\frac{1}{2}\mathrm{i}(\boldsymbol{Y}-\overline{\boldsymbol{Y}})\boldsymbol{\Delta}\boldsymbol{\Omega}^{-1}\end{array}\right].\end{aligned} \tag{6.59}$$

Due to (4.81), one gets by (6.51) the submatrix representation

$$\begin{aligned}\left[\begin{array}{c}\dot{\hat{\boldsymbol{x}}}_1(t)\\ \hdashline \dot{\hat{\boldsymbol{x}}}_2(t)\end{array}\right]&=\left[\begin{array}{c:c}\boldsymbol{E} & \boldsymbol{E}\\ \hdashline \mathrm{i}\boldsymbol{\Omega} & -\mathrm{i}\boldsymbol{\Omega}\end{array}\right]\left[\begin{array}{c:c}\boldsymbol{\Delta}+\mathrm{i}\boldsymbol{\Omega} & \boldsymbol{0}\\ \hdashline \boldsymbol{0} & \boldsymbol{\Delta}-\mathrm{i}\boldsymbol{\Omega}\end{array}\right]\left[\begin{array}{c:c}\frac{1}{2}\boldsymbol{E} & -\frac{1}{2}\mathrm{i}\boldsymbol{\Omega}^{-1}\\ \hdashline \frac{1}{2}\boldsymbol{E} & \frac{1}{2}\mathrm{i}\boldsymbol{\Omega}^{-1}\end{array}\right]\left[\begin{array}{c}\hat{\boldsymbol{x}}_1(t)\\ \hdashline \hat{\boldsymbol{x}}_2(t)\end{array}\right]\\ &=\left[\begin{array}{c:c}\boldsymbol{\Delta} & \boldsymbol{E}\\ \hdashline -\boldsymbol{\Omega}^2 & \boldsymbol{\Delta}\end{array}\right]\left[\begin{array}{c}\hat{\boldsymbol{x}}_1(t)\\ \hdashline \hat{\boldsymbol{x}}_2(t)\end{array}\right]\end{aligned} \tag{6.60}$$

or

$$\left.\begin{aligned}\ddot{\hat{x}}_1(t)-2\boldsymbol{\Delta}\dot{\hat{x}}_1(t)+(\boldsymbol{\Delta}^2+\boldsymbol{\Omega}^2)\hat{x}_1(t)=\mathbf{0},\\ \hat{x}_2(t)=\dot{\hat{x}}_1(t)-\boldsymbol{\Delta}\hat{x}_1(t).\end{aligned}\right\}\tag{6.61}$$

This is again a system of f second order differential equations, but the coordinates $\hat{x}_{1j}$, $j=1(1)f$, are related by the similarity transformation with matrix (6.59) to the position and velocity coordinates y_j and $\dot{y}_j$, $j=1(1)f$, in contrast to the coordinates $\hat{y}_j$, $j=1(1)f$, in (6.21) which were related by a congruence transformation (6.20) to the position coordinates y_j, $j=1(1)f$. Thus the coordinates in (6.61) have lost their physical meaning and must simply be viewed as mathematically convenient coordinates. Moreover one should observe that suitable numerical procedures are not known for solving (6.53) and (6.55). Thus it is recommended to solve the eigenvalue problem in the state space representation directly and to use (6.58) for controlling the result.

By (6.58) and (4.74), the general solution of the vibrating system (6.47) can be written as

$$y(t)=\boldsymbol{Y}\mathrm{e}^{(\boldsymbol{\Delta}+\mathrm{i}\boldsymbol{\Omega})t}\boldsymbol{c}_1+\bar{\boldsymbol{Y}}\mathrm{e}^{(\boldsymbol{\Delta}-\mathrm{i}\boldsymbol{\Omega})t}\boldsymbol{c}_2.\tag{6.62}$$

Should the j-th natural mode be excited, i.e., the partial vibration with frequency ω_j, then one has to substitute (6.26) in (6.62). It follows that the initial conditions for the j-th natural mode are given by

$$\left.\begin{aligned}y_0&=(\alpha\boldsymbol{Y}+\beta\bar{\boldsymbol{Y}})\boldsymbol{e}_j=\alpha\tilde{y}_j+\beta\bar{\tilde{y}}_j,\\ \dot{y}_0&=(\alpha\boldsymbol{Y}+\beta\bar{\boldsymbol{Y}})\boldsymbol{\Delta}\boldsymbol{e}_j+\mathrm{i}(\alpha\boldsymbol{Y}-\beta\bar{\boldsymbol{Y}})\boldsymbol{\Omega}\boldsymbol{e}_j\\ &=\left(\alpha\tilde{y}_j+\beta\bar{\tilde{y}}_j\right)\delta_j+\mathrm{i}\left(\alpha\tilde{y}_j-\beta\bar{\tilde{y}}_j\right)\omega_j.\end{aligned}\right\}\tag{6.63}$$

Substituting $\alpha=\beta=\frac{1}{2}$, one gets for $\boldsymbol{Y}=\boldsymbol{U}+\mathrm{i}\boldsymbol{V}$ the real initial conditions

$$y_0=\boldsymbol{U}\boldsymbol{e}_j,\ \dot{y}_0=(\boldsymbol{U}\boldsymbol{\Delta}-\boldsymbol{V}\boldsymbol{\Omega})\boldsymbol{e}_j.\tag{6.64}$$

This means that for damped systems the natural modes are, as a rule, excited by the initial displacements and initial velocities. Hence the j-th natural mode has the solution

$$\begin{aligned}y_j(t)&=\tfrac{1}{2}\left[\mathrm{e}^{(\delta_j+\mathrm{i}\omega_j)t}\tilde{y}_j+\mathrm{e}^{(\delta_j-\mathrm{i}\omega_j)t}\bar{\tilde{y}}_j\right]\\ &=\mathrm{e}^{\delta_j t}\left[\cos\omega_j t\,\boldsymbol{U}\boldsymbol{e}_j-\sin\omega_j t\,\boldsymbol{V}\boldsymbol{e}_j\right]\\ &=\mathrm{e}^{\delta_j t}\cos\left(\omega_j t\boldsymbol{E}+\boldsymbol{\varphi}_j\right)\boldsymbol{a}_j,\end{aligned}\tag{6.65}$$

where $\boldsymbol{a}_j$ and $\boldsymbol{\varphi}_j$ are defined by

$$\boldsymbol{a}_j = \left[(a_j)_k\right], \qquad (a_j)_k = \sqrt{U_{kj}^2 + V_{kj}^2}, \tag{6.66}$$

$$\boldsymbol{\varphi}_j = \operatorname{diag}\left\{(\varphi_j)_k\right\}, \qquad (\varphi_j)_k = \arctan\frac{V_{kj}}{U_{kj}}, \qquad k = 1(1)f. \tag{6.67}$$

Consequently the real and imaginary part of the j-th eigenvector $\tilde{\boldsymbol{y}}_j$ determine the amplitudes $(a_j)_k$ and phase angles $(\varphi_j)_k$ for the individual coordinates y_k, $k = 1(1)f$, of the j-th natural mode. A characteristic plot for the coordinate y_k of the j-th natural mode can be seen in Fig. 6.3 as a $(y_k, \dot{y}_k)$ representation, and in Fig. 6.4 as a (y_k, t)-representation.

The essential difference between the determination of the natural modes for nongyroscopic conservative systems (6.1) and for lightly damped systems (6.47) can be formulated as follows: the decoupling of the system can be obtained by a congruence transformation in the first case, and in the second case, in general, only by a similarity transformation. The question arises if there exists mechanical systems (6.47) whose

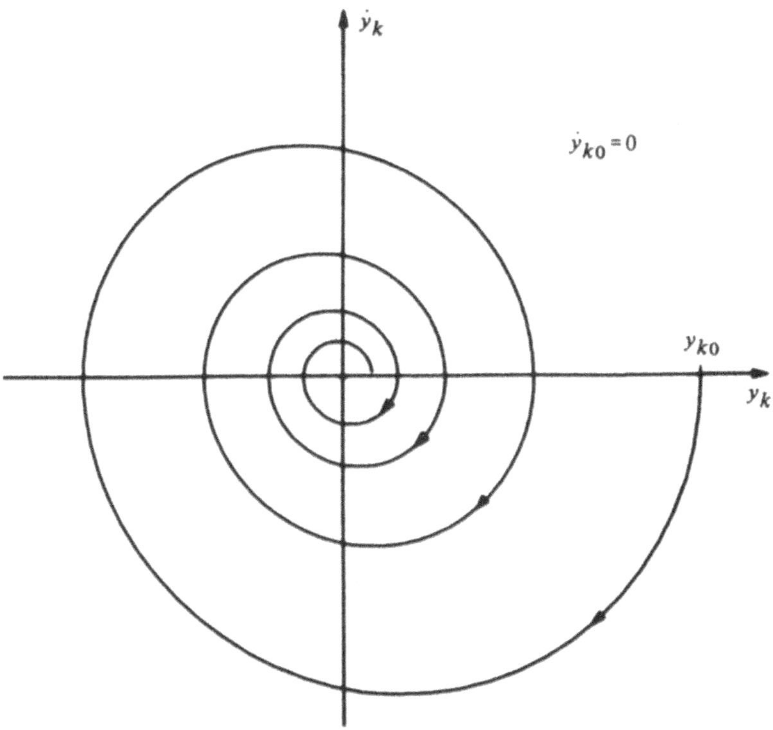

Fig. 6.3. Lightly damped natural mode in the state plane

differential equations can be decoupled by a congruence transformation.

A congruence transformation preserves the properties of symmetry and skewsymmetry of the matrices appearing in (6.47). However, since the equations are always coupled by skewsymmetric matrices the above problems can be investigated with success only for systems of the form

$$\boldsymbol{M}\ddot{\boldsymbol{y}}(t)+\boldsymbol{D}\dot{\boldsymbol{y}}(t)+\boldsymbol{K}\boldsymbol{y}(t)=\boldsymbol{0}. \tag{6.68}$$

Consequently, one assertains that the system (6.68) can be decoupled by a congruence transformation if and only if

$$\boldsymbol{K}\boldsymbol{M}^{-1}\boldsymbol{D}=\boldsymbol{D}\boldsymbol{M}^{-1}\boldsymbol{K} \tag{6.69}$$

holds. In this case (6.68) is taken by the transformation (6.20) into

$$\ddot{\hat{\boldsymbol{y}}}(t)-2\boldsymbol{\Delta}\dot{\hat{\boldsymbol{y}}}(t)+(\boldsymbol{\Delta}^2+\boldsymbol{\Omega}^2)\hat{\boldsymbol{y}}(t)=\boldsymbol{0} \tag{6.70}$$

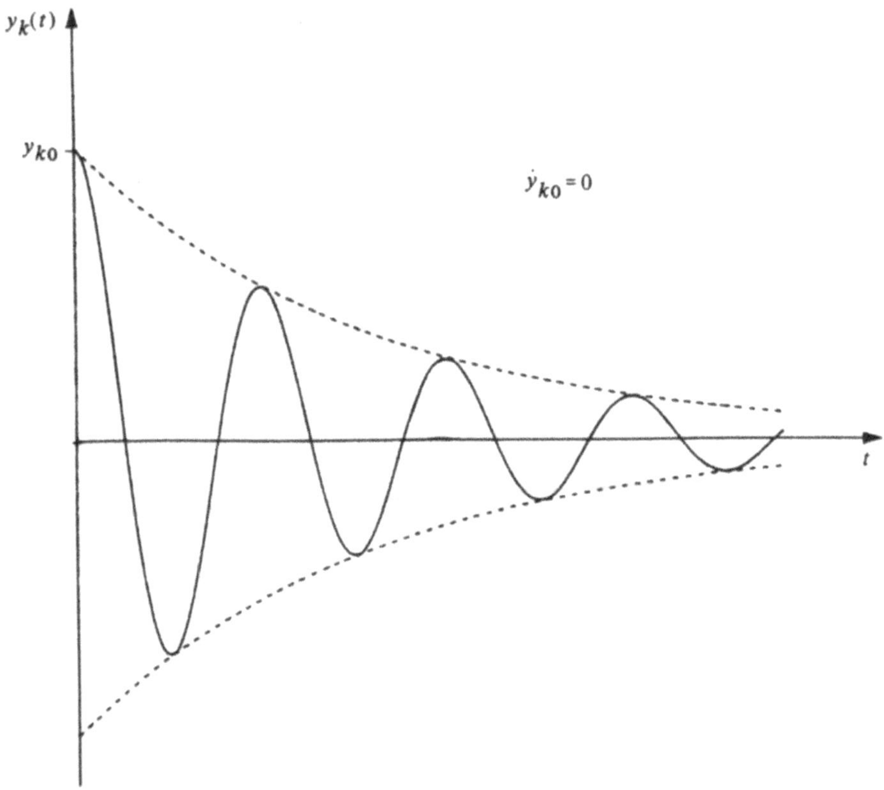

Fig. 6.4. Amplitude-time diagram of a lightly damped natural mode

with

$$\left.\begin{aligned}\hat{\boldsymbol{Y}}^T\boldsymbol{M}\hat{\boldsymbol{Y}}=\boldsymbol{E}, \quad \hat{\boldsymbol{Y}}^T\boldsymbol{D}\hat{\boldsymbol{Y}}= -2\boldsymbol{\Delta},\\ \hat{\boldsymbol{Y}}^T\boldsymbol{K}\hat{\boldsymbol{Y}}=\hat{\boldsymbol{\Omega}}^2=\boldsymbol{\Delta}^2+\boldsymbol{\Omega}^2.\end{aligned}\right\} \tag{6.71}$$

The general solution of (6.68) with (6.69), generalizing (6.25), is now given by

$$y(t)=\boldsymbol{Y}\mathrm{e}^{\boldsymbol{\Delta}t}\{(\cos\boldsymbol{\Omega}t-\boldsymbol{\Delta}\boldsymbol{\Omega}^{-1}\sin\boldsymbol{\Omega}t)\boldsymbol{Y}^{-1}y_0+\boldsymbol{\Omega}^{-1}\sin\boldsymbol{\Omega}t\boldsymbol{Y}^{-1}\dot{y}_0\}. \tag{6.72}$$

It can be seen from (6.71) that the natural frequencies ($\boldsymbol{\Omega}^2$) are smaller for damped systems (6.68) than for undamped systems (6.1).

Condition (6.69) is satisfied for instance when $\boldsymbol{D}$ is a linear combination of the inertia and stiffness matrix,

$$\boldsymbol{D}=\alpha\boldsymbol{M}+\beta\boldsymbol{K}. \tag{6.73}$$

For $\alpha=0$ one obtains the known relation often used in structural mechanics for viscous material damping. Because of the *simultaneous diagonalization of* the matrices $\boldsymbol{M}$, $\boldsymbol{D}$, $\boldsymbol{K}$ one also speaks of *modal damping.*

Example 6.2. Vertical vibrations of an automobile

The equation of motion (4.99) of the vertical vibrations of an automobile has the form (6.68). Thus it may be firstly checked whether the system can be decoupled by a congruence transformation. If (6.69) is written out in detail,

$$\boldsymbol{K}\boldsymbol{M}^{-1}\boldsymbol{D}=\begin{bmatrix} k_1d_1\left(\frac{1}{m_1}+\frac{1}{m_2}\right) & -k_1d_1\left(\frac{1}{m_1}+\frac{1}{m_2}\right)\\ -k_1d_1\left(\frac{1}{m_1}+\frac{1}{m_2}\right)-\frac{k_2d_1}{m_2} & k_1d_1\left(\frac{1}{m_1}+\frac{1}{m_2}\right)+\frac{k_2d_1}{m_2}\end{bmatrix}$$

$$\neq\begin{bmatrix} k_1d_1\left(\frac{1}{m_1}+\frac{1}{m_2}\right) & -k_1d_1\left(\frac{1}{m_1}+\frac{1}{m_2}\right)-\frac{k_2d_1}{m_2}\\ -k_1d_1\left(\frac{1}{m_1}+\frac{1}{m_2}\right) & k_1d_1\left(\frac{1}{m_1}+\frac{1}{m_2}\right)+\frac{k_2d_1}{m_2}\end{bmatrix}$$

$$=\boldsymbol{D}\boldsymbol{M}^{-1}\boldsymbol{K}, \tag{6.74}$$

then it can be seen that the similarity transformation (6.59) is needed for the determination of the natural modes. The transformation matrix $\hat{\boldsymbol{X}}$ formed from the eigenvectors $\tilde{\boldsymbol{x}}_i$, $i=1(1)4$, in (4.106) is

$$\hat{\boldsymbol{X}} = [\tilde{\boldsymbol{x}}_1 \,\vdots\, \tilde{\boldsymbol{x}}_2 \,\vdots\, \tilde{\boldsymbol{x}}_3 \,\vdots\, \tilde{\boldsymbol{x}}_4]\frac{1}{2}\left[\begin{array}{c|c} \boldsymbol{E} & -\mathrm{i}\boldsymbol{\Omega}^{-1} \\ \hline \boldsymbol{E} & \mathrm{i}\boldsymbol{\Omega}^{-1} \end{array}\right] \tag{6.75}$$

where

$$\boldsymbol{\Omega} = \begin{bmatrix} 57.003 & 0 \\ 0 & 4.578 \end{bmatrix}. \tag{6.76}$$

Therefore

$$\hat{\boldsymbol{X}} = \left[\tfrac{1}{2}(\tilde{\boldsymbol{x}}_1 + \tilde{\boldsymbol{x}}_2) \,\vdots\, \tfrac{1}{2}(\tilde{\boldsymbol{x}}_3 + \tilde{\boldsymbol{x}}_4) \,\vdots\, -\frac{\mathrm{i}}{2\omega_1}(\tilde{\boldsymbol{x}}_1 - \tilde{\boldsymbol{x}}_2) \,\vdots\, -\frac{\mathrm{i}}{2\omega_2}(\tilde{\boldsymbol{x}}_3 - \tilde{\boldsymbol{x}}_4)\right]$$

$$= \begin{bmatrix} -0.031 & 8.2 & 0.001 & -1.64 \\ 1 & 1 & 0 & 0 \\ 4.01 & 20 & 0.002 & 11.08 \\ -30.25 & -1.75 & 1 & 1 \end{bmatrix}. \tag{6.77}$$

Then one has with $\hat{\boldsymbol{x}}_1 = [\hat{x}_1 \quad \hat{x}_2]^T$, $\hat{\boldsymbol{x}}_2 = [\hat{x}_3 \quad \hat{x}_4]^T$, by (6.61),

$$\left.\begin{aligned} &\ddot{\hat{\boldsymbol{x}}}_1(t) - 2\begin{bmatrix} -30.253 & 0 \\ 0 & -1.747 \end{bmatrix}\dot{\hat{\boldsymbol{x}}}_1(t) + \begin{bmatrix} 4168 & 0 \\ 0 & 24 \end{bmatrix}\hat{\boldsymbol{x}}_1(t) = \boldsymbol{0}, \\ &\hat{\boldsymbol{x}}_2(t) = \dot{\hat{\boldsymbol{x}}}_1(t) - \begin{bmatrix} -30.253 & 0 \\ 0 & -1.747 \end{bmatrix}\hat{\boldsymbol{x}}_1(t). \end{aligned}\right\} \tag{6.78}$$

Both natural modes are shown in Fig. 6.5. The results of Fig. 4.3 are obtained by a suitable superposition of both natural modes. ◆

6.1.3. Ordinary mechanical systems

For ordinary mechanical systems with differential equations (6.47) or (6.48) and the characteristic equation (6.49) the eigenvalues have the form

$$\left.\begin{aligned} &\lambda_i = \delta_j \pm \mathrm{i}\omega_j, \quad \omega_j \neq 0, \quad i = 1(1)2s, \quad j = 1(1)s, \\ &\lambda_i = \sigma_k, \qquad\qquad\qquad\quad i = 2s+1(1)n, \quad k = 1(1)n-2s, \end{aligned}\right\} \tag{6.79}$$

where for repeated eigenvalues the multiplicity is equal to the nullity of the characteristic matrix (Case I in Section 4.3). The corresponding Jordan matrix is then a diagonal matrix (4.72) of the form

$$\Lambda = \left[\begin{array}{c:c:c} \Delta + \mathrm{i}\Omega & \mathbf{0} & \mathbf{0} \\ \hdashline \mathbf{0} & \Delta - \mathrm{i}\Omega & \mathbf{0} \\ \hdashline \mathbf{0} & \mathbf{0} & \Sigma \end{array}\right], \quad \left.\begin{array}{l} \Delta = \operatorname{diag}\{\delta_j\}, \\ \Omega = \operatorname{diag}\{\omega_j\}, \\ \Sigma = \operatorname{diag}\{\sigma_k\}. \end{array}\right\} \tag{6.80}$$

From the eigenvalue problem (4.53) one obtains, by (6.48) and (6.79), the modal matrix

$$\boldsymbol{X} = \left[\begin{array}{c:c:c} \boldsymbol{Y}_1 & \overline{\boldsymbol{Y}}_1 & \boldsymbol{Y}_3 \\ \hdashline \boldsymbol{Y}_1(\Delta + \mathrm{i}\Omega) & \overline{\boldsymbol{Y}}_1(\Delta - \mathrm{i}\Omega) & \boldsymbol{Y}_3\Sigma \end{array}\right], \tag{6.81}$$

where $\boldsymbol{Y}_1$ follows from the eigenvalue problem (6.53) and $\boldsymbol{Y}_3$ is formed from the eigenvectors of the equation

$$\left[\boldsymbol{M}\sigma_k^2 + (\boldsymbol{D} + \boldsymbol{G})\sigma_k + (\boldsymbol{K} + \boldsymbol{N})\right]\tilde{\boldsymbol{y}}_k = \boldsymbol{0}, \quad k = 1(1)n - 2s. \tag{6.82}$$

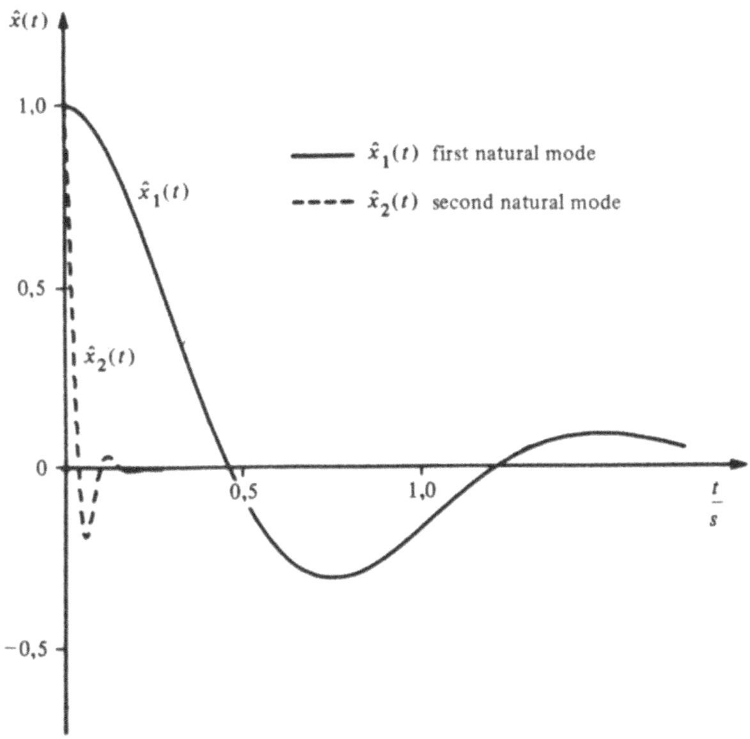

Fig. 6.5. Natural modes $\hat{x}_1(t)$ and $\hat{x}_2(t)$ for the vertical vibrations of a motor car

If one extends the similarity transformation (6.59) to

$$\hat{X} = X \begin{bmatrix} \frac{1}{2}E & -\frac{1}{2}\mathrm{i}\Omega^{-1} & 0 \\ \frac{1}{2}E & \frac{1}{2}\mathrm{i}\Omega^{-1} & 0 \\ 0 & 0 & E \end{bmatrix}, \tag{6.83}$$

one gets in submatrix notation

$$\begin{bmatrix} \dot{\hat{x}}_1(t) \\ \dot{\hat{x}}_2(t) \\ \dot{\hat{x}}_3(t) \end{bmatrix} = \begin{bmatrix} \Delta & E & 0 \\ -\Omega^2 & \Delta & 0 \\ 0 & 0 & \Sigma \end{bmatrix} \begin{bmatrix} \hat{x}_1(t) \\ \hat{x}_2(t) \\ \hat{x}_3(t) \end{bmatrix} \tag{6.84}$$

with the $s \times 1$ vectors $\hat{x}_1$, $\hat{x}_2$ and the $(n-2s)\times 1$ vector $\hat{x}_3$. Finally (6.84) yields the decoupled system

$$\left.\begin{aligned} &\ddot{\hat{x}}_1(t) - 2\Delta\dot{\hat{x}}_1(t) + (\Delta^2 + \Omega^2)\hat{x}_1(t) = 0, \\ &\hat{x}_2(t) = \dot{\hat{x}}_1(t) - \Delta\hat{x}_1(t), \\ &\dot{\hat{x}}_3(t) - \Sigma\hat{x}_3(t) = 0. \end{aligned}\right\} \tag{6.85}$$

The general solution of the vibrating system (6.47) is obtained by superimposing the various natural modes, and by (6.81) and (4.74) it can be written as

$$y(t) = Y_1\,\mathrm{e}^{(\Delta+\mathrm{i}\Omega)t}c_1 + \bar{Y}_1\,\mathrm{e}^{(\Delta-\mathrm{i}\Omega)t}c_2 + Y_3\,\mathrm{e}^{\Sigma t}c_3. \tag{6.86}$$

To get a real solution, one must take $c_2 = \bar{c}_1$ and c_3 real. Should the j-th natural mode be excited, $j = 1(1)s$, i.e., the individual vibration with the frequency ω_j, then one has to set $c_3 = 0$, and the results (6.63 to 67) hold without alterations. Should the i-th natural mode be excited, $i = 2s + 1(1)n$, i.e., the individual motion with the eigenvalue σ_k, $k = 1(1)n - 2s$, then one has to set $c_1 = c_2 = 0$ and $c_3 = e_k$. The corresponding initial conditions read as

$$y_0 = Y_3 e_k, \quad \dot{y}_0 = Y_3 \Sigma e_k. \tag{6.87}$$

Therefore the i-th natural mode, $i = 2s + 1(1)n$, has the solution

$$y_i(t) = \mathrm{e}^{\sigma_k t} Y_3 e_k = \mathrm{e}^{\sigma_k t}\tilde{y}_i, \quad i = 2s + 1(1)n, \quad k = 1(1)n - 2s. \tag{6.88}$$

The i-th eigenvector $\tilde{y}_i$ characterizes the amplitudes of the individual coordinates $(y_i)_j$, $j = 1(1)f$, of the i-th natural mode. Fig. 6.6 shows the time variation of a coordinate $(y_i(t))_j$ of an i-th natural mode (6.88) with $\sigma_k < 0$. It is seen that the motion is typically aperiodic.

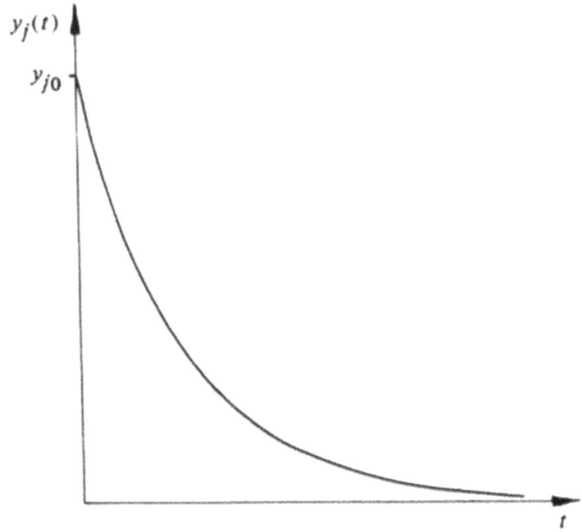

Fig. 6.6. Amplitude-time diagram of a strongly damped natural mode

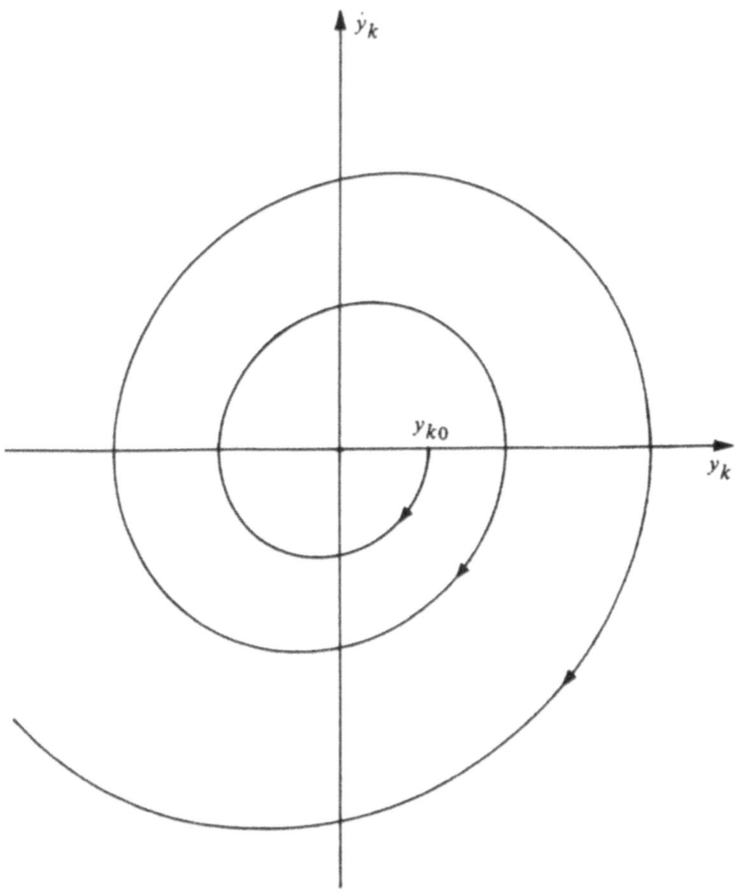

Fig. 6.7. Oscillatory unstable natural mode in the state plane

Figures 6.1 to 6.6 illustrate various stable or asymptotically stable natural modes. However in the vibrations of ordinary mechanical systems unstable natural modes can also occur. Thus the diagrams are supplemented by those on Figures 6.7 to 6.10 exhibiting typical vibrations for $\delta_j > 0$, $\sigma_k = 0$ and $\sigma_k > 0$.

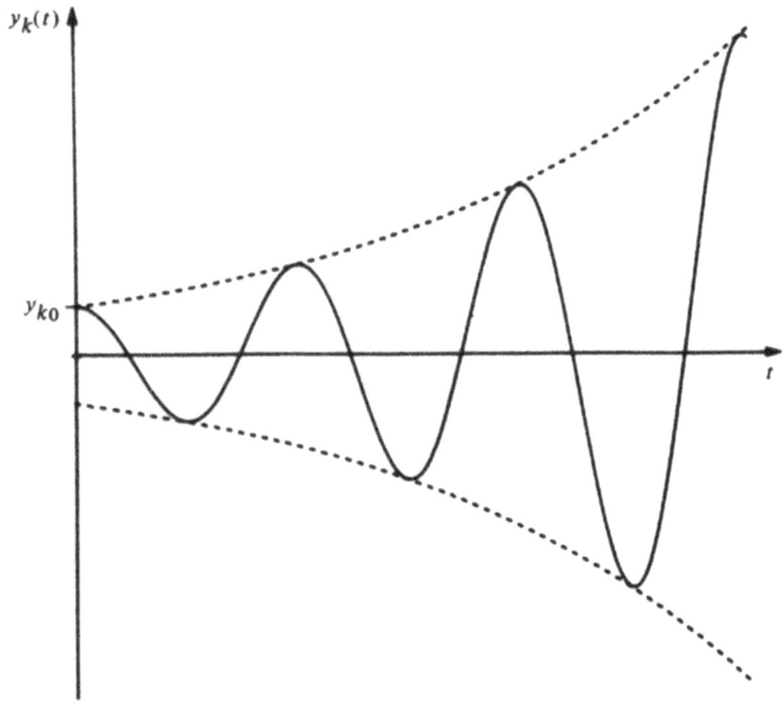

Fig. 6.8. Amplitude-time diagram of a oscillatory unstable natural mode

Fig. 6.9. Amplitude-time diagram of a monotone marginally stable natural mode

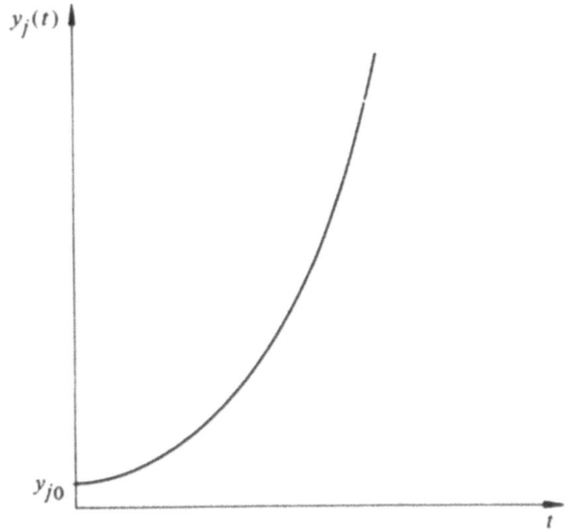

Fig. 6.10. Amplitude-time diagram of a monotone unstable natural mode

6.2. Optimal natural vibrations

The parameters determining the behavior of mechanical vibrating systems, such as spring or damping constants, often do not have fixed values, the paramaters can vary within a certain range. One may then attempt to choose the *parameters* within these ranges so that the system becomes asymptotically stable and moreover exhibits an optimum behavior in some specified sense. Here three problems turn up. One has to decide on a *cost functional* rating the behavior of the system; one has to calculate the values of this functional for the parameter combinations of interest; finally the optimization of the functional has to take place. These problems are summarized as

1. Choice of the cost functional,
2. Calculation of the cost functional,
3. Optimization with respect to the cost functional.

6.2.1. Cost functionals

The choice of a meaningful cost functional is a proper engineering problem which can often be tackled only after a detailed analysis of the proposed problem. Subsequently some functionals will be explained which are frequently used for the estimation of natural vibrations.

Let again the state equation of the vibrating system be given by

$$\dot{\boldsymbol{x}}(t) = \boldsymbol{A}\boldsymbol{x}(t) = \boldsymbol{A}(p_j)\boldsymbol{x}(t), \quad \boldsymbol{x}(0) = \boldsymbol{x}_0, \tag{6.89}$$

where the $n \times n$ system matrix $\boldsymbol{A}$ depends on certain system parameters p_j, $j = 1(1)l$, varying within specified ranges. Assuming a continuous dependence $\boldsymbol{A}(p_j)$, one sees that the whole system as well as the eigenvalues continuously depend on the parameters. If the vibrating system (6.89) is asymptotically stable for some admissible parameter combination, $\operatorname{Re} \lambda_i < 0$, $i = 1(1)n$, then it follows from the assumption of continuity that the system (6.89) will remain asymptotically stable when the parameters are varied within certain ranges. Thus it is meaningful to investigate asymptotically stable systems by means of cost functionals.

The first cost functional to be considered is the *degree of stability* (absolute stability reserve) defined as the distance d_0 from the imaginary axis of the nearest eigenvalue (Fig. 6.11)

$$d_0 = -\max_{i=1(1)n} \{\operatorname{Re} \lambda_i\}. \tag{6.90}$$

The degree of stability characterizes the rate of decaying of the transition process. By (4.33), the natural vibration $\boldsymbol{x}(t) = \boldsymbol{\Phi}(t)\boldsymbol{x}_0$ can be estimated approximately by

$$\| \boldsymbol{x}(t) \| < c\, \mathrm{e}^{-d_0 t}. \tag{6.91}$$

If the degree of stability is large, then the natural vibrations will decay fast.

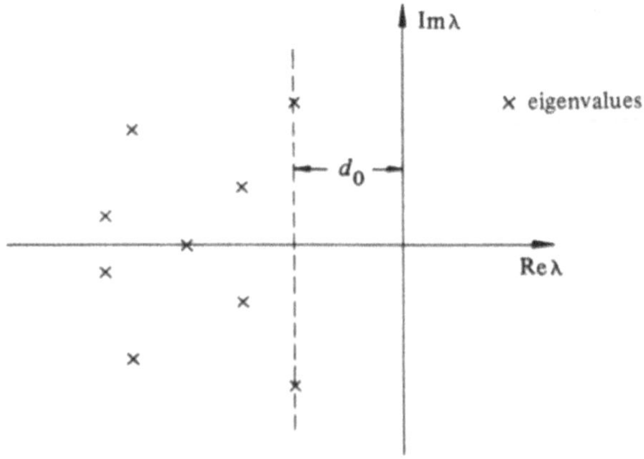

Fig. 6.11. Degree of stability d_0

For lightly damped systems with complex conjugate eigenvalues $\lambda_i = \delta_i \pm i\omega_i$, $\delta_i < 0$ one can define the *degree of damping* (relative stability reserve), Fig. 6.12,

$$D_0 = -\max_{i=1(1)n} \left\{ \frac{\operatorname{Re} \lambda_i}{|\lambda_i|} \right\}. \tag{6.92}$$

The degree of damping is the smallest *damping factor*

$$D_i = \frac{|\delta_i|}{\sqrt{\delta_i^2 + \omega_i^2}}, \qquad i = 1(1)n, \tag{6.93}$$

of the various natural vibrations, and it indicates the reduction of an initial displacement to a certain ratio with respect to the number of periods. Thus a large degree of damping implies that the natural vibration will decay in just a few periods. For $D_0 = 1$, one gets an aperiodic transition. A large degree of damping also guarantees small amplitudes for harmonically excited vibrating systems (Section 8) and hence a good disturbance behavior.

An essential characteristic of the degree of stability and the degree of damping is that they do not depend on the initial conditions $\boldsymbol{x}_0$ of the vibrating system (6.89).

An especially important place among the cost functionals optimizing the parameters is occupied by the *integrals of error*. For asymptotically stable systems these are defined as improper converging integrals over nonnegative functionals of the solution vector $\boldsymbol{x}(t)$. They yield measures for the displacement $\boldsymbol{x}(t)$ from the zero position over the total duration

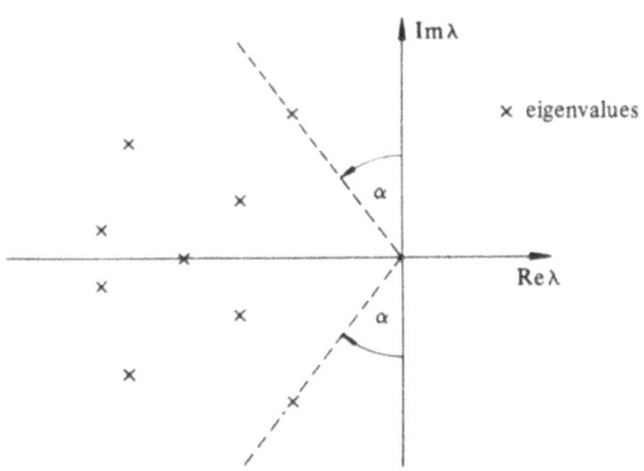

Fig. 6.12. Degree of damping $D_0 = \sin \alpha$

of the process, $0 \leqslant t < \infty$. The best known integral cost functional is the *integral of squared error*

$$IE^2 = \int_0^\infty \boldsymbol{x}^T(\tau)\boldsymbol{S}\boldsymbol{x}(\tau)\,\mathrm{d}\tau \tag{6.94}$$

with a symmetric positive semidefinite $n \times n$ weighting matrix $\boldsymbol{S} = \boldsymbol{S}^T \geqslant 0$ (IE^2 means Integral of $(\text{Error})^2$). The integral of squared error characterizes the transition behavior of the vibration and it strongly evaluates the large amplitudes because the integrand in (6.94) depends on the square of $\boldsymbol{x}(t)$. Small cost IE^2 implies small amplitudes at the transition of the vibration. The integral of squared error depends on the initial conditions $\boldsymbol{x}_0$ and on the weighting matrix $\boldsymbol{S}$. These quantities have to be chosen according to the engineering requirements.

Example 6.3. Integral of squared error for a single-dof-vibrating system
Consider the single-dof-vibrating system

$$m\ddot{y}(t) + d\dot{y}(t) + ky(t) = 0 \tag{6.95}$$

where for fixed m and k the damping parameter d undergoes variation. For

$$\boldsymbol{x} = \begin{bmatrix} y \\ \dot{y} \end{bmatrix}, \qquad \boldsymbol{S} = \begin{bmatrix} 1 & 0 \\ 0 & 0 \end{bmatrix}, \qquad \boldsymbol{x}_0 = \begin{bmatrix} 1 \\ 0 \end{bmatrix} \tag{6.96}$$

the integral of squared error is

$$IE^2 = \int_0^\infty y^2(\tau)\mathrm{d}\tau. \tag{6.97}$$

The time variation of $y(t)$ and $\int_0^t y^2(\tau)\mathrm{d}\tau$ is shown in Fig. 6.13. One can recognize that the minimum of the integral of squared error is to be expected between light and strong damping for a middle value of the damping constant. ◆

Besides the integral of squared error (6.94), other important integrals used for parameter optimization are the *time weighted integrals of squared error*

$$IT^kE^2 = \int_0^\infty \tau^k \boldsymbol{x}^T(\tau)\boldsymbol{S}\boldsymbol{x}(\tau)\mathrm{d}\tau \tag{6.98}$$

with $\boldsymbol{S} = \boldsymbol{S}^T \geqslant 0$, and the *time weighted integrals of absolute value of error*

$$IT^kAE = \int_0^\infty \tau^k \|\boldsymbol{x}(\tau)\|\mathrm{d}\tau \tag{6.99}$$

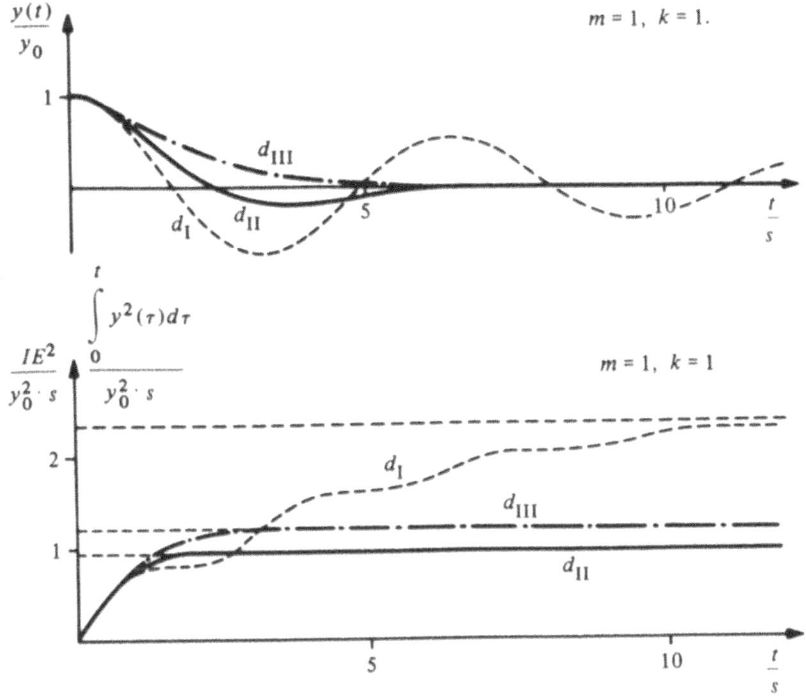

Fig. 6.13. Integral of square error for a single-dof vibrating system: d_{I} light, d_{II} middle, d_{III} strong damping.

with one of the vector norms (5.2 to 5.5). (Here IT^kE^2 stands for Integral of (Time)^k times (Error)^2, and IT^kAE stands for Integral of (Time)^k times Absolute Error). An important feature of the cost functionals (6.98) and (6.99) is that due to the presence of the time factor the later appearing amplitudes are evaluated more strongly. Besides the dependence on $\boldsymbol{x}_0$ and $\boldsymbol{S}$ there is also present the dependence on the time power k.

For asymptotically stable systems one can further define a *absolute value criterion* based on the maximum amplitudes of individual or all state coordinates,

$$AE = \max_i \left\{ \max_{0 \leqslant t < \infty} |x_i(t)| \right\}, \tag{6.100}$$

where i characterizes the coordinates which are of interest (AE stands for Absolute Error). The absolute value criterion depends on the initial conditions $\boldsymbol{x}_0$ and the choice of the state coordinates x_i.

6.2.2. Calculation of the cost functionals

The calculation procedures for the cost functionals introduced in Section

6.2.1 are quite different from each other. The calculation of the degree of stability d_0 and the degree of damping D_0 is based on the definitions (6.90) and (6.92) and on the knowledge of the eigenvalues which, in general, have to be determined numerically.

The integral of squared error (6.94) has the special property that it can be calculated without integration applying the *algebraic Ljapunov matrix* equation. From the proposition

$$IE^2(t) = \int_0^t \boldsymbol{x}^T(\tau)\boldsymbol{S}\boldsymbol{x}(\tau)\mathrm{d}\tau = \boldsymbol{x}_0^T\boldsymbol{R}\boldsymbol{x}_0 - \boldsymbol{x}^T(t)\boldsymbol{R}\boldsymbol{x}(t) \tag{6.101}$$

one obtains by differentiating and taking into account (6.89)

$$\begin{aligned}\boldsymbol{x}^T(t)\boldsymbol{S}\boldsymbol{x}(t) &= -\dot{\boldsymbol{x}}^T(t)\boldsymbol{R}\boldsymbol{x}(t) - \boldsymbol{x}^T(t)\boldsymbol{R}\dot{\boldsymbol{x}}(t)\\ &= -\boldsymbol{x}^T(t)[\boldsymbol{A}^T\boldsymbol{R} + \boldsymbol{R}\boldsymbol{A}]\boldsymbol{x}(t),\end{aligned} \tag{6.102}$$

where $\boldsymbol{R}$ is a symmetric $n \times n$ matrix. Since equation (6.102) has to hold for arbitrary solutions $\boldsymbol{x}(t)$, (6.102) implies the algebraic Ljapunov matrix equation

$$\boldsymbol{A}^T\boldsymbol{R} + \boldsymbol{R}\boldsymbol{A} = -\boldsymbol{S}, \tag{6.103}$$

which occured already in (5.59) during the discussion of stability problems. Since $\boldsymbol{x}(t) \to \boldsymbol{0}$ for $t \to \infty$, (6.101) implies

$$IE^2 = \boldsymbol{x}_0^T\boldsymbol{R}\boldsymbol{x}_0. \tag{6.104}$$

Thereby the calculation of the integral of squared error (6.94) is reduced to finding a solution $\boldsymbol{R}$ of Ljapunov's equation (6.103). For asymptotically stable systems the matrix $\boldsymbol{R}$ is uniquely determined by (6.103) and for $\boldsymbol{S} = \boldsymbol{S}^T \geqslant 0$ it is at least positive semidefinite. The methods of solving (6.103) will be dealt with in Sections 13.2 and 14.2.

The calculation of the time weighted integrals of squared errors (6.98) proceeds by partial integration and $(k+1)$ applications of the above shown solution method for the quadratic cost functional IE^2. Using the notation

$$\boldsymbol{S}_0 = \boldsymbol{S}, \qquad IT^{k-l}E^2(\boldsymbol{S}_l) = \int_0^\infty \boldsymbol{x}^T(\tau)\boldsymbol{S}_l\boldsymbol{x}(\tau)\mathrm{d}\tau \tag{6.105}$$

one obtains the recursion algorithm

$$\begin{aligned}IT^kE^2(\boldsymbol{S}_0) &= l!\binom{k}{l} IT^{k-l}E^2(\boldsymbol{S}_l)\\ &= k!IE^2(\boldsymbol{S}_k) = k!\boldsymbol{x}_0^T\boldsymbol{S}_{k+1}\boldsymbol{x}_0\end{aligned} \tag{6.106}$$

with

$$A^T S_l + S_l A = -S_{l-1}, \qquad l = 1(1)k+1. \tag{6.107}$$

In general, it is not possible to explicitly calculate with algebraic methods the cost functionals (6.99) and the absolute value criterion (6.100). As a rule, one has to carry out numerical evaluations of the defining equations (6.99) and (6.100), where the trajectories have to be determined from the general solution given in Section 4.

6.2.3. Optimal parameters

The free system parameters p_j, $j = 1(1)l$, can be used to optimize the behaviour of the vibrating system in the sense of a given cost functional

$$J = J(p_j). \tag{6.108}$$

The optimal behavior takes place when the cost functional attains its minimum or maximum value, respectively. The corresponding parameters are called *optimal parameters.*

In case of unbounded parameters a necessary condition for the existence of an extremum of the cost functional (6.108) is the vanishing of all partial derivatives

$$\frac{\partial J(p_j)}{\partial p_k} = 0, \qquad k = 1(1)l. \tag{6.109}$$

However condition (6.109) is in general not very helpful for the determination of the optimal parameters, since the analytic calculation of (6.109) for the cost functionals presented in Section 6.2.1 is usually not possible. Besides, the parameters are mostly subject to certain restrictions, hence boundary extrema are possible. But the latter cannot be obtained from (6.109). Hence mostly a numerical *search procedure* has to be applied in order to calculate the optimal parameters. Such a search

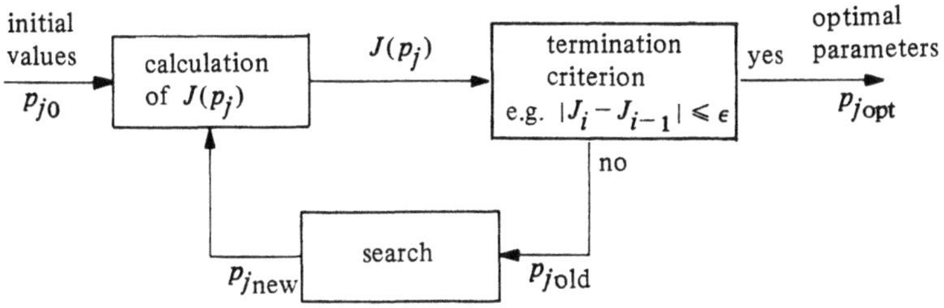

Fig. 6.14. Schema of a search procedure

procedure consists of rules for discrete steps within the space of the free parameters bounded by the additional conditions resulting in better values of the cost functionals. Thus one arrives in a close vicinity of the searched optimum. A principal scheme of a search procedure is shown in Fig. 6.14. Among the various types of numerical search procedures the following should be mentioned:

1. Search parallel to axes;
2. Gradient procedure;
3. Direct search procedure.

Detailed descriptions of all three procedures can be found in the literature, see e.g. Drenick. The direct search procedure is especially recommended when the additional restrictions on the free parameters are complicated.

Example 6.4. Centrifuge

Consider the centrifuge discussed in Problem 4.1 (see Fig. 3.4) with the angular displacements α and β of the axis of symmetry of the centrifuge from the vertical. The corresponding equations of motion (4.149) are

$$\ddot{\boldsymbol{y}}(t)+(\vartheta\boldsymbol{E}+g\Omega\boldsymbol{S})\,\dot{\boldsymbol{y}}(t)+\kappa\boldsymbol{y}(t)=\boldsymbol{0}. \tag{6.110}$$

Suppose that for fixed values of κ and $g\Omega$ one has to choose the damping coefficient ϑ optimally with respect to the integral of squared error

$$IE^2=\int_0^\infty \boldsymbol{y}^T(\tau)\,\boldsymbol{y}(\tau)\mathrm{d}\tau \tag{6.111}$$

for a given initial condition $\boldsymbol{y}_0\neq\boldsymbol{0}$, $\dot{\boldsymbol{y}}_0=\boldsymbol{0}$. In the state space representation

$$\boldsymbol{x}=\left[\begin{array}{c}\boldsymbol{y}\\ \hline \dot{\boldsymbol{y}}\end{array}\right],\quad \dot{\boldsymbol{x}}(t)=\boldsymbol{A}\boldsymbol{x}(t),\quad \boldsymbol{A}=\left[\begin{array}{c|c}\boldsymbol{0} & \boldsymbol{E}\\ \hline -\kappa\boldsymbol{E} & -\vartheta\boldsymbol{E}-g\Omega\boldsymbol{S}\end{array}\right], \tag{6.112}$$

one can write (6.111) as

$$IE^2=\int_0^\infty \boldsymbol{x}^T(\tau)\boldsymbol{S}'\boldsymbol{x}(\tau)\mathrm{d}\tau \quad\text{with}\quad \boldsymbol{S}'=\left[\begin{array}{c|c}\boldsymbol{E} & \boldsymbol{0}\\ \hline \boldsymbol{0} & \boldsymbol{0}\end{array}\right]. \tag{6.113}$$

The solution matrix $\boldsymbol{R}$ of the Ljapunov equation (6.103) can be written in the form

$$\boldsymbol{R}=\left[\begin{array}{c|c}\boldsymbol{R}_{11} & \boldsymbol{R}_{12}\\ \hline \boldsymbol{R}_{12}^T & \boldsymbol{R}_{22}\end{array}\right],\qquad \boldsymbol{R}_{11}=\boldsymbol{R}_{11}^T,\quad \boldsymbol{R}_{22}=\boldsymbol{R}_{22}^T. \tag{6.114}$$

From

$$\left[\begin{array}{c|c} \mathbf{0} & -\kappa \boldsymbol{E} \\ \hline \boldsymbol{E} & -\vartheta \boldsymbol{E}+g\Omega \boldsymbol{S} \end{array}\right]\left[\begin{array}{c|c} \boldsymbol{R}_{11} & \boldsymbol{R}_{12} \\ \hline \boldsymbol{R}_{12}^T & \boldsymbol{R}_{22} \end{array}\right]+\left[\begin{array}{c|c} \boldsymbol{R}_{11} & \boldsymbol{R}_{12} \\ \hline \boldsymbol{R}_{12}^T & \boldsymbol{R}_{22} \end{array}\right]\left[\begin{array}{c|c} \mathbf{0} & \boldsymbol{E} \\ \hline -\kappa \boldsymbol{E} & -\vartheta \boldsymbol{E}-g\Omega \boldsymbol{S} \end{array}\right]$$

$$= -\left[\begin{array}{c|c} \boldsymbol{E} & \mathbf{0} \\ \hline \mathbf{0} & \mathbf{0} \end{array}\right] \tag{6.115}$$

follow three equations for the submatrices:

$$\left.\begin{aligned} &\kappa\left(\boldsymbol{R}_{12}+\boldsymbol{R}_{12}^T\right)=\boldsymbol{E}, \\ &\boldsymbol{R}_{12}+\boldsymbol{R}_{12}^T-2\vartheta \boldsymbol{R}_{22}+g\Omega(\boldsymbol{S}\boldsymbol{R}_{22}-\boldsymbol{R}_{22}\boldsymbol{S})=\mathbf{0}, \\ &\boldsymbol{R}_{11}-\vartheta \boldsymbol{R}_{12}-g\Omega \boldsymbol{R}_{12}\boldsymbol{S}-\kappa \boldsymbol{R}_{22}=\mathbf{0}. \end{aligned}\right\} \tag{6.116}$$

These imply

$$\left.\begin{aligned} &\boldsymbol{R}_{12}+\boldsymbol{R}_{12}^T=\frac{1}{\kappa}\boldsymbol{E}, \qquad \boldsymbol{R}_{22}=\frac{1}{2\vartheta\kappa}\boldsymbol{E}, \\ &\boldsymbol{R}_{11}=\vartheta \boldsymbol{R}_{12}+g\Omega \boldsymbol{R}_{12}\boldsymbol{S}+\frac{1}{2\vartheta}\boldsymbol{E}. \end{aligned}\right\} \tag{6.117}$$

The unknown matrix $\boldsymbol{R}_{12}$ is determined from the conditions

$$\boldsymbol{R}_{12}+\boldsymbol{R}_{12}^T=\frac{1}{\kappa}\boldsymbol{E}, \qquad \boldsymbol{R}_{11}=\boldsymbol{R}_{11}^T. \tag{6.118}$$

One obtains

$$\boldsymbol{R}_{12}=\frac{1}{2\kappa}\boldsymbol{E}-\frac{g\Omega}{2\vartheta\kappa}\boldsymbol{S}. \tag{6.119}$$

This leads to

$$\boldsymbol{R}_{11}=\frac{1}{2\kappa}\left[\vartheta+\frac{1}{\vartheta}\left(\kappa+g^2\Omega^2\right)\right]\boldsymbol{E}. \tag{6.120}$$

By (6.104), the integral of squared error (6.111) for $\dot{\boldsymbol{y}}_0=\mathbf{0}$ is

$$\begin{aligned} IE^2 &= \boldsymbol{x}_0^T\boldsymbol{R}\boldsymbol{x}_0=\boldsymbol{y}_0^T\boldsymbol{R}_{11}\boldsymbol{y}_0 \\ &= \frac{1}{2\kappa}\left[\vartheta+\frac{1}{\vartheta}\left(\kappa+g^2\Omega^2\right)\right]\boldsymbol{y}_0^T\boldsymbol{y}_0. \end{aligned} \tag{6.121}$$

A search for the minimum of IE^2 by the method (6.109) of the differential calculus leads to the condition

$$\frac{\partial IE^2}{\partial \vartheta} = \frac{1}{2\kappa}\left[1 - \frac{1}{\vartheta^2}\left(\kappa + g^2\Omega^2\right)\right] y_0^T y_0 = 0, \tag{6.122}$$

i.e.,

$$\vartheta_{\text{opt}}^2 = \kappa + g^2\Omega^2. \tag{6.123}$$

The minimum of the integral of squared error is

$$IE_{\min}^2 = \frac{\vartheta_{\text{opt}}}{\kappa} y_0^T y_0 = \frac{1}{\kappa}\sqrt{\kappa + g^2\Omega^2}\, y_0^T y_0 . \tag{6.124}$$

A graph of the integral of squared error $IE^2(\vartheta)$ is plotted in Fig. 6.15. ◆

Example 6.5. Vibrating system with elastic damper

According to Problem 4.3, the vibrating system with elastic damper shown in Fig. 3.1 has the eigenvalues

$$\lambda_1 = -\frac{k}{d}, \qquad \lambda_{2,3} = -\frac{1}{2}\frac{d}{m} \pm \mathrm{i}\sqrt{\frac{k}{m} + \frac{3}{4}\frac{d^2}{m^2}}, \tag{6.125}$$

provided $b = k + d^2/m$ holds for the damper elasticity. One may try to determine an optimal damping parameter d for fixed values of the mass m and the spring constant k. Since $m > 0$, $k > 0$, $d > 0$, it follows from (6.125) that the degree of stability (6.90) is

$$d_0 = \min\left\{\frac{k}{d}, \frac{1}{2}\frac{d}{m}\right\}. \tag{6.126}$$

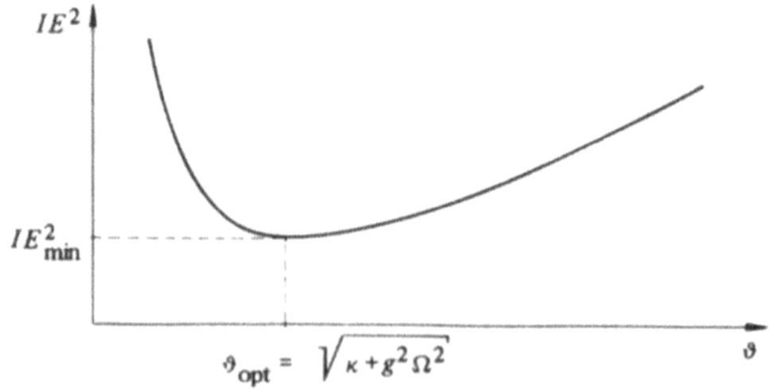

Fig. 6.15. Integral of squared error $IE^2(\vartheta)$

If one tries, by varying d, to maximize the degree of stability (6.126) one obtains

$$\frac{k}{d_{opt}} = \frac{1}{2} \frac{d_{opt}}{m},$$

i.e.,

$$d_{opt} = \sqrt{2km}. \tag{6.127}$$

For the maximum degree of stability

$$d_{0\max} = \sqrt{\frac{1}{2} \frac{k}{m}} \tag{6.128}$$

the real parts of the eigenvalues (6.125) are all equal to $-d_{0\max}$ and the eigenvalues are

$$\lambda_1 = -\sqrt{\frac{1}{2} \frac{k}{m}}, \qquad \lambda_{2,3} = -\sqrt{\frac{1}{2} \frac{k}{m}} \pm \mathrm{i} \sqrt{\frac{5}{2} \frac{k}{m}}. \tag{6.129}$$

If optimization should be attained with respect to the degree of damping (6.92) one has to calculate first D_0, using (6.125):

$$D_0 = -\max\left\{-1, \frac{-\frac{1}{2}\frac{d}{m}}{\sqrt{\frac{k}{m} + \frac{d^2}{m^2}}}\right\} = \frac{1}{2} \frac{\frac{d}{m}}{\sqrt{\frac{k}{m} + \frac{d^2}{m^2}}}. \tag{6.130}$$

According to (6.109) a maximum value of (6.130) is obtained by equating to zero the derivative of D_0 with respect to d:

$$\frac{\partial D_0}{\partial d} = \frac{1}{2m} \frac{\frac{k}{m}}{\sqrt{\frac{k}{m} + \frac{d^2}{m^2}}} = 0. \tag{6.131}$$

Since (6.131) can hold only for $d \to \infty$, it follows that (6.130) does not attain a proper maximum relative to d. It follows

$$D_0 \to D_{0\max} = \tfrac{1}{2} \quad \text{for} \quad d \to \infty \tag{6.132}$$

and

$$\lambda_1 \to 0, \quad \lambda_{2,3} \to \tfrac{1}{2}(-1 \pm \mathrm{i}\sqrt{3})\frac{d}{m} \quad \text{for} \quad d \to \infty. \tag{6.133}$$

Now $\lambda_1 \to 0$ implies that asymptotic stability is lost in this case. However in real problems the parameter range is always bounded. If, for instance $0 \leqslant d \leqslant 2\sqrt{km}$ is adopted, then the maximum degree of damping is attained as a boundary maximum for

$$d_{\text{opt}} = 2\sqrt{km} \quad \text{with} \quad D_{0\,\max} = \frac{1}{\sqrt{5}}. \tag{6.134}$$

The corresponding eigenvalues are

$$\lambda_1 = -\frac{1}{2}\sqrt{\frac{k}{m}}, \qquad \lambda_{2,3} = -\sqrt{\frac{k}{m}} \pm 2\sqrt{\frac{k}{m}}\,\mathrm{i}. \tag{6.135}$$

If one compares in this example the optimum with respect to the degree of stability and the degree of damping, one sees that often an optimization has only a meaningful solution if the parameters are bounded by technical reasons. Moreover optimizations with respect to different cost functionals yield different values for the optimal parameters. Then, the decision has to be made by engineering requirements or a compromise may be necessary. In the present case the parameter (6.127) which optimizes the degree of stability leads to the following values:

$$d_{0\,\max} = 0.707\sqrt{\frac{k}{m}}, \qquad D_0 = \frac{1}{\sqrt{6}} = 0.408. \tag{6.136}$$

On the other hand the optimization (6.134) with respect to the degree of damping yields the values

$$d_0 = 0.5\sqrt{\frac{k}{m}}, \qquad D_{0\,\max} = \frac{1}{\sqrt{5}} = 0.447. \tag{6.137}$$

A possible compromise between (6.136) and (6.137) is given by

$$d = 1.667\sqrt{km} \quad \text{with} \quad d_0 = 0.6\sqrt{\frac{k}{m}}, \quad D_0 = 0.428. \quad \blacklozenge \tag{6.138}$$

Problem 6.1. Undamped vertical vibrations of an automobile

Consider the vertical vibrations of an automobile (Fig. 2.12, Problem

2.2, Example 4.3) without shock absorbers and without tire damping, $d_{1,2} = 0$. The corresponding equation of motion is

$$\boldsymbol{M}\ddot{\boldsymbol{y}}(t) + \boldsymbol{K}\boldsymbol{y}(t) = \boldsymbol{0} \tag{6.139}$$

with the system matrices

$$\boldsymbol{M} = \begin{bmatrix} m_1 & 0 \\ 0 & m_2 \end{bmatrix}, \qquad \boldsymbol{K} = \begin{bmatrix} k_1 & -k_1 \\ -k_1 & k_1 + k_2 \end{bmatrix}. \tag{6.140}$$

The following numerical data are adopted:

Car body mass $m_1 = 1200$ kg,

Mass of the axles $m_2 = 80$ kg,

Car body spring constant $k_1 = 300 \dfrac{N}{\text{cm}}$,

Tire spring constant $k_2 = 3200 \dfrac{N}{\text{cm}}$.

a) What are the coefficients of the characteristic equation

$$p(\lambda) \equiv \frac{1}{\det \boldsymbol{M}} \det(\boldsymbol{M}\lambda^2 + \boldsymbol{K}) \equiv \lambda^4 + a_2\lambda^2 + a_4 = 0?$$

□ $a_2 = m_1(k_1 + k_2) + m_2 k_1, \quad a_4 = k_1 k_2;$

□ $a_2 = \dfrac{k_1}{m_1} + \dfrac{k_1 + k_2}{m_2}, \quad a_4 = \dfrac{k_1 k_2}{m_1 m_2};$

□ $a_2 = \dfrac{m_1(k_1 + k_2) + m_2 k_1}{m_1 m_2}, \quad a_4 = \dfrac{k_1 k_2}{m_1 m_2};$

□ $a_2 = \dfrac{k_1 + k_2}{m_1} + \dfrac{k_2}{m_2}, \quad a_4 = \dfrac{k_1 k_2}{m_1 m_2}.$

b) What is the characteristic equation of the vibrating system?

□ $\lambda^4 + 4025\,\lambda^2 + 9\,063 = 0;$

□ $\lambda^4 + 4400\,\lambda^2 + 109\,375 = 0;$

□ $\lambda^4 + 4400\,\lambda^2 + 100\,000 = 0;$

□ $\lambda^4 + 4025\,\lambda^2 + 109\,375 = 0.$

c) What are the natural frequencies of the system?

☐ $\omega_1 = 0.7608$ Hz, $\omega_2 = 10.53$ Hz;
☐ $\omega_1 = 4.78\ \text{s}^{-1}$, $\omega_2 = 66.16\ \text{s}^{-1}$;
☐ $\omega_1 = 4.58\ \text{s}^{-1}$, $\omega_2 = 57.00\ \text{s}^{-1}$;
☐ $\omega_1 = 0.729$ Hz, $\omega_2 = 9.08$ Hz.

d) Find the eigenvector for the first natural frequency.

☐ $\tilde{y}_1 = \begin{bmatrix} 11.606 \\ 1 \end{bmatrix}$, ☐ $\tilde{y}_1 = \begin{bmatrix} 1 \\ 0.08616 \end{bmatrix}$,

☐ $\tilde{y}_1 = \begin{bmatrix} 348\,172 \\ 30\,000 \end{bmatrix}$, ☐ $\tilde{y}_1 = \begin{bmatrix} -30\,000 \\ 348\,172 \end{bmatrix}$.

e) Find the eigenvector for the second natural frequency.

☐ $\tilde{y}_2 = \begin{bmatrix} -0.00574 \\ 1 \end{bmatrix}$, ☐ $\tilde{y}_2 = \begin{bmatrix} 1 \\ -174 \end{bmatrix}$,

☐ $\tilde{y}_2 = \begin{bmatrix} -172 \\ 30\,000 \end{bmatrix}$, ☐ $\tilde{y}_2 = \begin{bmatrix} -30\,000 \\ -172 \end{bmatrix}$.

f) Show the orthogonality of the eigenvectors with respect to the inertia matrix.

$$\boldsymbol{MY} = \begin{bmatrix} 600 & 0 \\ 0 & 40 \end{bmatrix} \begin{bmatrix} 1 & \quad 1 \\ & \end{bmatrix},$$

$$\boldsymbol{Y}^T\boldsymbol{MY} = \begin{bmatrix} 1 & \\ 1 & \end{bmatrix} \begin{bmatrix} \quad & \quad \\ & \end{bmatrix} = \begin{bmatrix} 600.5 & 0 \\ 0 & 1\,211\,640 \end{bmatrix}.$$

g) What are the natural modes (6.30)?

$$y_1(t) = \cos(\quad)t \begin{bmatrix} \quad \\ \quad \end{bmatrix},$$

$$y_2(t) = \cos(\quad)t \begin{bmatrix} \quad \\ \quad \end{bmatrix}.$$

h) For given total energy, by which factor are the amplitudes of the second natural mode smaller than the amplitudes of the first natural mode?

☐ $\frac{\omega_1}{\omega_2} \leqslant 0.080$, ☐ $\frac{\omega_1}{\omega_2} \leqslant 0.052$,

☐ $\frac{\omega_1}{\omega_2} \leqslant 0.063$, ☐ $\frac{\omega_1}{\omega_2} \leqslant 0.072$. ◆

Problem 6.2. Single-dof-vibrating system

The various cost functionals are used to optimize the damping parameter d of a single-dof-vibrating system

$$m\ddot{y}(t)+d\dot{y}(t)+ky(t)=0 \tag{6.141}$$

with fixed m and k. In each case the corresponding value of the damping factor

$$D=\frac{d}{2\sqrt{km}} \tag{6.142}$$

should be calculated.

a) Which is the optimal value of D for a maximum degree of stability d_0?

$D=\frac{1}{2}$,	$D=\frac{1}{\sqrt{2}}$,	$D\geqslant\frac{1}{\sqrt{2}}$,	$D=1$,	$D\geqslant 1$.
□	□	□	□	□

b) Which is the optimal value of D for a maximum degree of damping D_0?

$D=\frac{1}{2}$,	$D=\frac{1}{\sqrt{2}}$,	$D\geqslant\frac{1}{\sqrt{2}}$,	$D=1$,	$D\geqslant 1$.
□	□	□	□	□

Consider now the problem of minimizing the integral of squared error

$$IE^2=\int_0^{\infty}\boldsymbol{x}^T(\tau)\boldsymbol{S}\boldsymbol{x}(\tau)\mathrm{d}\tau$$

with

$$\boldsymbol{x}=\begin{bmatrix} y \\ \dot{y} \end{bmatrix},\qquad \boldsymbol{S}=\begin{bmatrix} 1 & 0 \\ 0 & \epsilon \end{bmatrix} \tag{6.143}$$

and $y_0=1,\ \dot{y}_0=0$.

c) According to Ljapunov's equation (6.103), what are the determining equations for the elements R_{ij}?

□ $\frac{k}{m}R_{12}=1,\qquad \frac{d}{m}R_{22}-R_{12}=\epsilon,$

$$R_{11}=\frac{d}{m}R_{12}+\frac{k}{m}R_{22};$$

□ $2\dfrac{k}{m}R_{12}=1, \qquad 2\dfrac{d}{m}R_{22}-2R_{12}=\epsilon,$

$R_{11}=\dfrac{d}{m}R_{12}+\dfrac{k}{m}R_{22};$

□ $2\dfrac{k}{m}R_{12}=1, \qquad 2\dfrac{d}{m}R_{22}-2R_{11}=\epsilon,$

$R_{11}=\dfrac{k}{m}R_{22};$

□ $\dfrac{k}{m}R_{12}=1, \qquad 2\dfrac{d}{m}R_{22}-2R_{12}=\epsilon,$

$R_{11}=\dfrac{k}{m}R_{12}+\dfrac{d}{m}R_{22};$

□ $\dfrac{k}{m}R_{12}=1, \qquad \dfrac{d}{m}R_{22}-R_{12}=\epsilon, \qquad R_{11}=\dfrac{d}{m}R_{12}.$

d) Evaluate the integral of squared error IE^2.

□ $IE^2=\dfrac{d}{2k}+\dfrac{m}{2d}\left(\epsilon+\dfrac{m}{k}\right);$

□ $IE^2=\dfrac{k}{2d}\left(\epsilon+\dfrac{m}{k}\right);$

□ $IE^2=\dfrac{d}{2k};$

□ $IE^2=\dfrac{m}{2d}\left(\epsilon+\dfrac{m}{k}\right);$

□ $IE^2=\dfrac{d}{2k}+\dfrac{k}{2d}\left(\epsilon+\dfrac{m}{k}\right).$

e) Which values of the damping factor (6.142) correspond for $\epsilon=0$ and $\epsilon=m/k$ to the minimum of the integral of squared error?

	$D=\frac{1}{2}$,	$D=\dfrac{1}{\sqrt{2}}$,	$D\geqslant\dfrac{1}{\sqrt{2}}$,	$D=1$,	$D\geqslant 1$.
$\epsilon=0$	□	□	□	□	□
$\epsilon=\dfrac{m}{k}$	□	□	□	□	□

To complete these answers, the optimal damping factors D for some other cost functionals are presented using also the weighting matrix (6.143):

$$\alpha)\ ITE^2 = \int_0^\infty \tau \boldsymbol{x}^T(\tau)\boldsymbol{S}\boldsymbol{x}(\tau)\mathrm{d}\tau:$$

$$D_{\mathrm{opt}}^4 = \frac{1}{8}\left(1 + \epsilon\frac{k}{m}\right);$$

$$\epsilon = 0: \quad D_{\mathrm{opt}} = 0.5946; \quad \epsilon = \frac{m}{k}: \quad D_{\mathrm{opt}} = 0.7071.$$

$$\beta)\ IAE = \int_0^\infty |y(\tau)|\mathrm{d}\tau: \quad D_{\mathrm{opt}} = 0.663.$$

$$\gamma)\ ITAE = \int_0^\infty \tau|y(\tau)|\mathrm{d}\tau: \quad D_{\mathrm{opt}} = 0.7524.$$ ◆

7. Forced vibrations

The general solution for externally excited vibrating systems with an arbitrary excitation function was given by (4.42) in Chapter 4. However in technical vibrating systems one often deals with very special excitation functions, such as the impulse function, the step function, and periodic functions. Impulsive excitation forces are frequently encountered in vibrating vehicles, e.g., when a motor car travels along a road with a few potholes, or if an airplane encounters local air turbulence. Periodic excitation forces appear in vibrating systems incorporating unbalanced rotating subsystems, e.g., in unbalanced centrifuges and in cars with unbalanced wheels. Periodic forces are also responsible for nearly all disturbances in the dynamics of machines. Further, periodic excitation of a motor car can be caused by an undulating road surface and ships may be periodically excited by a swell. The multitude of applications justifies the investigation of vibrating systems with impulsive, step, and periodic excitation. Here one should also take into account the special properties of mechanical systems.

7.1. Impulsive excitation

Impulsive excitation forces can be represented by the *Dirac function* *) $\delta(t-t_I)$

$$\boldsymbol{b}(t)=\boldsymbol{b}_I\delta(t-t_I), \tag{7.1}$$

where $\boldsymbol{b}_I$ is a constant $n\times 1$ vector and t_I is the instant of the impulse. The Dirac function has the following properties:

$$\left.\begin{aligned}
&\delta(t-t_I)=0 \quad \text{for} \quad t\neq t_I,\\
&\delta(t-t_I)=\infty \quad \text{for} \quad t=t_I,\\
&\lim_{\epsilon\to 0}\int_{t_I-\epsilon}^{t_I+\epsilon}\delta(\tau-t_I)\mathrm{d}\tau=1,\\
&\int_a^b f(\tau)\delta(\tau-t_I)\mathrm{d}\tau=f(t_I) \quad \text{for} \quad a<t_I<b
\end{aligned}\right\} \tag{7.2}$$

*) According to the mathematical definition, $\delta(t-t_I)$ is not a function but a distribution. However, this distinction will not be made here and δ will be called a function, as is customary in physics.

and any continuous function f(t).

The time dependence of the Dirac function is shown in Fig. 7.1. It is evident that the Dirac function is an idealization of a real impulse or jerk. This function is always meaningful when the time duration of the impulse is very short in comparison with the time constants of the vibrating system, given for instance by the size of the eigenvalues. This idealization allows considerable simplification of the general solution of the considered vibrating system.

The state equation of a dynamic system with impulsive excitation is

$$\dot{\boldsymbol{x}}(t) = \boldsymbol{A}\boldsymbol{x}(t) + \boldsymbol{b}_{\mathrm{I}}\delta(t - t_{\mathrm{I}}), \quad \boldsymbol{x}(0) = \boldsymbol{x}_0 \tag{7.3}$$

by (4.2) and (7.1). The general solution implies now the *impulsive response*

$$\boldsymbol{x}_{\mathrm{I}}(t) = \boldsymbol{\Phi}(t)\boldsymbol{x}_0 + \int_0^t \boldsymbol{\Phi}(t - \tau)\boldsymbol{b}_{\mathrm{I}}\delta(\tau - t_{\mathrm{I}})\,d\tau, \tag{7.4}$$

i.e., $\boldsymbol{x}_{\mathrm{I}}(t)$ is the solution corresponding to impulsive excitation. To further evaluate the integral (7.4) the properties (7.2) of the Dirac function are used. One then obtains for the *impulsive response*

$$\boldsymbol{x}_{\mathrm{I}}(t) = \begin{cases} \boldsymbol{\Phi}(t)\boldsymbol{x}_0 & \text{for} \quad 0 \leqslant t < t_{\mathrm{I}}, \\ \boldsymbol{\Phi}(t)[\boldsymbol{x}_0 + \boldsymbol{\Phi}(-t_{\mathrm{I}})\boldsymbol{b}_{\mathrm{I}}] & \text{for} \quad t_{\mathrm{I}} < t. \end{cases} \tag{7.5}$$

This means that impulsive excitation forces cause a step change of the state variables at the instant of the impulse. The special structure of the state and excitation vectors for mechanical vibrating systems, Section 3.1, implies that for the time immediately after the impulse

$$\begin{bmatrix} \boldsymbol{y}_{\mathrm{I}} \\ \hline \dot{\boldsymbol{y}}_{\mathrm{I}} \end{bmatrix} = \boldsymbol{\Phi}(t_{\mathrm{I}}) \begin{bmatrix} \boldsymbol{y}_0 \\ \hline \dot{\boldsymbol{y}}_0 \end{bmatrix} + \begin{bmatrix} \boldsymbol{0} \\ \hline \boldsymbol{M}^{-1}\boldsymbol{h}_{\mathrm{I}} \end{bmatrix}, \tag{7.6}$$

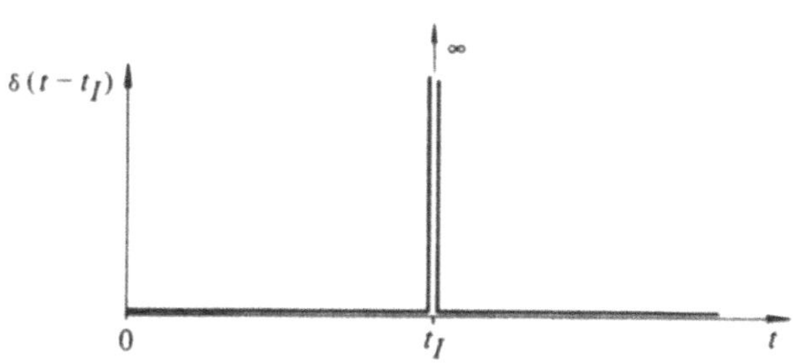

Fig. 7.1. Dirac function

i.e., the mechanical impulsive forces give rise to a stepwise change of the velocities $\dot{y}(t)$.

The steady state behaviour of the impulsive response, i.e., the behaviour of (7.5) for $t \to \infty$, depends on the stability of the vibrating system. Applying the stability definitions in Section 5.1, one obtains from (7.5) the following conclusions:

$\boldsymbol{x}_I(t \to \infty)$ vanishes if the homogeneous system is asymptotically stable;
$\boldsymbol{x}_I(t \to \infty)$ is bounded if the homogeneous system is marginally stable;
$\boldsymbol{x}_I(t \to \infty)$ is unbounded if the homogeneous system is unstable.

Thus the steady state behaviour of the impulsive response corresponds to the steady state behaviour of the vibrating system.

Example 7.1. Vertical vibrations of an automobile after a pothole

The simplest model for the vertical vibrations of a motor car with defective shock absorber is given by a sprung mass, as shown on Fig. 7.2. The equation of motion $m\ddot{y}(t) + ky(t) = 0$ implies the state equation

$$\dot{\boldsymbol{x}}(t) = \boldsymbol{A}\boldsymbol{x}(t), \quad \boldsymbol{A} = \begin{bmatrix} 0 & 1 \\ \omega^2 & 0 \end{bmatrix}, \quad \omega^2 = \frac{k}{m}. \tag{7.7}$$

Before arriving at the pothole the motor car is still in the equilibrium position $y_0 = 0$, $\dot{y}_0 = 0$. The impulsive response $\boldsymbol{x}_I(t)$ for $t > t_I = 0$, i.e., after the pothole is given by

$$\begin{bmatrix} y_I(t) \\ \dot{y}_I(t) \end{bmatrix} = \begin{bmatrix} \cos \omega t & \dfrac{1}{\omega} \sin \omega t \\ -\omega \sin \omega t & \cos \omega t \end{bmatrix} \begin{bmatrix} 0 \\ \dfrac{h_I}{m} \end{bmatrix} \tag{7.8}$$

or

$$y_I(t) = \frac{1}{m\omega} \sin \omega t \, h_I. \tag{7.9}$$

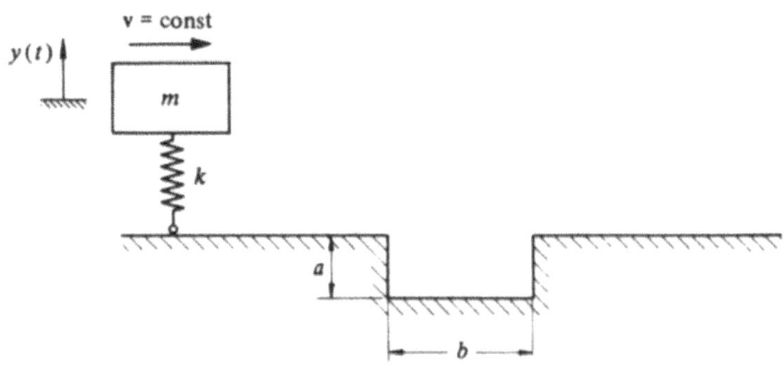

Fig. 7.2. Motor car before the pothole

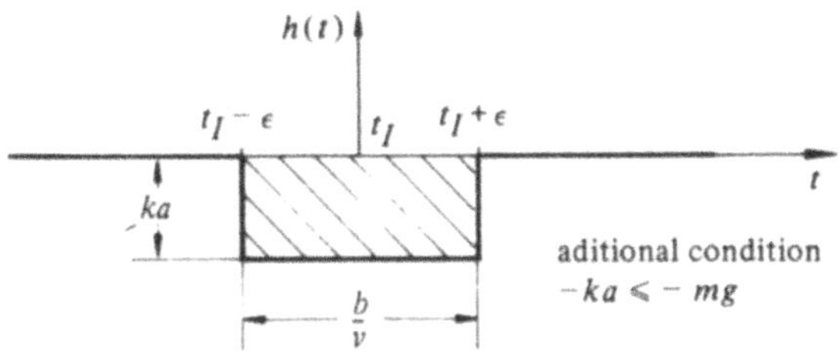

Fig. 7.3. Time duration of the impulse force

Here use was made of the fundamental matrix (4.50). Moreover it is assumed that the impulse duration b/v is small in comparison with the vibration period $T = 2\pi/\omega$,

$$\frac{b}{v} \ll \frac{2\pi}{\omega}. \tag{7.10}$$

The impulse strength h_{I} can be obtained from the time duration of the impulse force, see Fig. 7.3:

$$h_{\mathrm{I}} = h_{\mathrm{I}} \int_{t_{\mathrm{I}}-\epsilon}^{t_{\mathrm{I}}+\epsilon} \delta(\tau) \mathrm{d}\tau = -\int_{t_{\mathrm{I}}-\epsilon}^{t_{\mathrm{I}}+\epsilon} ka \, \mathrm{d}\tau = -ka\frac{b}{v}. \tag{7.11}$$

After substituting (7.11) in (7.9) one obtains

$$y_{\mathrm{I}}(t) = -\frac{ab\omega}{v} \sin \omega t. \tag{7.12}$$

The numerical values $m = 1200$ kg, $k = 300$ N/cm (hence $\omega = 5\mathrm{s}^{-1}$), $a = 0.2$ m, $b = 0.5$ m and $v = 72$ km/h $= 20$ m/s lead to the solution

$$y_I(t) = -0{,}025 \sin 5t \quad [\mathrm{m}]. \tag{7.13}$$

After passing the pothole the vehicle vibrates with an amplitude of 2,5 cm. ◆

7.2. Step excitation

Step excitation forces can be represented by the *step function* $s(t - t_{\mathrm{S}})$,

$$\boldsymbol{b}(t) = \boldsymbol{b}_{\mathrm{S}}\, s(t - t_{\mathrm{S}}), \tag{7.14}$$

where $\boldsymbol{b}_{\mathrm{S}}$ is a constant $n \times 1$ vector, and t_{S} is the instant of the step. The step function is defined by

$$s(t - t_{\mathrm{S}}) = \begin{cases} 0 & \text{for} \quad t < t_{\mathrm{S}}, \\ 1 & \text{for} \quad t > t_{\mathrm{S}} \end{cases} \tag{7.15}$$

and its graph is shown in Fig. 7.4. According to (4.2) and (7.14), the state equation of a step-excited system is

$$\dot{\boldsymbol{x}}(t)=\boldsymbol{A}\boldsymbol{x}(t)+\boldsymbol{b}_S s(t-t_S), \quad \boldsymbol{x}(0)=\boldsymbol{x}_0. \tag{7.16}$$

Hence (4.42) implies that the *step response* $\boldsymbol{x}_S(t)$ is given by

$$\boldsymbol{x}_S(t)=\boldsymbol{\Phi}(t)\boldsymbol{x}_0+\int_0^t \boldsymbol{\Phi}(t-\tau)\boldsymbol{b}_S s(\tau-t_S)\mathrm{d}\tau. \tag{7.17}$$

Assuming a regular system matrix $\boldsymbol{A}$ and applying the definition (7.15) of $s(t-t_S)$ and relation (4.24), one can evaluate the integral in (7.17). This yields

$$\boldsymbol{x}_S(t)=\begin{cases}\boldsymbol{\Phi}(t)\boldsymbol{x}_0 & \text{for} \quad 0 \leqslant t < t_S,\\ \boldsymbol{\Phi}(t)\boldsymbol{x}_0+(\boldsymbol{E}-\boldsymbol{\Phi}(t-t_S))\boldsymbol{x}_\infty & \text{for} \quad t \geqslant t_S\end{cases} \tag{7.18}$$

with

$$\boldsymbol{x}_\infty = -\boldsymbol{A}^{-1}\boldsymbol{b}_S. \tag{7.19}$$

In other words, the action of step excitation forces results in a change of the position of equilibrium at the instant when the step occurs. For vibrating systems with a regular matrix $\boldsymbol{A}$, det $\boldsymbol{A} \neq 0$, the motion consists of natural vibrations about the new position of equilibrium (7.19).

The evaluation of (7.17) is less simple if the system matrix $\boldsymbol{A}$ is singular, det $\boldsymbol{A}=0$. If that is so, one has to use the generalized inverse matrix $\boldsymbol{A}^+$ (Section 13.1) which also exists for singular matrices, to distinguish between the two cases. For

$$(\boldsymbol{E}-\boldsymbol{A}\boldsymbol{A}^+)\boldsymbol{b}_S=\boldsymbol{0} \tag{7.20}$$

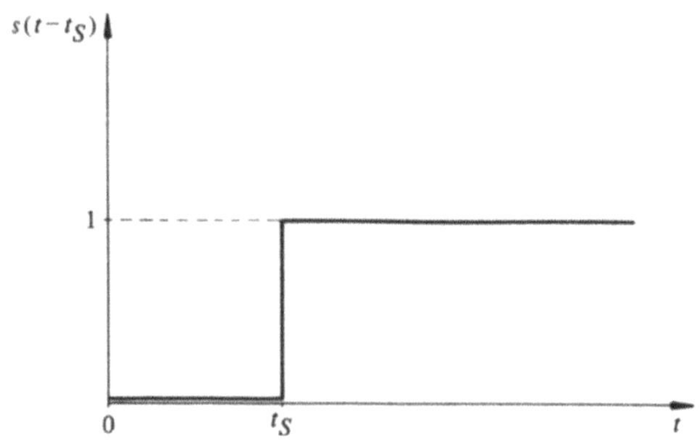

Fig. 7.4. Step function

one gets the step response (7.18) with

$$x_\infty = -A^+ b_S. \tag{7.21}$$

On the other hand, for

$$(E - AA^+) b_S \neq 0 \tag{7.22}$$

one obtains a linear solution growing unboundedly with the time $(t - t_S)$. Then, in the case when the multiplicity υ of the eigenvalue $\lambda = 0$ is equal to the nullity d of the matrix A (Section 4.3), the step response for $t \geqslant t_S$ can be written with the notation (13.31 to 33) as

$$x_S(t) = \Phi(t) x_0 - (E - \Phi(t - t_S)) T_2 \tilde{A}^{-1} T_4^* b_S + (t - t_S) T_1 T_3^* b_S. \tag{7.23}$$

In the case, when $\upsilon > d$ for the eigenvalue $\lambda = 0$, the homogeneous system is already unstable, hence the explicit determination of the step response is of no interest.

The steady state behaviour of the step response can be described with reference to the stability behaviour of the homogeneous vibrating system as follows:

$x_S(t \to \infty)$ is bounded when the homogeneous system is asymptotically stable

$$x_S(t \to \infty) = x_\infty; \tag{7.24}$$

$x_S(t \to \infty)$ is bounded or unbounded, respectively, when the homogeneous system is marginally stable and (7.20) or (7.22), respectively holds;

$x_S(t \to \infty)$ is unbounded when the inhomogeneous system is unstable.

Example 7.2. Vertical vibrations of an automobile after a pothole

In contrast to Example 7.1, the vertical vibrations of a motor car after a pothole will be now calculated with the help of the step response. Before encountering the pothole the motor car is in the equilibrium position $y_0 = 0$, $\dot{y}_0 = 0$. At the instant $t_{S1} = 0$ the step function $b_{S1} s(t)$ with

$$b_{S1} = \begin{bmatrix} 0 \\ -\dfrac{ka}{m} \end{bmatrix} \tag{7.25}$$

marks the beginning of the pothole and the function $b_{S2} s(t - b/\upsilon)$ with $b_{S2} = -b_{S1}$ marks its end. Here ka is the spring force which acts on the motor car as a result of passing the pothole of depth a, and b/v is the duration of the action. Thus the state equation is

$$\dot{\boldsymbol{x}}(t)=\boldsymbol{A}\boldsymbol{x}(t)+\boldsymbol{b}_{\mathrm{S1}}\left(s(t)-s\left(t-\frac{b}{v}\right)\right). \tag{7.26}$$

By (7.7), the system matrix $\boldsymbol{A}$ is regular for $\omega \neq 0$. Thus the step response (7.18) can be applied for solving (7.26). The two steps cause the superposition of the two particular solutions (7.18) and with $\boldsymbol{x}_0=\boldsymbol{0}$ one obtains the response

$$\begin{aligned}\boldsymbol{x}_{\mathrm{S}}(t)&=(\boldsymbol{E}-\boldsymbol{\Phi}(t))(-\boldsymbol{A}^{-1}\boldsymbol{b}_{\mathrm{S1}})+\left(\boldsymbol{E}-\boldsymbol{\Phi}\left(t-\frac{b}{v}\right)\right)(+\boldsymbol{A}^{-1}\boldsymbol{b}_{\mathrm{S1}})\\&=\left[\boldsymbol{\Phi}(t)-\boldsymbol{\Phi}\left(t-\frac{b}{v}\right)\right]\boldsymbol{A}^{-1}\boldsymbol{b}_{\mathrm{S1}}\quad\text{for}\quad t\geqslant\frac{b}{v}.\end{aligned}\tag{7.27}$$

From this solution together with the fundamental matrix (4.50) the position coordinate y of the vehicle is found as

$$y_{\mathrm{S}}(t)=a\left[\cos\omega t-\cos\omega\left(t-\frac{b}{v}\right)\right],\quad t\geqslant\frac{b}{v}. \tag{7.28}$$

In order to compare this result with (7.12), one has to write (7.28) in a different form and then linearize with respect to $\omega b/v$:

$$\begin{aligned}y_{\mathrm{S}}(t)&=a\left[\cos\omega t\left(1-\cos\frac{\omega b}{v}\right)-\sin\omega t\,\sin\frac{\omega b}{v}\right]\\&\approx-\frac{ab\omega}{v}\sin\omega t=y_I(t).\end{aligned}$$

For short times b/v, in first approximation, (7.12) and (7.28) are identical. But the calculation is simpler for (7.12) than for (7.28). Hence if the rectangular impulses as shown in Fig. 7.3 are of short duration, the impulsive response is of great practical importance.

7.3. Periodic excitation

Periodic excitation forces can be represented by a *periodic* $n\times 1$ vector function

$$\boldsymbol{b}_{\mathrm{P}}(t)=\boldsymbol{b}_{\mathrm{P}}(t+T) \tag{7.29}$$

where T is the period and

$$\Omega = \frac{2\pi}{T} \tag{7.30}$$

is the excitation frequency. For bounded and piecewise continuous functions (7.29) the periodic forces can be expanded in a convergent *Fourier series*

$$\boldsymbol{b}_P(t) = \tfrac{1}{2}\boldsymbol{b}_0 + \sum_{k=1}^{\infty} \left(\boldsymbol{b}_k^{(1)} \cos k\Omega t + \boldsymbol{b}_k^{(2)} \sin k\Omega t\right) \tag{7.31}$$

with the $n \times 1$ vectors of the Fourier-coefficients

$$\boldsymbol{b}_k^{(1)} = \frac{2}{T}\int_0^T \boldsymbol{b}_P(\tau) \cos k\Omega\tau \mathrm{d}\tau, \quad k = 0(1)\infty, \tag{7.32}$$

$$\boldsymbol{b}_k^{(2)} = \frac{2}{T}\int_0^T \boldsymbol{b}_P(\tau) \sin k\Omega\tau \mathrm{d}\tau, \quad k = 1(1)\infty. \tag{7.33}$$

The Fourier coefficients of all the important functions can be found in the corresponding textbooks. Fig. 7.5 shows a periodic function with mean value $\frac{1}{2}b_0$.

According to (4.2) and (4.7), the state equation of a periodically excited vibrating system is

$$\dot{\boldsymbol{x}}(t) = \boldsymbol{A}\boldsymbol{x}(t) + \boldsymbol{b}_P(t), \quad \boldsymbol{x}(0) = \boldsymbol{x}_0. \tag{7.34}$$

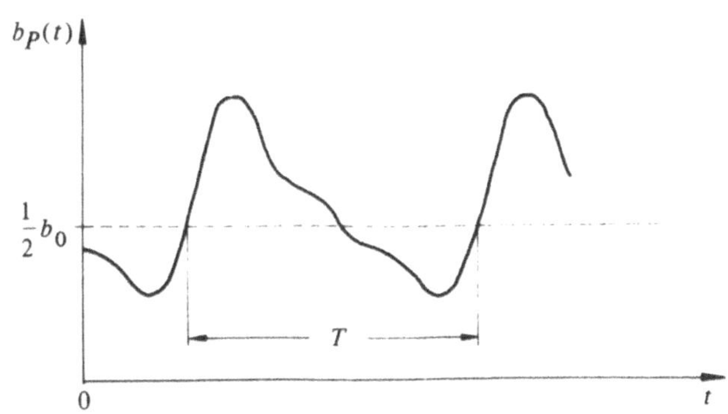

Fig. 7.5. Periodic function

From the general solution one now obtains the *response to periodic excitation*:

$$\boldsymbol{x}_\mathrm{P}(t) = \boldsymbol{\Phi}(t)\boldsymbol{x}_0 + \tfrac{1}{2}\int_0^t \boldsymbol{\Phi}(t-\tau)\boldsymbol{b}_0 \mathrm{d}\tau$$

$$+ \sum_{k=1}^{\infty} \int_0^t \boldsymbol{\Phi}(t-\tau)\left[\boldsymbol{b}_k^{(1)} \cos k\Omega\tau + \boldsymbol{b}_k^{(2)} \sin k\Omega\tau\right]\mathrm{d}\tau, \qquad (7.35)$$

where the Fourier series (7.31) can be integrated term by term provided it converges uniformly. To continue the evaluation, one has to determine the integrals in (7.35). To determine the basic behavior, it is sufficient to consider only the constant excitation and the first harmonic excitating function ($k = 1$). Corresponding results are found for the higher harmonic excitating functions.

The response to the constant excitation $\boldsymbol{b}(t) = \frac{1}{2}\boldsymbol{b}_0$ was derived in Section 7.2. For a regular system matrix $\boldsymbol{A}$, it follows from (7.18) and (7.19) with $t_\mathrm{S} = 0$ and $\boldsymbol{x}_0 = \boldsymbol{0}$:

$$\boldsymbol{x}_{\mathrm{P}0}(t) = (\boldsymbol{E} - \mathrm{e}^{\boldsymbol{A}t})\boldsymbol{x}_{\mathrm{P}0\infty}, \qquad (7.36)$$

$$\boldsymbol{x}_{\mathrm{P}0\infty} = -\tfrac{1}{2}\boldsymbol{A}^{-1}\boldsymbol{b}_0. \qquad (7.37)$$

If $\boldsymbol{A}$ is a singular matrix, one has to make use of (7.23) instead of (7.18) and (7.19).

To derive the response of the first harmonic excitation, it will be convenient to adopt an equivalent complex notation,

$$\left.\begin{aligned} &\boldsymbol{b}_{\mathrm{P}1}(t) = \boldsymbol{b}_1^{(1)} \cos \Omega t + \boldsymbol{b}_1^{(2)} \sin \Omega t = \boldsymbol{b}\,\mathrm{e}^{\mathrm{i}\Omega t} + \bar{\boldsymbol{b}}\,\mathrm{e}^{-\mathrm{i}\Omega t},\\ &\text{where } \boldsymbol{b} = \tfrac{1}{2}\left(\boldsymbol{b}_1^{(1)} - i\boldsymbol{b}_1^{(2)}\right) \end{aligned}\right\} \qquad (7.38)$$

is a complex $n \times 1$ vector. In (7.38) the formula of Euler $\mathrm{e}^{\mathrm{i}\Omega t} = \cos \Omega t + \mathrm{i} \sin \Omega t$ has been used. The complex conjugate of the vector $\boldsymbol{b}$ was denoted by $\bar{\boldsymbol{b}} = \frac{1}{2}(\boldsymbol{b}_1^{(1)} + \mathrm{i}\boldsymbol{b}_1^{(2)})$. While the excitation frequency Ω is the same for all coordinates of the harmonic vector functions, the amplitudes e_i and the phase angles φ_i are certainly different:

$$\left.\begin{aligned} &b_{\mathrm{P}1i}(t) = b_{1i}^{(1)} \cos \Omega t + b_{1i}^{(2)} \sin \Omega t = e_i \cos(\Omega t - \varphi_i),\\ &e_i = \sqrt{\left(b_{1i}^{(1)}\right)^2 + \left(b_{1i}^{(2)}\right)^2}, \quad \tan \varphi_i = \frac{b_{1i}^{(2)}}{b_{1i}^{(1)}}, \quad i = 1(1)n. \end{aligned}\right\} \qquad (7.39)$$

By (7.35), the response to the harmonic excitation (7.38) is

$$x_{P1}(t) = \int_0^t e^{A(t-\tau)}[b\, e^{i\Omega\tau} + \bar{b}\, e^{-i\Omega\tau}]d\tau$$

$$= e^{At}\left\{\int_0^t e^{(i\Omega E - A)\tau}d\tau \cdot b + \int_0^t e^{(-i\Omega E - A)\tau}d\tau \cdot \bar{b}\right\}. \tag{7.40}$$

If $\det(i\Omega E - A) \neq 0$, the integrals appearing in (7.40) can be evaluated according to (4.24):

$$x_{P1}(t) = e^{At}\left\{\left(e^{(i\Omega E - A)t} - E\right)(i\Omega E - A)^{-1} b\right.$$

$$\left. + \left(e^{(-i\Omega E - A)t} - E\right)(-i\Omega E - A)^{-1}\bar{b}\right\}$$

$$= (i\Omega E - A)^{-1} b\, e^{i\Omega t} + (-i\Omega E - A)^{-1}\bar{b}\, e^{-i\Omega t}$$

$$- e^{At}\left[(i\Omega E - A)^{-1} b + (-i\Omega E - A)^{-1}\bar{b}\right],$$

i.e., there remains

$$x_{P1}(t) = g\, e^{i\Omega t} + \bar{g}\, e^{-i\Omega t} - e^{At} g^{(1)} \tag{7.41}$$

with the complex $n \times 1$ *amplitude vector*

$$g = \tfrac{1}{2}\left(g^{(1)} - i g^{(2)}\right) = F b \tag{7.42}$$

and the complex $n \times n$ *frequency matrix*

$$F = (i\Omega E - A)^{-1}. \tag{7.43}$$

In the response (7.41) there appears the harmonic function

$$x_{P1\infty}(t) = g\, e^{i\Omega t} + \bar{g}\, e^{-i\Omega t} = g^{(1)} \cos \Omega t + g^{(2)} \sin \Omega t \tag{7.44}$$

which has the same period T as the excitation, but different *amplitudes* a_i and *phase angles* ψ_i:

$$a_i = \sqrt{\left(g_i^{(1)}\right)^2 + \left(g_i^{(2)}\right)^2}, \quad \tan \psi_i = \frac{g_i^{(2)}}{g_i^{(1)}}, \quad i = 1(1)n. \tag{7.45}$$

The complex vector $g(\Omega)$, as function of the excitation frequency, is called the *frequency response vector.*

After these considerations the response (7.35) for the constant and first harmonic excitating function, $k=0$ and $k=1$ ($\det \boldsymbol{A} \neq 0$, $\det(\mathrm{i}\Omega\boldsymbol{E} - \boldsymbol{A}) \neq 0$), is known

$$\boldsymbol{x}_{\mathrm{P}}(t) = \mathrm{e}^{\boldsymbol{A}t}\left[\boldsymbol{x}_0 - \boldsymbol{x}_{\mathrm{P}0\infty} - \boldsymbol{g}^{(1)}\right] + \boldsymbol{x}_{\mathrm{P}0\infty} + \boldsymbol{x}_{\mathrm{P}1\infty}(t). \tag{7.46}$$

The response (7.41) of the system (7.34) to a harmonic excitation was obtained under the assumption that $\det(\mathrm{i}\Omega\boldsymbol{E} - \boldsymbol{A}) \neq 0$. For applications in engineering problems this is the most important case. However, the integrals in (7.40) will also be evaluated in the case when the excitation frequency Ω coincides with an eigenvalue $\lambda_k = \mathrm{i}\Omega$ of the system matrix $\boldsymbol{A}$: $\det(\mathrm{i}\Omega\boldsymbol{E} - \boldsymbol{A}) = 0$. Equation (7.40) can also be written in the form

$$\boldsymbol{x}_{\mathrm{P}1}(t) = \mathrm{e}^{\mathrm{i}\Omega t}\int_0^t \mathrm{e}^{(\boldsymbol{A} - \mathrm{i}\Omega\boldsymbol{E})(t-\tau)}\boldsymbol{b}\,\mathrm{d}\tau + \mathrm{e}^{-\mathrm{i}\Omega t}\int_0^t \mathrm{e}^{(\boldsymbol{A} + \mathrm{i}\Omega\boldsymbol{E})(t-\tau)}\bar{\boldsymbol{b}}\,\mathrm{d}\tau. \tag{7.47}$$

To evaluate these integrals for $\det(\pm\mathrm{i}\Omega\boldsymbol{E} - \boldsymbol{A}) = 0$ one can choose the method described in Section 7.2. The integrals in (7.17) coincide for $t_{\mathrm{S}} = 0$, $\boldsymbol{b}_{\mathrm{S}} = \boldsymbol{b}$ and $\boldsymbol{A} \Rightarrow \boldsymbol{A} \mp \mathrm{i}\Omega\boldsymbol{E}$ with the integrals (7.47), thus for pairs of eigenvalues $\lambda = \pm\mathrm{i}\Omega$ and the case $v = d$ (Section 4.3) one can use the solution according to (7.23):

$$\begin{aligned}\boldsymbol{x}_{\mathrm{P}1}(t) = {} & \mathrm{e}^{\mathrm{i}\Omega t}\left\{(\boldsymbol{E} - \mathrm{e}^{(\boldsymbol{A} - \mathrm{i}\Omega\boldsymbol{E})t})\boldsymbol{T}_2(\Omega)\tilde{\boldsymbol{A}}^{-1}(\Omega)\boldsymbol{T}_4^*(\Omega)\boldsymbol{b} + t\boldsymbol{T}_1(\Omega)\boldsymbol{T}_3^*(\Omega)\boldsymbol{b}\right\} \\ & + \mathrm{e}^{-\mathrm{i}\Omega t}\left\{(\boldsymbol{E} - \mathrm{e}^{(\boldsymbol{A} + \mathrm{i}\Omega\boldsymbol{E})t})\bar{\boldsymbol{T}}_2(\Omega)\bar{\tilde{\boldsymbol{A}}}^{-1}(\Omega)\bar{\boldsymbol{T}}_4^*(\Omega)\bar{\boldsymbol{b}}\right. \\ & \left. + t\bar{\boldsymbol{T}}_1(\Omega)\bar{\boldsymbol{T}}_3^*(\Omega)\bar{\boldsymbol{b}}\right\},\end{aligned}$$

that is,

$$\boldsymbol{x}_{\mathrm{P}1}(t) = \boldsymbol{g}^+\,\mathrm{e}^{\mathrm{i}\Omega t} + \bar{\boldsymbol{g}}^+\,\mathrm{e}^{-\mathrm{i}\Omega t} - \mathrm{e}^{\boldsymbol{A}t}\boldsymbol{g}^{+(1)} + t\left[\tilde{\boldsymbol{g}}\,\mathrm{e}^{\mathrm{i}\Omega t} + \bar{\tilde{\boldsymbol{g}}}\,\mathrm{e}^{-\mathrm{i}\Omega t}\right] \tag{7.48}$$

with

$$\boldsymbol{g}^+ = \tfrac{1}{2}\left(\boldsymbol{g}^{+(1)} - i\boldsymbol{g}^{+(2)}\right) = \boldsymbol{T}_2(\Omega)\tilde{\boldsymbol{A}}^{-1}(\Omega)\boldsymbol{T}_4^*(\Omega)\boldsymbol{b} \tag{7.49}$$

and

$$\tilde{\boldsymbol{g}} = \boldsymbol{T}_1(\Omega)\boldsymbol{T}_3^*(\Omega)\boldsymbol{b}. \tag{7.50}$$

Here the matrices $\boldsymbol{T}_i(\Omega)$, $i = 1(1)4$, and $\tilde{\boldsymbol{A}}(\Omega)$ are defined for the complex matrices $(\boldsymbol{A} \mp \mathrm{i}\Omega\boldsymbol{E})$ in the same way as in (13.31 to 33) the matrices $\boldsymbol{T}_i$, $i = 1(1)4$, and $\tilde{\boldsymbol{A}}$ are defined for $\boldsymbol{A}$. Thus $\boldsymbol{T}_1(\pm\Omega)$ contains all eigenvectors of $\boldsymbol{A}$ for the eigenvalues $\pm\mathrm{i}\Omega$, and $\boldsymbol{T}_2(\pm\Omega)$ contains all eigen and principal vectors for the remaining eigenvalues.

It is significant for the solution (7.48) that the terms corresponding to (7.41) ($\boldsymbol{g}^+$ in place of $\boldsymbol{g}$, where for $\det(\mathrm{i}\Omega\boldsymbol{E}-\boldsymbol{A})\neq 0$ one has $\boldsymbol{g}^+=\boldsymbol{g}$) are completed by a vibration whose amplitudes grow linearly with time. Hence the response (7.48) is unbounded if $\det(\mathrm{i}\Omega\boldsymbol{E}-\boldsymbol{A})=0$ and $\tilde{\boldsymbol{g}}\neq\boldsymbol{0}$. To verify that $\tilde{\boldsymbol{g}}\neq\boldsymbol{0}$ one may use

$$\tilde{\boldsymbol{g}}\neq\boldsymbol{0}\Leftrightarrow\left[\boldsymbol{E}-(\mathrm{i}\Omega\boldsymbol{E}-\boldsymbol{A})(\mathrm{i}\Omega\boldsymbol{E}-\boldsymbol{A})^+\right]\boldsymbol{b}\neq\boldsymbol{0} \tag{7.51}$$

which is a consequence of (7.22).

The case $v>d$ for the eigenvalues $\lambda=\pm\mathrm{i}\Omega$ is not investigated in detail because the homogeneous solution is already unstable.

In conclusion, for the steady state behavior of periodically excitated systems (7.34) it is found:

$\boldsymbol{x}_P(t\to\infty)$ is bounded if the homogeneous system is asymptotically stable;

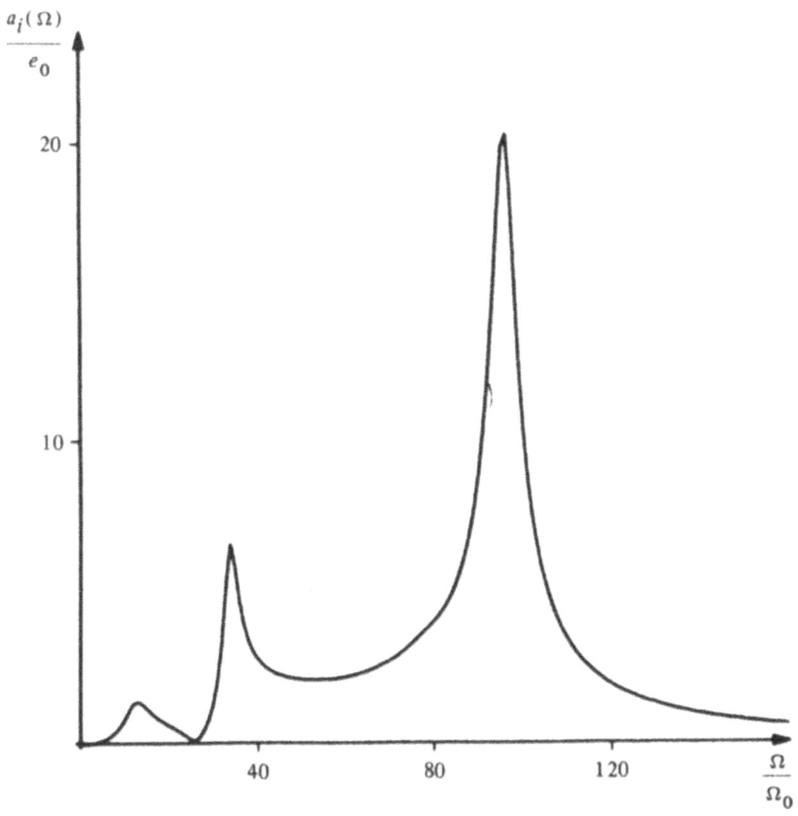

Fig. 7.6. Typical amplitude frequency response

$x_P(t \to \infty)$ is bounded or unbounded, respectively if the homogeneous system is marginally stable and (7.22) or (7.51) fails to hold or holds, respectively;

$x_P(t \to \infty)$ is unbounded if the homogeneous system is unstable.

According to (7.46), the steady state response of an asymptotically stable system is

$$x_P(t \to \infty) = x_{P0\infty} + x_{P1\infty}(t). \tag{7.52}$$

This shows that the frequency response (7.42) is the most important characteristic of the steady state response (7.52). For this reason the frequency response $\boldsymbol{g}(\Omega)$ and the frequency response matrix $\boldsymbol{F}(\Omega)$ will be extensively investigated in the next chapter. Moreover graphical representations of the coordinates of the frequency response vector $\boldsymbol{g}(\Omega)$ can be given if one uses the excitation frequency Ω as a variable parameter,

$$2g_i(\Omega) = a_i(\Omega)\, e^{-i\psi_i(\Omega)}, \quad i = 1(1)n, \tag{7.53}$$

where $a_i(\Omega)$ is the *amplitude frequency response,* $\psi_i(\Omega)$ is the phase frequency response, and $2g_i(\Omega)$ is the *polar plot* in the i-th coordinate.

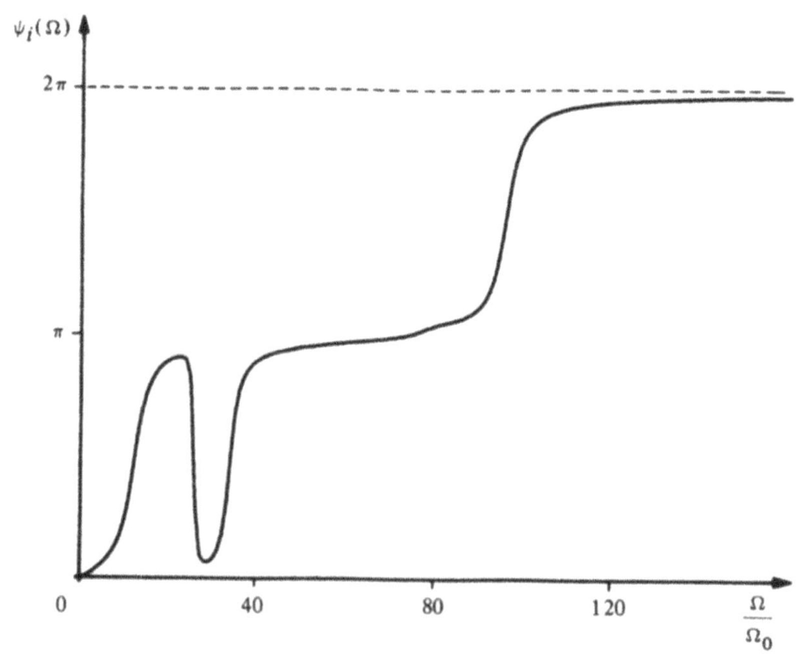

Fig. 7.7. Typical phase frequency response

Some typical graphs of such functions in normalized representation can be seen in Figures 7.6 to 7.8.

It has been very convenient to use complex notation for calculating forced vibrations with harmonic excitation forces since all results can be obtained by linear operations from the system matrix $\boldsymbol{A}$. In particular, the frequency response matrix (7.43) representing the complex transition relation between the excitation vector and the complex amplitude vector (7.42), is the inverse of the characteristic matrix $(\lambda \boldsymbol{E} - \boldsymbol{A})$ occuring in the eigenvalue problem for $\lambda = \mathrm{i}\Omega$. Nevertheless it can be quite useful to present also the real relations equivalent to the complex equation (7.42):

$$\left.\begin{aligned} \boldsymbol{g}^{(1)} &= -(\Omega^2 \boldsymbol{E} + \boldsymbol{A}^2)^{-1}(\boldsymbol{A}\boldsymbol{b}_1^{(1)} + \Omega \boldsymbol{b}_1^{(2)}), \\ \boldsymbol{g}^{(2)} &= -(\Omega^2 \boldsymbol{E} + \boldsymbol{A}^2)^{-1}(-\Omega \boldsymbol{b}_1^{(1)} + \boldsymbol{A}\boldsymbol{b}_1^{(2)}). \end{aligned}\right\} \tag{7.54}$$

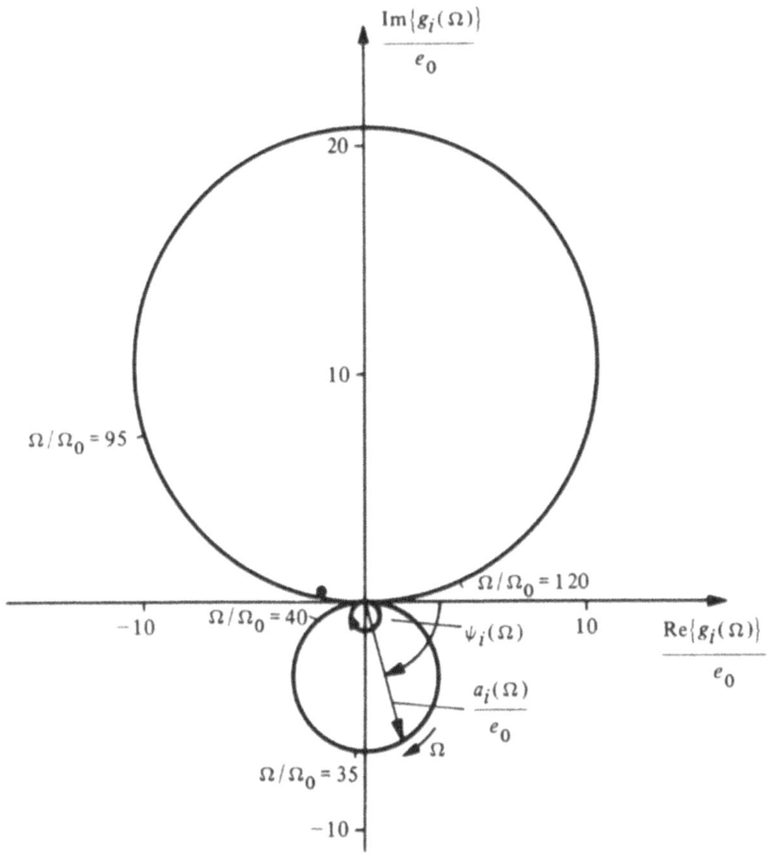

Fig. 7.8. Typical polar plot

The real calculation proves to be the more difficult one, since it requires forming the products A^2, $Ab_1^{(1)}$ and $Ab_1^{(2)}$.

7.4. Harmonically excited mechanical systems

Consider the equations of motion of a mechanical system

$$M\ddot{y}(t)+(D+G)\dot{y}(t)+(K+N)y(t)=h_F(t) \tag{7.55}$$

with the $f\times 1$ harmonic excitation vector

$$h_P(t)=h^{(1)}\cos\Omega t+h^{(2)}\sin\Omega t=h\,e^{i\Omega t}+\bar{h}\,e^{-i\Omega t}, \tag{7.56}$$

$$h=\tfrac{1}{2}(h^{(1)}-ih^{(2)}). \tag{7.57}$$

Assuming that the homogeneous mechanical system (7.55) is asymptotically stable, one obtains from the previous section the steady state response

$$y_{P1\infty}(t)=q^{(1)}\cos\Omega t+q^{(2)}\sin\Omega t=q\,e^{i\Omega t}+\bar{q}\,e^{-i\Omega t}, \tag{7.58}$$

$$q=\tfrac{1}{2}(q^{(1)}-iq^{(2)}). \tag{7.59}$$

Substituting (7.56) and (7.58) into (7.55), one gets the relation

$$q=F_M h \tag{7.60}$$

with the $f\times f$ frequency response matrix typical for mechanical systems

$$F_M=\left[-\Omega^2 M+i\Omega(D+G)+(K+N)\right]^{-1}. \tag{7.61}$$

Moreover, by (7.58), one has for the velocities $\dot{y}(t)$

$$\begin{aligned}\dot{y}_{P1\infty}(t)&=i\Omega\,q\,e^{i\Omega t}-i\Omega\,\bar{q}\,e^{-i\Omega t}\\&=\Omega\,q^{(2)}\cos\Omega t-\Omega\,q^{(1)}\sin\Omega t,\end{aligned} \tag{7.62}$$

i.e., one obtains the steady state response for the velocities from the steady state response for the position coordinates by multiplying the amplitudes with the frequency Ω, and a phase shift by $\pi/2$.

The connection between the $n\times n$ state frequency response matrix (7.43) and the $f\times f$ frequency response matrix (7.61) of the generalized coordinates is given by

$$F=\left[\begin{array}{c|c}F_M(i\Omega M+D+G) & F_M M\\ \hline -F_M(K+N) & i\Omega F_M M\end{array}\right] \tag{7.63}$$

and the connection between the vectors $\boldsymbol{g}$ and $\boldsymbol{q}$, by

$$\boldsymbol{g}=\left[\begin{array}{c}\boldsymbol{q}\\ \hline \mathrm{i}\Omega\boldsymbol{q}\end{array}\right]. \tag{7.64}$$

Thus, for mechanical systems (7.55) only the calculation of the $f\times f$ frequency response matrix $\boldsymbol{F}_{\mathrm{M}}$ (7.61) is necessary.

The real relations for $\boldsymbol{q}^{(1)}$ and $\boldsymbol{q}^{(2)}$ cannot be written as two separate equations, as was the case with (7.54) in the state space; they are given by a joint matrix relation as follows:

$$\left[\begin{array}{c}\boldsymbol{q}^{(1)}\\ \hline \boldsymbol{q}^{(2)}\end{array}\right]=\left[\begin{array}{c|c}-\Omega^2\boldsymbol{M}+(\boldsymbol{K}+\boldsymbol{N}) & \Omega(\boldsymbol{D}+\boldsymbol{G})\\ \hline -\Omega(\boldsymbol{D}+\boldsymbol{G}) & -\Omega^2\boldsymbol{M}+(\boldsymbol{K}+\boldsymbol{N})\end{array}\right]^{-1}\left[\begin{array}{c}\boldsymbol{h}^{(1)}\\ \hline \boldsymbol{h}^{(2)}\end{array}\right]. \tag{7.65}$$

In this real representation the mechanical systems don't offer any advantage. While, by (7.60), one has to solve a linear complex system of equations of order f to determine the frequency response, (7.65) is a real system with equations of order $2f$. This increase in dimension may create difficulties in numerical calculations. Thus in the next chapter the complex notation will always be preferred.

If the homogeneous mechanical system is not asymptotically stable but merely marginally stable, then the steady state solution (7.55) can be described with the aid of relations analogous to (7.48 to 51). Thus, according to (7.51) a vibration linearly increasing with time occurs exactly for

$$\begin{gathered}\left\{\boldsymbol{E}-\left[-\Omega^2\boldsymbol{M}+\mathrm{i}\Omega(\boldsymbol{D}+\boldsymbol{G})+(\boldsymbol{K}+\boldsymbol{N})\right]\right.\\ \left.\left[-\Omega^2\boldsymbol{M}+\mathrm{i}\Omega(\boldsymbol{D}+\boldsymbol{G})+(\boldsymbol{K}+\boldsymbol{N})\right]^{+}\right\}\boldsymbol{h}\neq\boldsymbol{0}.\end{gathered} \tag{7.66}$$

Problem 7.1. Impulsively excited double pendulum

Assume that the double pendulum considered in Example 4.6 is impulsively excited by hammer hits, Fig. 7.9. Taking (4.135), (7.3) with $t_{\mathrm{I}}=0$ one obtains the state equation

$$\left.\begin{aligned}&\dot{\boldsymbol{x}}(t)=\boldsymbol{A}\boldsymbol{x}(t)+\boldsymbol{b}_{\mathrm{I}}\delta(t),\quad \boldsymbol{x}_0=\boldsymbol{0},\\ &\boldsymbol{x}=\begin{bmatrix}y_1\\ y_2\\ \dot{y}_1\\ \dot{y}_2\end{bmatrix},\quad \boldsymbol{A}=\begin{bmatrix}0 & 0 & 1 & 0\\ 0 & 0 & 0 & 1\\ -3\dfrac{g}{L} & \dfrac{g}{L} & 0 & 0\\ \dfrac{g}{L} & -\dfrac{g}{L} & 0 & 0\end{bmatrix},\quad \boldsymbol{b}_I=\frac{1}{m}\begin{bmatrix}0\\ 0\\ h_1\\ h_2\end{bmatrix}.\end{aligned}\right\} \tag{7.67}$$

In Example 4.6 the corresponding fundamental matrix $\Phi(t)$ was calculated (4.142):

$$\Phi(t)=\left[\begin{array}{c|c}\alpha_0(t)\boldsymbol{E}-\alpha_2(t)\boldsymbol{K} & \alpha_1(t)\boldsymbol{E}-\alpha_3(t)\boldsymbol{K}\\ \hline -\alpha_1(t)\boldsymbol{K}+\alpha_3(t)\boldsymbol{K}^2 & \alpha_0(t)\boldsymbol{E}-\alpha_2(t)\boldsymbol{K}\end{array}\right].$$

Here $\alpha_i(t)$, $i=0(1)3$, are the trigonometric functions (4.141) and $\boldsymbol{K}$ is the normalized stiffness matrix $\boldsymbol{K}=g/L\begin{bmatrix}3 & -1\\ -1 & 1\end{bmatrix}$.

a) Which vibration is introduced by a unit hit on the first pendulum mass, $h_1=1$, $h_2=0$?

☐ $y_1(t)=\frac{1}{m}\frac{g}{L}\alpha_3(t),\quad y_2(t)=\frac{1}{m}\frac{g}{L}\alpha_3(t);$

☐ $y_1(t)=\frac{1}{m}\frac{g}{L}\alpha_3(t),\quad y_2(t)=\frac{1}{m}\left(\alpha_1(t)-\frac{g}{L}\alpha_3(t)\right);$

☐ $y_1(t)=\frac{1}{m}\left(\alpha_1(t)-3\frac{g}{L}\alpha_3(t)\right),\quad y_2(t)=\frac{1}{m}\left(\alpha_1(t)-\frac{g}{L}\alpha_3(t)\right);$

☐ $y_1(t)=\frac{1}{m}\left(\alpha_1(t)-3\frac{g}{L}\alpha_3(t)\right),\quad y_2(t)=\frac{1}{m}\frac{g}{L}\alpha_3(t).$

b) Which vibration is introduced by a unit hit on the second pendulum mass, $h_1=0$, $h_2=1$?

☐ $y_1(t)=\frac{1}{m}\frac{g}{L}\alpha_3(t),\quad y_2(t)=\frac{1}{m}\frac{g}{L}\alpha_3(t);$

☐ $y_1(t)=\frac{1}{m}\frac{g}{L}\alpha_3(t),\quad y_2(t)=\frac{1}{m}\left(\alpha_1(t)-\frac{g}{L}\alpha_3(t)\right),$

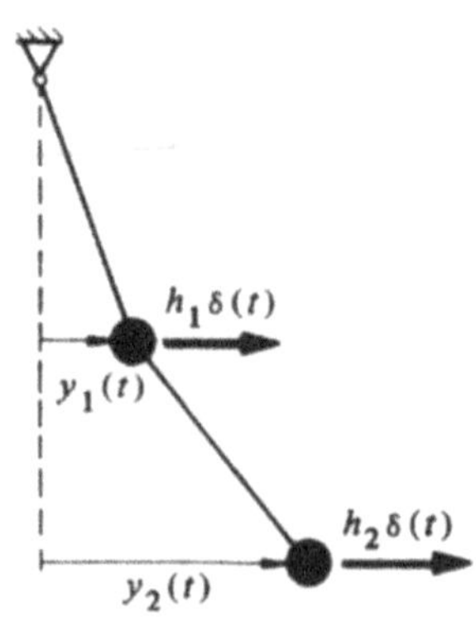

Fig. 7.9. Impulsively excited pendulum

□ $y_1(t)=\frac{1}{m}\left(\alpha_1(t)-3\frac{g}{L}\alpha_3(t)\right),\quad y_2(t)=\frac{1}{m}\left(\alpha_1(t)-\frac{g}{L}\alpha_3(t)\right);$

□ $y_1(t)=\frac{1}{m}\left(\alpha_1(t)-3\frac{g}{L}\alpha_3(t)\right),\quad y_2(t)=\frac{1}{m}\frac{g}{L}\alpha_3(t).$

c) Does a hit upon one of the pendulum masses initiate for small times $0<t\ll\sqrt{L/g}$ the motion of
□ the excited pendulum mass?
□ both pendulum masses?

d) Which vibration is initiated by unit hits upon both pendulum masses, $h_1=1,\ h_2=0$?

□ $y_1(t)=\frac{1}{m}\left(\alpha_1(t)-4\frac{g}{L}\alpha_3(t)\right),\quad y_2(t)=\frac{1}{m}\left(-\alpha_1(t)+2\frac{g}{L}\alpha_3(t)\right);$

□ $y_1(t)=\frac{1}{m}\left(\alpha_1(t)-3\frac{g}{L}\alpha_3(t)\right),\quad y_2(t)=\frac{1}{m}\frac{g}{L}\alpha_3(t);$

□ $y_1(t)=\frac{1}{m}\left(\alpha_1(t)-2\frac{g}{L}\alpha_3(t)\right),\quad y_2(t)=\frac{1}{m}\alpha_1(t);$

□ $y_1(t)=\frac{1}{m}\left(\alpha_1(t)-\frac{g}{L}\alpha_3(t)\right),\quad y_2(t)=\frac{1}{m}\left(2\alpha_1(t)-\frac{g}{L}\alpha_3(t)\right).$ ◆

Problem 7.2. Ramp excitation

Ramp excitation forces can be represented by the ramp function

$$\boldsymbol{b}(t)=\boldsymbol{b}_{\mathrm{R}}r(t-t_{\mathrm{R}}), \tag{7.68}$$

where $(t-t_{\mathrm{R}})$ is defined by

$$r(t-t_{\mathrm{R}})=\begin{cases}0 & \text{for} \quad t<t_{\mathrm{R}},\\ t-t_{\mathrm{R}} & \text{for} \quad t\geqslant t_{\mathrm{R}}.\end{cases} \tag{7.69}$$

Fig. 7.10. By (4.2) and (7.68), the state equation of a vibrating system with ramp excitation is

$$\dot{\boldsymbol{x}}(t)=\boldsymbol{A}\boldsymbol{x}(t)+\boldsymbol{b}_{\mathrm{R}}r(t-t_{\mathrm{R}}),\quad \boldsymbol{x}(0)=\boldsymbol{x}_0. \tag{7.70}$$

The *ramp response* $\boldsymbol{x}_{\mathrm{R}}(t)$ is thus, according to (4.42)

$$\boldsymbol{x}_{\mathrm{R}}(t)=\boldsymbol{\Phi}(t)\boldsymbol{x}_0+\int_0^t\boldsymbol{\Phi}(t-\tau)\boldsymbol{b}_{\mathrm{R}}r(\tau-\tau_{\mathrm{R}})\mathrm{d}\tau. \tag{7.71}$$

a) Which integral has to be calculated, so that (7.71) can be further evaluated?

□ $\int_0^t \tau\, e^{A(t_R-\tau)} d\tau$;

□ $\int_{t_R}^t (\tau - t_R)\, e^{-A\tau} d\tau$;

□ $\int_0^t \tau\, e^{-A\tau} b_R\, d\tau$,

□ $\int_{t_R}^t (\tau - t_R)\, e^{A(t-\tau)} b_R\, d\tau$.

b) What is the value of $\int_{t_R}^t \Phi(t-\tau) b_R r(\tau - t_R) d\tau$ for a regular system matrix A?

□ $-(E - e^{A(t-t_R)}) A^{-1} b_R$;

□ $-(E - e^{A(t-t_R)}) A^{-2} b_R$;

□ $-(t - t_R) A^{-1} b_R - (E - e^{A(t-t_R)}) A^{-2} b_R$;

□ $-(t - t_R) b_R - (E - e^{A(t-t_R)}) A^{-1} b_R$.

c) What is the ramp response obtained for the result in b)?

$$x_R(t) = \begin{cases} \Phi(t) x_0 & \text{for} \quad 0 \leqslant t < t_R, \\ \Phi(t) x_0 + \ldots\ldots\ldots\ldots\ldots & \text{for} \quad t \geqslant t_R. \end{cases} \tag{7.72}$$

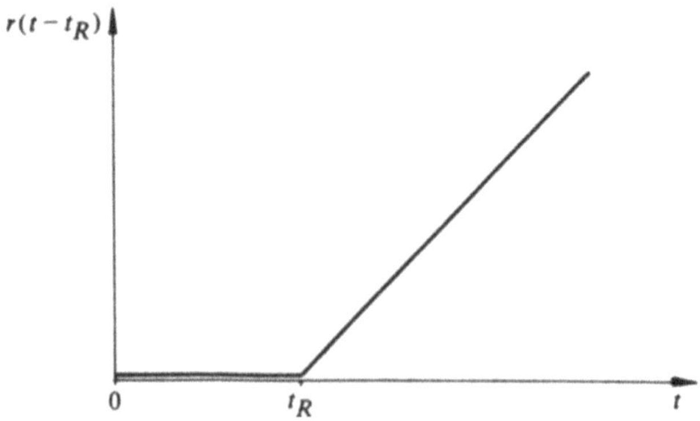

Fig. 7.10. Ramp function

d) What is the steady state behaviour of the ramp response, as characterized by the steady state behaviour of the homogeneous vibrating system?

- ☐ $\boldsymbol{x}_R(t \to \infty)$ is bounded or unbounded, respectively if $\dot{\boldsymbol{x}}(t) = \boldsymbol{A}\boldsymbol{x}(t)$ is stable or unstable, respectively
- ☐ $\boldsymbol{x}_R(t \to \infty)$ is unbounded, independently of the stability behaviour of $\dot{\boldsymbol{x}}(t) = \boldsymbol{A}\boldsymbol{x}(t)$;
- ☐ $\boldsymbol{x}_R(t \to \infty)$ is bounded for asymptotically stable systems $\dot{\boldsymbol{x}}(t) = \boldsymbol{A}\boldsymbol{x}(t)$, and otherwise unbounded. ◆

8. Resonance and absorption

In harmonically excited vibrating systems one can observe for certain exciting frequencies a strong growth of the vibration amplitudes. This phenomenon is usually called resonance and it is characteristic for undamped and lightly damped vibrating systems. For multi-dof-systems there may exist not only one but possibly more resonance frequencies. Since the occurance of resonance in vibrating systems is undesirable, due to the large amplitudes, the phenomenon should be more closely investigated. This investigation can be limited to stable vibrating systems, since for an unstable system the natural vibrations are already unbounded so that these systems are of no interest for engineering applications. The formulation of the conditions for resonance will be preceded by a discussion of the *elementary frequency response matrix* based on the idea of a *magnification function* for the principal or normal vibrations. This formulation implies necessary conditions for *strict resonance* and conditions for *resonance phenomena*. In multi-dof-vibrating systems there may also appear *pseudoresonances* with finite amplitudes beside strict resonances with infinite amplitudes. Conditions for both cases will be derived. Moreover, one can find excitation vectors which certainly lead to pseudoresonance. Also, in multi-dof-vibrating systems the amplitudes for individual state variables may vanish for certain excitation frequencies. One then speaks of *vibration absorption*. Conditions will be given for absorption and there will also be listed the corresponding requirements on excitation vectors useful in the design of absorption devices. To conclude this section, an *optimization* relative to the *disturbance behavior* of the vibrating system in order to determine the system parameters, such as spring or damping constants, according to the criterion of small resonance amplitudes will be undertaken.

8.1. Elementary frequency response matrix

By (7.34) and (7.38), the state equation of a harmonically excited vibrating system is

$$\dot{\boldsymbol{x}}(t)=\boldsymbol{A}\boldsymbol{x}(t)+\boldsymbol{b}\,\mathrm{e}^{\mathrm{i}\Omega t}+\bar{\boldsymbol{b}}\,\mathrm{e}^{-\mathrm{i}\Omega t}, \tag{8.1}$$

where Ω is the excitation frequency. According to (7.44),

$$\left.\begin{aligned} \boldsymbol{x}_{\mathrm{P1}\infty}(t) &= \boldsymbol{g}\,\mathrm{e}^{\mathrm{i}\Omega t} + \bar{\boldsymbol{g}}\,\mathrm{e}^{-\mathrm{i}\Omega t}, \\ \boldsymbol{g} = \boldsymbol{F}\boldsymbol{b}, &\quad \boldsymbol{F} = (\mathrm{i}\Omega\boldsymbol{E} - \boldsymbol{A})^{-1} \end{aligned}\right\} \tag{8.2}$$

is a particular solution of the system (8.1), provided $(\mathrm{i}\Omega\boldsymbol{E} - \boldsymbol{A})$ is a regular matrix. The $n \times n$ frequency response matrix is of great importance here.

The particular solution (8.2) shall now be written in principal or normal coordinates. Assuming that the multiplicity is equal to the nullity for every eigenvalue of $\boldsymbol{A}(d_i = v_i)$,

$$\left.\begin{aligned} &\lambda_i = \delta_j \pm \mathrm{i}\omega_j, \quad \omega_j \neq 0, \quad i = 1(1)2s, \qquad j = 1(1)s, \\ &\lambda_i = \sigma_k, \qquad\qquad\qquad\quad\; i = 2s+1(1)n, \quad k = 1(1)n-2s, \end{aligned}\right\} \tag{8.3}$$

the system (8.1) may be subjected to a real similarity transformation

$$\boldsymbol{x}(t) = \hat{\boldsymbol{X}}\hat{\boldsymbol{x}}(t), \quad \boldsymbol{b} = \hat{\boldsymbol{X}}\hat{\boldsymbol{b}} \tag{8.4}$$

yielding

$$\dot{\hat{\boldsymbol{x}}}(t) = \left[\begin{array}{c|c|c} \boldsymbol{\Delta} & \boldsymbol{E} & \boldsymbol{0} \\ \hline -\boldsymbol{\Omega}^2 & \boldsymbol{\Delta} & \boldsymbol{0} \\ \hline \boldsymbol{0} & \boldsymbol{0} & \boldsymbol{\Sigma} \end{array}\right] \hat{\boldsymbol{x}}(t) + \hat{\boldsymbol{b}}\,\mathrm{e}^{\mathrm{i}\Omega t} + \bar{\hat{\boldsymbol{b}}}\,\mathrm{e}^{-\mathrm{i}\Omega t} \tag{8.5}$$

with the diagonal matrices

$$\boldsymbol{\Delta} = \mathrm{diag}\{\delta_j\}, \quad \boldsymbol{\Omega} = \mathrm{diag}\{\omega_j\}, \quad \boldsymbol{\Sigma} = \mathrm{diag}\{\sigma_k\}. \tag{8.6}$$

For ordinary mechanical systems the similarity transformation (8.4) is given by (6.83) in Section 6.1.3. The corresponding transformation for the frequency response vector $\boldsymbol{g}$ is

$$\boldsymbol{g} = \hat{\boldsymbol{X}}\hat{\boldsymbol{g}}, \tag{8.7}$$

where the frequency response vector $\hat{\boldsymbol{g}}$ of the normal coordinates can be calculated by solving the linear system of equations

$$\left[\begin{array}{c|c|c} \mathrm{i}\Omega\boldsymbol{E} - \boldsymbol{\Delta} & -\boldsymbol{E} & \boldsymbol{0} \\ \hline \boldsymbol{\Omega}^2 & \mathrm{i}\Omega\boldsymbol{E} - \boldsymbol{\Delta} & \boldsymbol{0} \\ \hline \boldsymbol{0} & \boldsymbol{0} & \mathrm{i}\Omega\boldsymbol{E} - \boldsymbol{\Sigma} \end{array}\right] \hat{\boldsymbol{g}} = \hat{\boldsymbol{b}}. \tag{8.8}$$

One obtains

$$\hat{g} = \left[\begin{array}{c|c|c} Z_2 & 0 & 0 \\ \hline 0 & Z_2 & 0 \\ \hline 0 & 0 & Z_1 \end{array}\right] \left[\begin{array}{c|c|c} \mathrm{i}\Omega E - \Delta & E & 0 \\ \hline -\Omega^2 & \mathrm{i}\Omega E - \Delta & 0 \\ \hline 0 & 0 & E \end{array}\right] \hat{b} \tag{8.9}$$

where

$$Z_1 = (\mathrm{i}\Omega E - \Sigma)^{-1} = \mathrm{diag}\left\{\frac{1}{\mathrm{i}\Omega - \sigma_k}\right\}, \tag{8.10}$$

$$Z_2 = \left[(\Delta^2 + \Omega^2 - \Omega^2 E) - \mathrm{i}(2\Omega\Delta)\right]^{-1}$$

$$= \mathrm{diag}\left\{\frac{1}{\left(\delta_j^2 + \omega_j^2 - \Omega^2\right) - 2\delta_j\Omega\mathrm{i}}\right\} \tag{8.11}$$

are the $(n-2s)\times(n-2s)$ and $s\times s$ *elementary frequency response matrices* of the first and second kind which are related to the real and complex eigenvalues. The elementary frequency response matrices are diagonal. Their elements are elementary frequency responses whose structure is either that of a *first order frequency response*

$$Z_{1kk} = V_{1k}\,\mathrm{e}^{-\mathrm{i}\alpha_{1k}} = \frac{1}{\mathrm{i}\Omega - \sigma_k}, \qquad k = 1(1)n-2s, \tag{8.12}$$

or that of a *second order frequency response*

$$Z_{2jj} = V_{2j}\,\mathrm{e}^{-\mathrm{i}\alpha_{2j}} = \frac{1}{\delta_j^2 + \omega_j^2 - \Omega^2 - 2\delta_j\Omega\mathrm{i}}, \qquad j = 1(1)s, \tag{8.13}$$

The elementary frequency responses (8.12) and (8.13) have an essential influence upon the amplitudes and phase frequency responses (7.53) of the vibrating system. However the elementary frequency responses do not totally determine the amplitudes and phase angles since an additional influence can be due to the modal matrix $\hat{X}$, and the excitation vector $b = b(\Omega)$ may also depend on the frequency.

8.2. Elementary frequency responses

The magnification functions $V_i(\Omega)$ and phase lags $\alpha_i(\Omega)$ will now be calculated from the complex frequency responses (8.12) and (8.13) as functions of the excitation frequency. Here the indexing corresponding to the eigenvalues will be abandoned and a distinction will be made between the quantities V_1, α_1 for the first order frequency response (8.12) and V_2, α_2 for the second order frequency response (8.13).

The first order frequency response (8.12)

$$V_1\,\mathrm{e}^{-\mathrm{i}\alpha_1} = \frac{1}{\mathrm{i}\Omega - \sigma} \tag{8.14}$$

and Euler's formula

$$V_1(\cos\alpha_1 - \mathrm{i}\sin\alpha_1) = \frac{1}{\sqrt{\sigma^2+\Omega^2}}\left(\frac{-\sigma}{\sqrt{\sigma^2+\Omega^2}} - \mathrm{i}\frac{\Omega}{\sqrt{\sigma^2+\Omega^2}}\right) \tag{8.15}$$

yield the following expression for the *magnification function (amplitude function, amplitude frequency response)*

$$V_1 = V_1(\Omega) = \frac{1}{\sqrt{\sigma^2+\Omega^2}} \tag{8.16}$$

and the following for the *phase lag (phase function, phase frequency response)*

$$\alpha_1 = \alpha_1(\Omega) = \arctan\frac{\Omega}{-\sigma}. \tag{8.17}$$

In Figures 8.1 to 8.3 one can find the *polar plots* of the first order

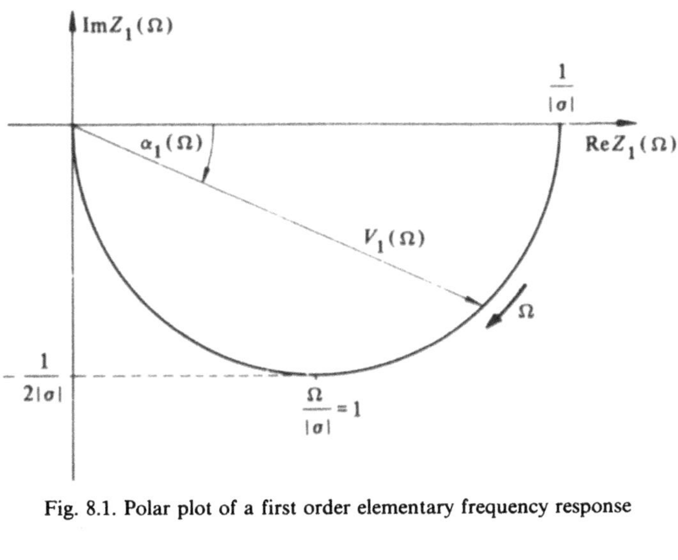

Fig. 8.1. Polar plot of a first order elementary frequency response

elementary frequency response (8.14), the corresponding magnification function (8.16), and the phase lag (8.17) for asymptotically stable systems, $\sigma < 0$. The following properties of the first order elementary frequency response now become apparent:

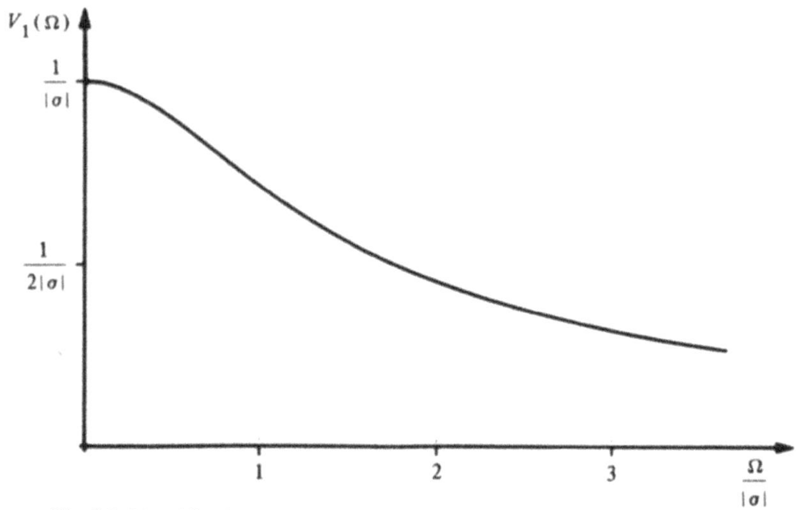

Fig. 8.2. Magnification function of a first order elementary frequency response

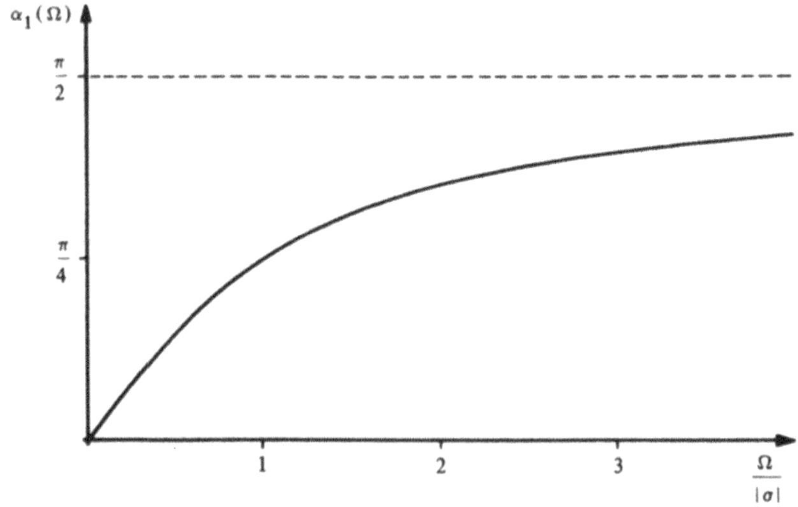

Fig. 8.3. Phase lag of a first order elementary frequency response

a) The polar plot is a semicircle.
b) The static amplitude magnification is

$$V_1(\Omega=0)=\frac{1}{|\sigma|} \tag{8.18}$$

which is also the maximum amplitude magnification.
c) The amplitude vanishes for $\Omega \to \infty$,

$$V_1(\Omega \to \infty)=0. \tag{8.19}$$

d) The static phase lag vanishes:

$$\alpha_1(\Omega=0)=0. \tag{8.20}$$

e) The maximum phase lag is

$$\alpha_1(\Omega \to \infty)=90^\circ. \tag{8.21}$$

It follows from (8.16) that under harmonic excitation with a constant amplitude one obtains smaller response amplitudes for a stronger damped eigenvalue than for a weaker damped eigenvalue.

Carrying out the corresponding calculation for the second order elementary frequency response (8.13),

$$V_2\,\mathrm{e}^{-\mathrm{i}\alpha_2}=\frac{1}{\delta^2+\omega^2-\Omega^2-2\delta\Omega\mathrm{i}}, \tag{8.22}$$

and introducing the notation

$$\eta=\frac{\Omega}{\sqrt{\delta^2+\omega^2}},\quad D=\frac{-\delta}{\sqrt{\delta^2+\omega^2}}\leqslant 1, \tag{8.23}$$

where η is an excitation frequency normalized by the absolute value of the eigenvalue and D is the damping factor (6.93), one obtains the relation

$$V_2(\cos\alpha_2-\mathrm{i}\sin\alpha_2)=\frac{1}{\delta^2+\omega^2}\,\frac{(1-\eta^2)-2D\eta\mathrm{i}}{(1-\eta^2)^2+4D^2\eta^2}. \tag{8.24}$$

Hence the *magnification function* is

$$V_2=V_2(\eta)=\frac{1}{\delta^2+\omega^2}\,\frac{1}{\sqrt{(1-\eta^2)^2+4D^2\eta^2}} \tag{8.25}$$

and the *phase lag* reads as

$$\alpha_2 = \alpha_2(\eta) = \arctan \frac{2D\eta}{1-\eta^2}. \tag{8.26}$$

In Figures 8.4 to 8.6 one can see the polar plots of the second order complex elementary frequency response (8.22), the magnification function (8.25), and the phase lag (8.26) for $0 \leqslant D \leqslant 1$. Note that in Figures 8.5 and 8.6 the abscissa is the frequency function

$$\zeta = \begin{cases} \eta & \text{for} \quad 0 \leqslant \eta \leqslant 1, \\ 2 - \dfrac{1}{\eta} & \text{for} \quad 1 \leqslant \eta < \infty \end{cases} \tag{8.27}$$

instead of the previously used frequency ratio η. This also makes it possible to represent very large frequency ratios by bounded quantities; in particular one gets for $\eta \to \infty$ the finite value $\zeta = 2$.

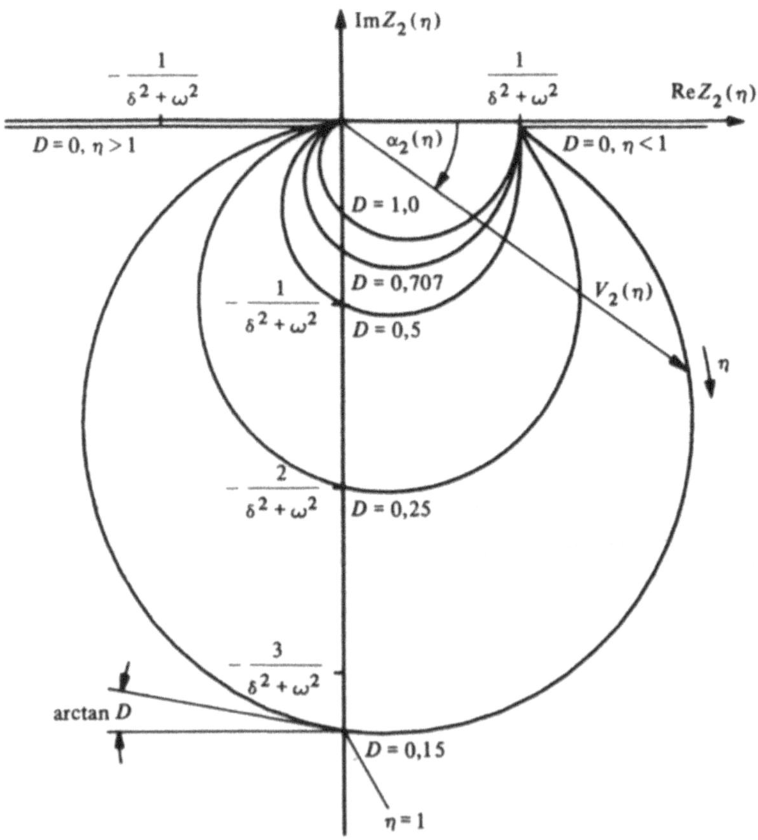

Fig. 8.4. Polar plot of a second order elementary frequency response

The following properties of the second order elementary frequency response follow from Figures 8.4 to 8.6:

a) The static amplitude magnification is

$$V_2(\eta = 0) = \frac{1}{\delta^2 + \omega^2}. \tag{8.28}$$

b) The maximum amplitude magnification is

$$V_{2\max} = \begin{cases} V_2(\eta = 0) = \dfrac{1}{\delta^2 + \omega^2} & \text{for} \quad D \geqslant \dfrac{1}{\sqrt{2}}, \\ V_2\left(\eta = \sqrt{1 - 2D^2}\right) = \dfrac{1}{\delta^2 + \omega^2} \dfrac{1}{2D\sqrt{1 - D^2}} & \\ \text{for } 0 < D \leqslant \dfrac{1}{\sqrt{2}}. & \end{cases} \tag{8.29}$$

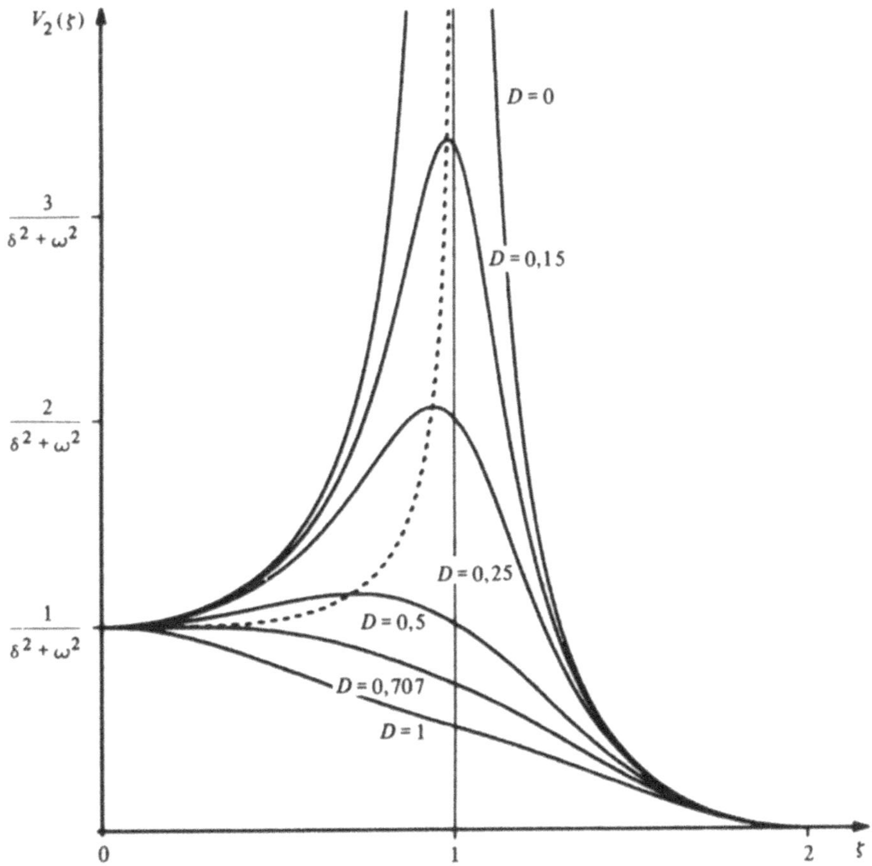

Fig. 8.5. Magnification function of a second order elementary frequency response

The dotted backbone curve in Fig. 8.5 is determined by

$$V_{2\max}(\eta) = \frac{1}{\delta^2 + \omega^2} \frac{1}{\sqrt{1-\eta^4}} \quad \text{for} \quad 0 \leqslant \eta < 1. \tag{8.30}$$

c) The amplitude vanishes for

$$V_2(\eta \to \infty) = 0. \tag{8.31}$$

d) The static phase lag vanishes,

$$\alpha_2(\eta = 0) = 0. \tag{8.32}$$

e) The middle phase lag

$$\alpha_2(\eta = 1) = 90° \tag{8.33}$$

does not depend on the damping factor D.

f) The maximal phase lag is

$$\alpha_2(\eta \to \infty) = 180°.$$

By (8.25), the amplitudes are proportional to the static displacement. Thus, if the system is subject to excitation with constant amplitude, and

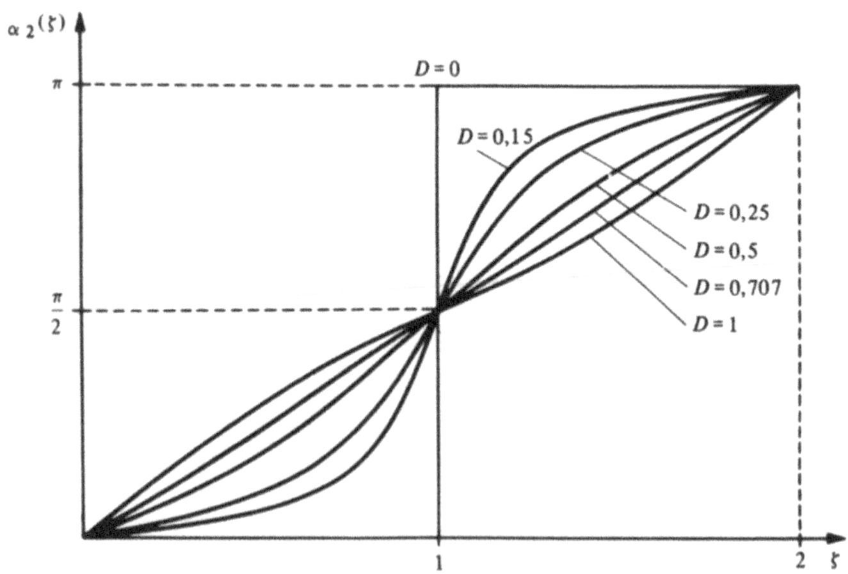

Fig. 8.6. Phase lag of a second order elementary frequency response

the damping factor is constant, then the higher frequency natural modes exhibit smaller amplitudes than the basic natural mode. This can be seen from (8.29) which implies for the damping factor (8.23) that

$$V_{2\max} = \frac{1}{2|\delta\omega|} = \frac{1}{2\omega^2}\sqrt{\frac{1-D^2}{D^2}} \quad \text{for} \quad 0 < D \leqslant \frac{1}{\sqrt{2}}. \tag{8.34}$$

8.3. Resonance and pseudoresonance

It can be seen from Figures 8.2 or 8.5 that for small or vanishing damping there may occur very large or infinite amplitudes. In the first case one speaks of the resonance phenomenon, the second case is called strict resonance. The following necessary conditions for resonance can be formulated.

For *strict resonance* to occur it is necessary that at least one of the elementary frequency responses does approach infinity for $0 \leqslant \Omega < \infty$, i.e., there exists an imaginary eigenvalue whose value is equal to the excitation frequency:

$$\left.\begin{array}{ll} \sigma_k = 0, & \Omega_R = 0 \\ \text{or} & \\ \delta_j = 0, & \omega_j = \Omega = \Omega_R, \end{array}\right\} \tag{8.35}$$

where Ω_R is the *resonance frequency*. The maximum number of resonance conditions correspond to the number of undamped natural modes. For instance, if a mechanical system (6.47) is totally undamped so that $\boldsymbol{\Delta} = \mathbf{0}$ in (6.61), then there exist $f = n/2$ resonance conditions. Condition (8.35) states that if $\Omega_R = \omega_j$, then, according to (8.9 to 11) the elementary frequency response matrix does not remain finite, and thus there does not exist a frequency response matrix (7.43). This in turn implies the singularity of the inverse frequency response matrix, so that the necessary condition for resonance (8.35) can also be stated as

$$p(\mathrm{i}\Omega_R) \equiv \det(\mathrm{i}\Omega_R \boldsymbol{E} - \boldsymbol{A}) = 0. \tag{8.36}$$

The frequency response continuously depends on the system parameters, provided the latter appears as continuous functions in the system matrices. Thus lightly damped systems certainly do not exhibit infinite values for the amplitude frequency response (8.25), nevertheless the amplitudes can become very large. According to (8.29), for $0 < D \leqslant \sqrt{\frac{1}{2}}$ or $0 < \delta_j^2 \leqslant \omega_j^2$ there occurs an amplitude much larger than the static displacement, and it attains its maximum value for the excitating frequency $\Omega = \sqrt{\omega_j^2 - \delta_j^2}$. In such case one speaks of a *resonance phenomenon,* given by the conditions

$$0 < \delta_j^2 \leqslant \omega_j^2, \quad \Omega = \sqrt{\omega_j^2 - \delta_j^2}. \tag{8.37}$$

Here again the number of possible distinct resonance phenomena depends on the number of eigenvalues which satisfy (8.37). For a lightly damped mechanical system (6.47) with $\boldsymbol{\Delta} \geqslant \mathbf{0}$ in (6.61) there can occur at most $f = n/2$ resonance peaks. The existence of the frequency response matrix is in this case always guaranteed; the inverse frequency response matrix remains regular.

A comparison of the conditions (8.35) and (8.37) for a strict resonance and for a resonance phenomenon leads to the conclusion that for dynamic systems whose eigenvalues do not depend on the excitation frequency (for a counterexample see Example 8.2) the resonance frequency Ω is shifted towards the smaller value. This deviation, however, is getting smaller as the damping, and thus δ_j decrease. This is utilized in the calculation of resonances. Since the verification of (8.37) is very laborious, and the resonance phenomena continuously depend on the damping parameters, one may investigate condition (8.35) for an undamped system instead of condition (8.37) for a lightly damped system. This technique will also be applied below.

By (8.35) and (8.37), in case of single eigenvalues, resonance takes place only in one elementary frequency response, and thus in only one natural mode or one normal coordinate $\hat{x}_j$. But this does not mean that resonance must occur in only one state coordinate x_i. Resonance in one normal coordinate can be carried over by the transformation (8.7) to more or all state coordinates.

Further, it can be seen from (8.9) that the requirements (8.35 to 37) are necessary conditions only. A strict resonance can only really occur if the essential coordinates of the excitation vector $\hat{\boldsymbol{b}}$ do not vanish. If this is the case, then it is still possible that the amplitudes remain finite. One then speaks of *pseudoresonance.* Just so the real appearance of resonance phenomena depends on the excitation vector. Since the amplitudes in resonance phenomena always remain finite, the pseudoresonance phenomena are of little importance. They are treated as a limiting case of pseudoresonances due to the above mentioned reasons of continuity.

As can be seen from the necessary conditions (8.35) or (8.36), strict resonance and pseudoresonance can occur only in partially undamped systems, i.e., in systems containing some purely imaginary eigenvalues. To distinguish additional conditions between strict resonance and pseudoresonance, one considers with $\sigma_k = 0$ or $\delta_j = 0$ for the j-th normal coordinate the particular solution (8.9):

$$\hat{g}_k = \frac{1}{i\Omega}\hat{b}_k \tag{8.38}$$

or

$$\hat{g}_j = \frac{i\Omega\hat{b}_j + \hat{b}_{j+s}}{\omega_j^2 - \Omega^2} \tag{8.39}$$

and

$$\hat{g}_{j+s} = \frac{-\hat{b}_j + i\Omega\hat{b}_{j+s}}{\omega_j^2 - \Omega^2} = \omega_j^2. \tag{8.40}$$

Moreover, in view of (8.35), one has in this case $\hat{g}_{j+s} = i\Omega\hat{g}_j$. It should also be noted that the excitation constants $\hat{b}_k$, $\hat{b}_j$, $\hat{b}_{j+s}$ may well depend on the excitation frequency,

$$\hat{b}_k = \hat{b}_k(\Omega), \quad \hat{b}_j = \hat{b}_j(\Omega), \quad \hat{b}_{j+s} = \hat{b}_{j+s}(\Omega). \tag{8.41}$$

Thus strict resonance occurs in (8.38) for $\Omega = 0$ precisely when

$$\hat{b}_k(0) \neq 0 \tag{8.42}$$

holds, and it occurs in (8.39, 8.40) for $\Omega = \omega_j$, if

$$i\omega_j\hat{b}_j(\omega_j) + \hat{b}_{j+s}(\omega_j) \neq 0 \tag{8.43}$$

holds. On the other hand, one obtains for pseudoresonance either the condition

$$\hat{b}_k(0) = 0 \tag{8.44}$$

for (8.38), or

$$i\omega_j\hat{b}_j(\omega_j) + \hat{b}_{j+s}(\omega_j) = 0 \tag{8.45}$$

for (8.39, 8.40). Indeed, if one applies to (8.38) and (8.39) De l'Hospital's rule for calculating limits, then one obtains

$$\lim_{\Omega \to 0} \hat{g}_k(\Omega) = \hat{g}_k(0) = \frac{\hat{b}'_k(0)}{i} \tag{8.46}$$

or

$$\lim_{\Omega \to \omega_j} \hat{g}_j(\Omega) = \hat{g}_j(\omega_j) = \frac{\mathrm{i}\hat{b}_j(\omega_j) + \mathrm{i}\omega_j \hat{b}_j'(\omega_j) + \hat{b}_{j+s}'(\omega_j)}{-2\omega_j}, \tag{8.47}$$

where (′) denotes the derivative with respect to the excitation frequency. This shows that although the necessary resonance condition (8.35) is satisfied for single eigenvalues under the assumption (8.44) or (8.45), the amplitudes of the particular solution remain finite. They can even vanish, e.g. when the excitation coordinates are identically zero, $\hat{b}_k \equiv 0$ or $\hat{b}_j \equiv 0$, $\hat{b}_{j+s} \equiv 0$.

Although the pseudoresonance conditions (8.44) and (8.45) are easy to understand due to the use of normal coordinates, better suited to practical applications is a representation of these conditions in terms of state coordinates. For this purpose the scalar conditions (8.44) and (8.45) have to be formulated as vector conditions for $\hat{\boldsymbol{b}}$, and next the similarity transformation (8.4) has to be applied to these conditions to obtain equivalent statements in the state space.

Consider first the pseudoresonance condition (8.44) for $\Omega_R = 0$ and the eigenvalue $\lambda_k = 0$. The excitation vectors $\hat{\boldsymbol{b}}(0)$ which satisfy (8.44) are described by

$$\hat{\boldsymbol{b}}(0) = \left[\begin{array}{c|c|c} -\boldsymbol{\Delta} & -\boldsymbol{E} & \boldsymbol{0} \\ \hline \boldsymbol{\Omega}^2 & -\boldsymbol{\Delta} & \boldsymbol{0} \\ \hline \boldsymbol{0} & \boldsymbol{0} & -\boldsymbol{\Sigma} \end{array}\right] \hat{\boldsymbol{d}}. \tag{8.48}$$

Since $\lambda_k = 0$, the k-th row of the matrix (8.48) vanishes and every $n \times 1$ vector $\hat{\boldsymbol{d}}$ yields $\hat{b}_k(0) = 0$ and thus with (8.48) also satisfies (8.44). Setting $\boldsymbol{b} = \hat{\boldsymbol{X}}\hat{\boldsymbol{b}}$ and $\boldsymbol{d} = \hat{\boldsymbol{X}}\hat{\boldsymbol{d}}$ and using (8.4), one obtains from (8.48) the equation

$$\boldsymbol{b}(0) = -\boldsymbol{A}\boldsymbol{d} = (\mathrm{i}\Omega\boldsymbol{E} - \boldsymbol{A})|_{\Omega=0}\boldsymbol{d}. \tag{8.49}$$

Thus for pseudoresonance with eigenvalue $\lambda_k = 0$, the excitation vector $\boldsymbol{b}(0)$ is an arbitrary linear combination of the columns of the inverse frequency response matrix (7.43) singular for $\Omega = 0$.

The same considerations will now be carried out for the pseudoresonance condition (8.45) for an excitation frequency (8.35). Using (8.45) and proceeding analogously to (8.48), one finds that the general excitation vector $\hat{\boldsymbol{b}}(\omega_j)$ as

$$\hat{\boldsymbol{b}}(\omega_j) = \left[\begin{array}{c|c|c} \mathrm{i}\omega_j\boldsymbol{E} - \boldsymbol{\Delta} & -\boldsymbol{E} & \boldsymbol{0} \\ \hline \boldsymbol{\Omega}^2 & \mathrm{i}\omega_j\boldsymbol{E} - \boldsymbol{\Delta} & \boldsymbol{0} \\ \hline \boldsymbol{0} & \boldsymbol{0} & \mathrm{i}\omega_j\boldsymbol{E} - \boldsymbol{\Sigma} \end{array}\right] \hat{\boldsymbol{d}}, \tag{8.50}$$

where $\hat{\boldsymbol{d}}$ is an arbitrary $n \times 1$ vector. One should observe here that, by (8.35), the j-th diagonal element in $\boldsymbol{\Delta}$ is zero, and the j-th diagonal element in $\boldsymbol{\Omega}$ is ω_j. Thus (8.50) implies

$$\hat{b}_j(\omega_j) = \mathrm{i}\omega_j \hat{d}_j - \hat{d}_{j+s}, \quad \hat{b}_{j+s}(\omega_j) = \omega_j^2 \hat{d}_j + \mathrm{i}\omega_1 \hat{d}_{j+s} \tag{8.51}$$

for arbitrary values of $\hat{d}_j$ and $\hat{d}_{j+s}$. They always satisfy condition (8.45). A coordinate transformation (8.4) back to the original state coordinates finally takes (8.50) into

$$\boldsymbol{b}(\omega_j) = (\mathrm{i}\omega_j \boldsymbol{E} - \boldsymbol{A})\boldsymbol{d} = (\mathrm{i}\Omega \boldsymbol{E} - \boldsymbol{A})|_{\Omega = \omega_j} \boldsymbol{d}. \tag{8.52}$$

Thus for pseudoresonance with eigenvalue $\lambda_j = \mathrm{i}\omega_j$ the excitation vector $\boldsymbol{b}(\omega_j)$ is an arbitrary linear combination of the columns of the inverse frequency response matrix (7.43) singular for $\Omega = \omega_j$.

Until now, the requirements (8.49) and (8.52) for pseudoresonance have been verified only for vibrating systems with single eigenvalues. But they are quite generally valid. If the necessary resonance condition (8.36) is satisfied, then pseudoresonance occurs iff the solution trajectories for the stable vibrating system (8.1) remain bounded. For vanishing initial conditions these trajectories were determined by (7.48). They are bounded iff the vector defined by (7.51) vanishes,

$$\left[\boldsymbol{E} - (\mathrm{i}\Omega_{\mathrm{R}}\boldsymbol{E} - \boldsymbol{A})(\mathrm{i}\Omega_{\mathrm{R}}\boldsymbol{E} - \boldsymbol{A})^{+}\right]\boldsymbol{b} = \boldsymbol{0}. \tag{8.53}$$

By (13.19c) and (13.35), this condition is equivalent to the requirement

$$\boldsymbol{b}(\Omega_{\mathrm{R}}) = (\mathrm{i}\Omega_{\mathrm{R}}\boldsymbol{E} - \boldsymbol{A})\boldsymbol{d} \tag{8.54}$$

for an arbitrary vector $\boldsymbol{d}$ and for a resonance frequency satisfying $\det(\mathrm{i}\Omega_{\mathrm{R}}\boldsymbol{E} - \boldsymbol{A}) = 0$. Thus for the resonance frequency Ω_{R} satisfying (8.36), pseudoresonance and strict resonance differ in the excitation

Table 8.1. Behavior of a harmonically excited vibrating system

$$\mathbf{b} = [\mathbf{E} - (i\Omega\mathbf{E} - \mathbf{A})(i\Omega\mathbf{E} - \mathbf{A})^{+}]\,\mathbf{b}$$

$\det(i\Omega\mathbf{E} - \mathbf{A})$ \ $\tilde{\mathbf{b}}$	$= 0$	$\neq 0$
$= 0$	pseudo resonance	strict resonance
$\neq 0$	vibration	

vector $\boldsymbol{b}(\Omega_R)$ given by (8.54). Taking into consideration (7.50, 7.51) and (8.53), one can describe by Table 8.1 the frequency response behaviour of the vibrating system (8.1).

The practical verification whether pseudoresonance or strict resonance takes place for a given resonance frequency (8.36) can be carried out either by verifying (8.54) or by a precise checking of (8.53). However this procedure is often quite laborious. If the resonance frequency corresponds to a single eigenvalue $\lambda_j = i\omega_j$ of the system matrix $\boldsymbol{A}$, which is the most important case for applications, a further condition for pseudoresonance can be given. Indeed, in this case one has a nonvanishing adjoint matrix to the singular inverse frequency response matrix (Section 13.1),

$$\operatorname{adj}(i\Omega_R \boldsymbol{E} - \boldsymbol{A}) \neq \boldsymbol{0}, \quad \det(i\Omega_R \boldsymbol{E} - \boldsymbol{A}) = 0. \tag{8.55}$$

By definition (13.13) it yields

$$[\operatorname{adj}(i\Omega_R \boldsymbol{E} - \boldsymbol{A})](i\Omega_R \boldsymbol{E} - \boldsymbol{A}) = \boldsymbol{0}. \tag{8.56}$$

If one multiplies (8.56) from the right by an arbitrary vector $\boldsymbol{d}$, then one gets by (8.54) the explicit condition for the resonance frequency Ω_R:

$$\operatorname{adj}(i\Omega_R \boldsymbol{E} - \boldsymbol{A})\boldsymbol{b}(\Omega_R) = \boldsymbol{0}. \tag{8.57}$$

This relation, however, is applicable only in the case of single eigenvalues, since the adjoint matrix vanishes for a multiple eigenvalue.

According to (7.49), the amplitude frequency response vector in the case of pseudoresonance is given by

$$\boldsymbol{g}(\Omega_R) = \boldsymbol{g}^+(\Omega_R) = \boldsymbol{T}_2(\Omega_R)\tilde{\boldsymbol{A}}^{-1}(\Omega_R)\boldsymbol{T}_4^*(\Omega_R)\boldsymbol{b}(\Omega_R). \tag{8.58}$$

However for a single eigenvalue the complex amplitudes can be also calculated from

$$\boldsymbol{g}(\Omega_R) = \lim_{\Omega \to \Omega_R} \boldsymbol{g}(\Omega) = \left. \frac{\dfrac{d}{d\Omega}[\operatorname{adj}(i\Omega \boldsymbol{E} - \boldsymbol{A})]\boldsymbol{b}(\Omega_R)}{\dfrac{d}{d\Omega}\det(i\Omega \boldsymbol{E} - \boldsymbol{A})} \right|_{\Omega = \Omega_R} \tag{8.59}$$

obtained by applying De l'Hospital's rule. According to the definition of pseudoresonance, this vector is finite. It is even possible that some of the coordinates $g_i(\Omega_R)$ are vanishing. Then, besides pseudoresonance there is also absorption occuring. Absorption will be discussed in greater detail in the next section.

In case of strict resonance, the vibration of (8.1) is not described by the particular solution (8.2) with the amplitude vector (8.58), but by the solution (7.48). Accordingly, the trajectories are oscillations increasing with time t:

$$\boldsymbol{x}_{P1}(t) \approx t\left[\tilde{\boldsymbol{g}}\, e^{i\Omega_R t} + \bar{\tilde{\boldsymbol{g}}}\, e^{-i\Omega_R t}\right] \quad \text{for great } t. \tag{8.60}$$

Thus the natural mode is determined by $\tilde{\boldsymbol{g}}$, as given in (7.50). Since the matrix $\boldsymbol{T}_1(\Omega_R)$ in (7.50) contains the eigenvectors of $\boldsymbol{A}$ for the eigenvalue $\lambda = i\Omega_R$, one has

$$(\boldsymbol{A} - i\Omega_R \boldsymbol{E})\tilde{\boldsymbol{g}} = \boldsymbol{0}. \tag{8.61}$$

Therefore at strict resonance the motion of the vibrating system is precisely determined by the natural modes with infinitely growing amplitudes belonging to the eigenvalue $\lambda = i\Omega_R$. Therefore in the neighbourhoods of resonances a dynamic system (8.1) vibrates precisely in its corresponding natural modes.

Example 8.1. Single-dof-vibrating system

Consider a harmonically excited single-dof-vibrating system

$$m\ddot{y}(t) + d\dot{y}(t) + ky(t) = h \cos \Omega t. \tag{8.62}$$

For this system there will be determined with the help of the elementary frequency response the amplitude and phase functions as well as the conditions for strict resonance. The state equation of the vibrating system is

$$\begin{bmatrix} \dot{y}(t) \\ \ddot{y}(t) \end{bmatrix} = \begin{bmatrix} 0 & 1 \\ -\dfrac{k}{m} & -\dfrac{d}{m} \end{bmatrix} \begin{bmatrix} y(t) \\ \dot{y}(t) \end{bmatrix} + \begin{bmatrix} 0 \\ \dfrac{h}{2m} \end{bmatrix} (e^{i\Omega t} + e^{-i\Omega t}). \tag{8.63}$$

The eigenvalues of the characteristic equation

$$\det(\lambda \boldsymbol{E} - \boldsymbol{A}) \equiv \det \begin{bmatrix} \lambda & -1 \\ \dfrac{k}{m} & \lambda + \dfrac{d}{m} \end{bmatrix} \equiv \lambda^2 + \frac{d}{m}\lambda + \frac{k}{m} = 0 \tag{8.64}$$

for light damping, $d^2 \leqslant 4km$, can be written as

$$\lambda_{1,2} = \delta \pm i\omega = -\frac{d}{2m} \pm \sqrt{\left(\frac{d}{2m}\right)^2 - \frac{k}{m}} \tag{8.65a}$$

with the notation

$$\delta = -\frac{d}{2m}, \qquad \omega^2 = \frac{k}{m} - \left(\frac{d}{2m}\right)^2. \tag{8.65b}$$

Since $\boldsymbol{A} = \boldsymbol{A}_{\mathrm{F}}$, (4.116) implies that the modal matrix is

$$\boldsymbol{X} = \begin{bmatrix} 1 & 1 \\ \delta + \mathrm{i}\omega & \delta - \mathrm{i}\omega \end{bmatrix}. \tag{8.66}$$

Thus, according to (6.83), the modified real modal matrix is

$$\hat{\boldsymbol{X}} = \begin{bmatrix} 1 & 1 \\ \delta + \mathrm{i}\omega & \delta - \mathrm{i}\omega \end{bmatrix} \begin{bmatrix} \frac{1}{2} & -\frac{\mathrm{i}}{2}\frac{1}{\omega} \\ \frac{1}{2} & \frac{\mathrm{i}}{2}\frac{1}{\omega} \end{bmatrix} = \begin{bmatrix} 1 & 0 \\ \delta & 1 \end{bmatrix}. \tag{8.67}$$

The modal matrix together with its inverse

$$\hat{\boldsymbol{X}}^{-1} = \begin{bmatrix} 1 & 0 \\ -\delta & 1 \end{bmatrix} \tag{8.68}$$

applied to the similarity transformation (8.4) results in the equation of motion in normal coordinates

$$\dot{\hat{\boldsymbol{x}}}(t) = \begin{bmatrix} \delta & 1 \\ -\omega^2 & \delta \end{bmatrix} \hat{\boldsymbol{x}}(t) + \begin{bmatrix} 0 \\ \frac{h}{2m} \end{bmatrix} (\mathrm{e}^{\mathrm{i}\Omega t} + \mathrm{e}^{-\mathrm{i}\Omega t}). \tag{8.69}$$

According to (8.9), the complex frequency response vector is

$$\hat{\boldsymbol{g}} = \begin{bmatrix} Z_2 & 0 \\ 0 & Z_2 \end{bmatrix} \begin{bmatrix} \mathrm{i}\Omega - \delta & 1 \\ -\omega^2 & \mathrm{i}\Omega - \delta \end{bmatrix} \begin{bmatrix} 0 \\ \frac{h}{2m} \end{bmatrix} = Z_2 \begin{bmatrix} 1 \\ \mathrm{i}\Omega - \delta \end{bmatrix} \frac{h}{2m}, \tag{8.70}$$

and it contains the second order frequency response

$$Z_2 = \frac{1}{\delta^2 + \omega^2 - \Omega^2 - 2\delta\Omega \mathrm{i}}. \tag{8.71}$$

The frequency response vector (8.7) can now be written in state coordinates:

$$\boldsymbol{g} = \hat{\boldsymbol{X}}\hat{\boldsymbol{g}} = \begin{bmatrix} 1 & 0 \\ \delta & 1 \end{bmatrix} Z_2 \begin{bmatrix} 1 \\ \mathrm{i}\Omega - \delta \end{bmatrix} \frac{h}{2m} = Z_2 \begin{bmatrix} 1 \\ \mathrm{i}\Omega \end{bmatrix} \frac{h}{2m}. \tag{8.72}$$

This immediately implies that the polar plot $g_1(\Omega)$ of the position coordinate of a simple vibrating system coincides, up to the factor $h/2m$ with the elementary frequency response Z_2,

$$g_1(\Omega)=\frac{h}{2m}Z_2(\Omega)=\frac{h}{2m}V_2(\Omega)\,\mathrm{e}^{-\mathrm{i}\alpha_2(\Omega)} \tag{8.73}$$

(see Figures 8.4 to 8.6). By (7.44), the particular solution $y_{\mathrm{P}1\infty}(t)$ of (8.62) is

$$\begin{aligned} y_{\mathrm{P}1\infty}(t) &= g_1\,\mathrm{e}^{\mathrm{i}\Omega t}+\bar{g}_1\,\mathrm{e}^{-\mathrm{i}\Omega t} \\ &= \frac{h}{2m}V_2\left[\mathrm{e}^{\mathrm{i}(\Omega t-\alpha_2)}+\mathrm{e}^{-\mathrm{i}(\Omega t-\alpha_2)}\right] \\ &= \frac{h}{m}V_2\cos(\Omega t-\alpha_2). \end{aligned} \tag{8.74}$$

For the amplitude function $a_1(\Omega)$ one obtains from (8.25)

$$a_1(\Omega)=\frac{h}{\sqrt{(k-m\Omega^2)^2+(d\Omega)^2}}. \tag{8.75}$$

Further, (8.26) gives the phase function $\psi_1(\Omega)=\alpha_2(\Omega)$, i.e.,

$$\tan\psi_1(\Omega)=\frac{d\Omega}{k-m\Omega^2}. \tag{8.76}$$

Moreover one has

$$\cos\psi_1(\Omega)=\frac{a_1(\Omega)}{h}(k-m\Omega^2),\qquad \sin\psi_1(\Omega)=\frac{a_1(\Omega)}{h}d\Omega. \tag{8.77}$$

One can read from Fig. 8.5 that strict resonance occurs for $d=0$ and $\Omega^2=\Omega_{\mathrm{R}}^2=k/m$, and a resonance phenomenon takes place for $0<d\leqslant\sqrt{2km}$ and $\Omega^2=k/m-2(d/2m)^2$. On the other hand, pseudoresonance cannot occur in a single-dof vibrating system.

Considering still the amplitude and phase functions for the velocity coordinate $\dot{y}$, one finds

$$a_2(\Omega)=\Omega a_1(\Omega),\qquad \psi_2(\Omega)=\psi_1(\Omega)-\frac{\pi}{2}. \tag{8.78}$$

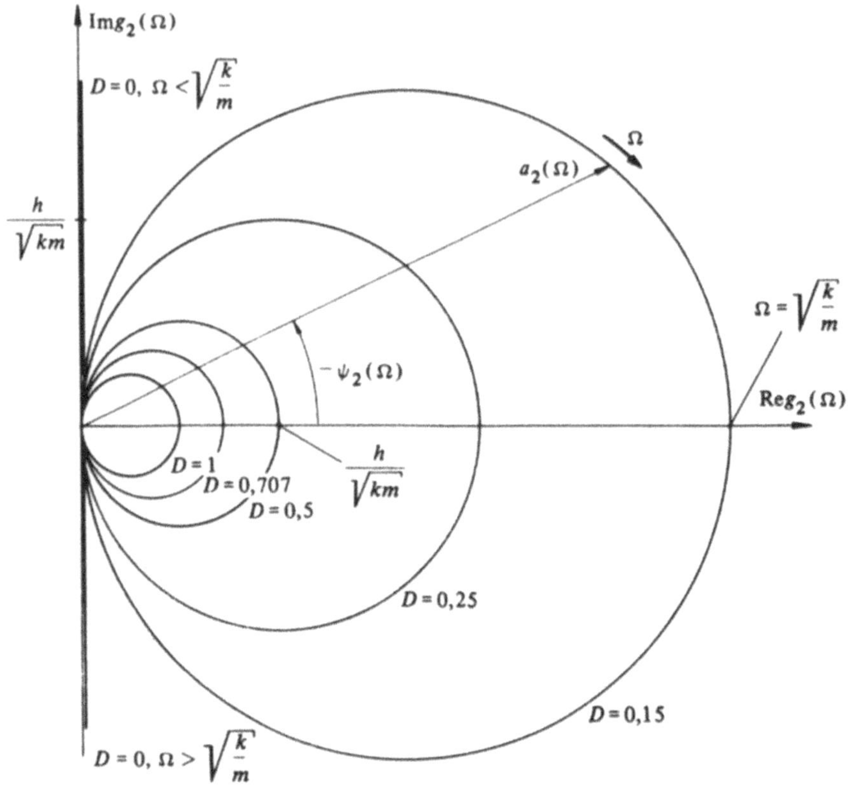

Fig. 8.7. Polar plot of the velocity frequency response of a single-dof-vibrating system

The corresponding polar plot

$$g_2(\Omega)=\frac{a_2}{2}(\Omega)\ \mathrm{e}^{-\mathrm{i}\psi_2(\Omega)}=\frac{h}{2m}\mathrm{i}\Omega V_2(\Omega)\ \mathrm{e}^{-\mathrm{i}\alpha_2(\Omega)} \tag{8.79}$$

is shown in Fig. 8.7 for various values of the damping factor $D=-\delta/\sqrt{\delta^2+\omega^2}=d/2\sqrt{km}$. One obtains a bundle of circles, all tangent to the imaginary axis.

Example 8.2. Centrifuge

In Problem 3.1 the following equations of motion for a centrifuge with a lightly unbalanced rotor and with position vector $y=[\alpha\ \beta]^{\mathrm{T}}$ was obtained:

$$\ddot{y}(t)+(\vartheta \boldsymbol{E}+g\Omega \boldsymbol{S})\,\dot{y}(t)+\kappa y(t)=\mathrm{e}\Omega^2\begin{bmatrix}\cos\Omega t\\ \sin\Omega t\end{bmatrix}. \tag{8.80}$$

The harmonic excitation vector

$$\boldsymbol{h}_{\mathrm{P}}(t)=\mathrm{e}\Omega^2\begin{bmatrix}\cos\Omega t\\ \sin\Omega t\end{bmatrix}=\tfrac{1}{2}\mathrm{e}\Omega^2\left\{\begin{bmatrix}1\\ -\mathrm{i}\end{bmatrix}\mathrm{e}^{\mathrm{i}\Omega t}+\begin{bmatrix}1\\ \mathrm{i}\end{bmatrix}\mathrm{e}^{-\mathrm{i}\Omega t}\right\} \tag{8.81}$$

depends on the square of the excitation frequency Ω. Applying (7.60, 7.61) one gets the frequency response vector of the position variables

$$\begin{aligned}\boldsymbol{q}&=\left[-\Omega^2\boldsymbol{E}+\mathrm{i}\Omega(\vartheta\boldsymbol{E}+g\Omega\boldsymbol{S})+\kappa\boldsymbol{E}\right]^{-1}\begin{bmatrix}1\\ -\mathrm{i}\end{bmatrix}\frac{\mathrm{e}\Omega^2}{2}\\ &=\frac{\frac{\mathrm{e}}{2}\Omega^2}{\left[\kappa-\Omega^2(1+g)+\mathrm{i}\vartheta\Omega\right]\left[\kappa-\Omega^2(1-g)+\mathrm{i}\vartheta\Omega\right]}\\ &\quad\times\begin{bmatrix}\kappa-\Omega^2(1+g)+\mathrm{i}\vartheta\Omega\\ +\mathrm{i}\left[\kappa-\Omega^2(1+g)+\mathrm{i}\vartheta\Omega\right]\end{bmatrix}\end{aligned} \tag{8.82}$$

or

$$\boldsymbol{q}=\frac{\frac{\mathrm{e}}{2}\Omega^2}{\kappa-\Omega^2(1-g)+\mathrm{i}\vartheta\Omega}\begin{bmatrix}1\\ -i\end{bmatrix}\quad\text{for}\quad\kappa-\Omega^2(1+g)+\mathrm{i}\vartheta\Omega\neq 0. \tag{8.83}$$

It follows from (8.82) that for $\vartheta=0$ strict resonance may occur provided the excitation frequencies are such that

$$\Omega^2_{\mathrm{R}_1}=\frac{\kappa}{1+g},\qquad\Omega^2_{\mathrm{R}_2}=\frac{\kappa}{1-g}. \tag{8.84}$$

However, for Ω_{R_1} the numerator vector is also vanishing, hence pseudo-resonance occurs for the excitation frequency Ω_{R_1} if $\vartheta=0$. This is a verification of the known phenomenon that it is not possible to excite by unbalances the so called backward whirl of a centrifuge so as to cause resonance. On the other hand, one obtains strict resonance for $\Omega=\Omega_{\mathrm{R}_2}$ when $\vartheta=0$ respresenting the critical speed or forward whirl, respectively. In particular, (8.83) implies

$$\lim_{\vartheta\to 0}\boldsymbol{q}(\Omega_{\mathrm{R}_2})=\lim_{\vartheta\to 0}\frac{\frac{\mathrm{e}}{2}\Omega^2_{\mathrm{R}_2}}{\kappa-\Omega^2_{\mathrm{R}_2}(1-g)+\mathrm{i}\vartheta\Omega_{\mathrm{R}_2}}\begin{bmatrix}1\\ -\mathrm{i}\end{bmatrix}=\infty \tag{8.85}$$

showing that at the resonance the centrifuge oscillates in the critical direction of its natural vibration $\tilde{y}_2=[1\ \ -i]^{\mathrm{T}}$.

For $\vartheta > 0$ there don't occur strict resonances. For $\vartheta > 0$, $\kappa > 0$ the centrifuge is always lightly damped, and thus a resonance phenomenon takes place. In Fig. 8.8 some graphs are shown of the amplitude frequency response

$$a_\alpha(\Omega) = \alpha_\beta(\Omega) = \frac{e\Omega^2}{\sqrt{\left(\kappa - \Omega^2(1-g)\right)^2 + \vartheta^2\Omega^2}} \tag{8.86}$$

for various values of the damping factor $D = \vartheta / 2\sqrt{\kappa(1-g)}$ $(g < 1)$. One

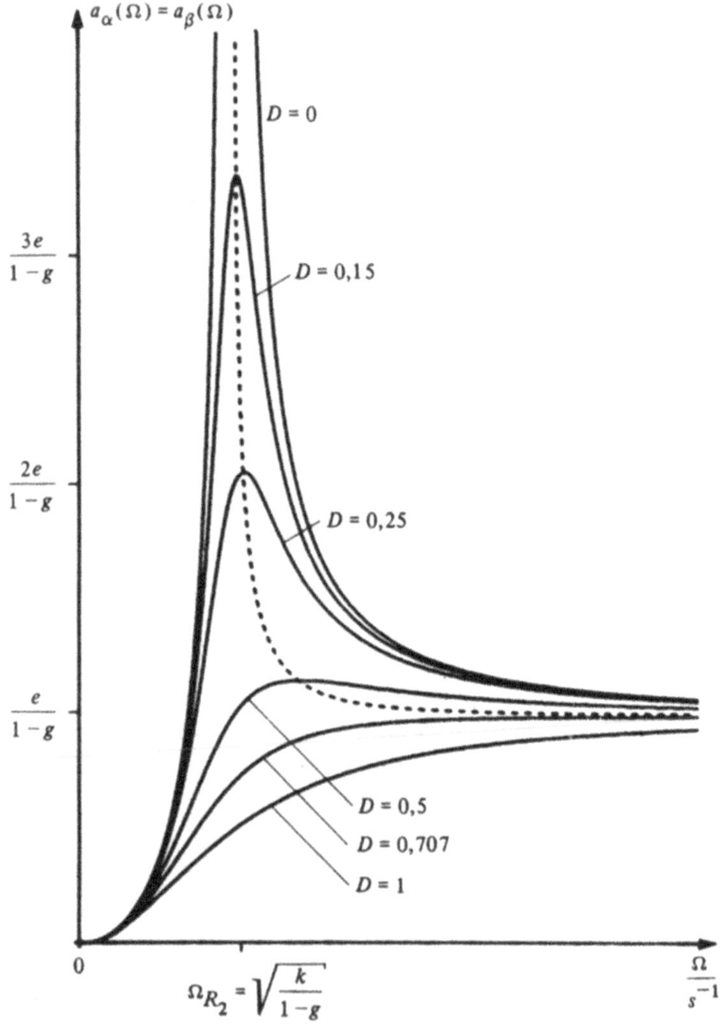

Fig. 8.8. Amplitude frequency response of a centrifuge

can also see from the dotted backbone curve that with growing D the frequency of the maximum amplitudes shifts towards higher values. It isn't necessary to discuss further the graphical representation of the amplitude frequency response $a_\alpha = a_\beta$ since the transformation

$$\eta = \frac{1}{\Omega}\sqrt{\frac{\kappa}{1-g}} \tag{8.87}$$

takes (8.86) into

$$a_\alpha(\eta) = a_\beta(\eta) = \frac{\mathrm{e}}{1-g}\frac{1}{\sqrt{(1-\eta^2)^2 + 4D^2\eta^2}} \tag{8.88}$$

which, up to a constant multiplier, coincides with the magnification function $V_2(\eta)$ (8.25). Thus, considering (8.29, 8.30) and (8.87), the frequency of the maximum amplitudes is obtained

$$\Omega^2 = \frac{\dfrac{\kappa}{1-g}}{1 - \dfrac{\vartheta^2}{2\kappa(1-g)}} \geqslant \frac{\kappa}{1-g} = \Omega^2_{\mathrm{R}_2} \tag{8.89}$$

and the corresponding amplitudes are given by

$$a_{\alpha\max} = a_{\beta\max} =$$

$$\begin{cases} \dfrac{e}{1-g}\dfrac{2\kappa(1-g)}{\sqrt{4\kappa(1-g)-\vartheta^2}} & \text{for } 0 < \vartheta^2 \leqslant 2\kappa(1-g), \\ \dfrac{e}{1-g} & \text{for } 2\kappa(1-g) \leqslant \vartheta^2. \quad \blacklozenge \end{cases} \tag{8.90}$$

Example 8.3. Vertical vibrations of an automobile

According to Problem 2.2, the vertical vibrations of a motor car riding on an profiled road can be described by the equation of motion

$$\begin{bmatrix} m_1 & 0 \\ 0 & m_2 \end{bmatrix}\ddot{y}(t) + \begin{bmatrix} d_1 & -d_1 \\ -d_1 & d_1 \end{bmatrix}\dot{y}(t) + \begin{bmatrix} k_1 & -k_1 \\ -k_1 & k_1 + k_2 \end{bmatrix} y(t)$$

$$= \begin{bmatrix} 0 \\ k_2 \end{bmatrix} y_e(t) \tag{8.91}$$

provided the tire damping is neglected. Assuming a harmonic excitation function, $y_e(t)=e\cos\Omega t$, for the road undulations, one can write the following complex representation of the excitation vector, using the notation of (7.55 to 57):

$$\boldsymbol{h}_{\mathrm{P}}(t)=\begin{bmatrix}0\\ k_2e\end{bmatrix}\cos\Omega t=\begin{bmatrix}0\\ \dfrac{k_2e}{2}\end{bmatrix}(\mathrm{e}^{\mathrm{i}\Omega t}+\mathrm{e}^{-\mathrm{i}\Omega t}). \tag{8.92}$$

Introducing the abbrevations (4.101), one can calculate the frequency matrix (7.61) making use of (13.14) as:

$$(-\Omega^2\boldsymbol{M}+\mathrm{i}\Omega\boldsymbol{D}+\boldsymbol{K})^{-1}=\frac{\operatorname{adj}(-\Omega^2\boldsymbol{M}+\mathrm{i}\Omega\boldsymbol{D}+\boldsymbol{K})}{\det(-\Omega^2\boldsymbol{M}+\mathrm{i}\Omega\boldsymbol{D}+\boldsymbol{K})} \tag{8.93}$$

where

$$\operatorname{adj}(-\Omega^2\boldsymbol{M}+\mathrm{i}\Omega\boldsymbol{D}+\boldsymbol{K})=\begin{bmatrix}\kappa_{22}+\kappa_{12}-\Omega^2+\mathrm{i}\Omega\vartheta_2 & \kappa_{11}+\mathrm{i}\Omega\vartheta_1\\ \kappa_{12}+\mathrm{i}\Omega\vartheta_2 & \kappa_{11}-\Omega^2+\mathrm{i}\Omega\vartheta_1\end{bmatrix}\times\begin{bmatrix}m_2 & 0\\ 0 & m_1\end{bmatrix} \tag{8.94}$$

and

$$\det(-\Omega^2\boldsymbol{M}+\mathrm{i}\Omega\boldsymbol{D}+\boldsymbol{K})=m_1m_2\{(\kappa_{11}-\Omega^2)(\kappa_{22}+\kappa_{12}-\Omega^2)-\kappa_{11}\kappa_{12}+ +\mathrm{i}\Omega[-(\vartheta_1+\vartheta_2)\Omega^2+\vartheta_1\kappa_{22}]\}. \tag{8.95}$$

Thus the frequency response vector (7.60) for the position coordinate of the vehicle is

$$\boldsymbol{q}=\tfrac{1}{2}(\boldsymbol{q}^{(1)}-\mathrm{i}\boldsymbol{q}^{(2)})$$

$$=\frac{1}{2}\,\frac{e\kappa_{22}}{(\kappa_{11}-\Omega^2)(\kappa_{22}+\kappa_{12}-\Omega^2)-\kappa_{11}\kappa_{12}+\mathrm{i}\Omega[-(\vartheta_1+\vartheta_2)\Omega^2+\vartheta_1\kappa_{22}]}\cdot\begin{bmatrix}\kappa_{11}+\mathrm{i}\Omega\vartheta_1\\ \kappa_{11}-\Omega^2+\mathrm{i}\Omega\vartheta_1\end{bmatrix}. \tag{8.96}$$

This leads to the following formula for the amplitude frequency responses:

$$a_1^2(\Omega) = e^2\left(\kappa_{22}^2\left(\kappa_{11}^2 + \vartheta_1^2\Omega^2\right)\right)$$
$$\times\left(\left[\left(\kappa_{11} - \Omega^2\right)\left(\kappa_{22} + \kappa_{12} - \Omega^2\right) - \kappa_{11}\kappa_{12}\right]^2\right.$$
$$\left. + \Omega^2\left[-\left(\vartheta_1 + \vartheta_2\right)\Omega^2 + \vartheta_1\kappa_{22}\right]^2\right)^{-1} \tag{8.97}$$

$$a_2^2(\Omega) = e^2\left(\kappa_{22}^2\left[\left(\kappa_{11} - \Omega^2\right)^2 + \vartheta_1^2\Omega^2\right]\right)$$
$$\times\left(\left[\left(\kappa_{11} - \Omega^2\right)\left(\kappa_{22} + \kappa_{12} - \Omega^2\right) - \kappa_{11}\kappa_{12}\right]^2\right.$$
$$\left. + \Omega^2\left[-\left(\vartheta_1 + \vartheta_2\right)\Omega^2 + \vartheta_1\kappa_{22}\right]^2\right)^{-1} \tag{8.98}$$

Strict resonance can occur only if there is no damping, $\vartheta_1 = \vartheta_2 = 0$. The resonance frequencies are then determined by

$$\left(\kappa_{11} - \Omega_{\mathrm{R}}^2\right)\left(\kappa_{22} + \kappa_{12} - \Omega_{\mathrm{R}}^2\right) - \kappa_{11}\kappa_{12} = 0. \tag{8.99}$$

One gets

$$\Omega_{R_{1,2}}^2 = \tfrac{1}{2}\left(\kappa_{11} + \kappa_{12} + \kappa_{22}\right) \mp \sqrt{\tfrac{1}{4}\left(\kappa_{11} + \kappa_{12} + \kappa_{22}\right)^2 - \kappa_{11}\kappa_{22}}\,, \tag{8.100}$$

where

$$\Omega_{\mathrm{R}_1} < \sqrt{\kappa_{11}}\,, \qquad \Omega_{\mathrm{R}_2} > \sqrt{\kappa_{12} + \kappa_{22}} \tag{8.101}$$

due to $\kappa_{11} < \kappa_{12} + \kappa_{22}$. The numerical data used in Problem 6.1 yield the resonance frequencies

$$\Omega_{\mathrm{R}_1} = 4.78\ s^{-1}, \qquad \Omega_{\mathrm{R}_2} = 66.16\ s^{-1}. \tag{8.102}$$

Since $\Omega_{\mathrm{R}_{1,2}} \neq \sqrt{\kappa_{11}}$ holds, the numerators in (8.97, 8.97) do not vanish at the resonance frequencies. If there is no damping, $\Omega = \Omega_{\mathrm{R}_1}$ and $\Omega = \Omega_{\mathrm{R}_2}$ lead to strict resonance. Thus for a lightly damped motor car one may expect resonance phenomena in the neighbourhoods of Ω_{R_1} and Ω_{R_2}.

In Fig. 8.9 the amplitude frequency response is recorded for the acceleration $\ddot{y}(t)$ of the car body. For judgement of riding comfort the frequency response of the acceleration of the car body is of greater interest than that of its position. Indeed, knowledge of the acceleration allows one to draw essential conclusions concerning the forces acting on

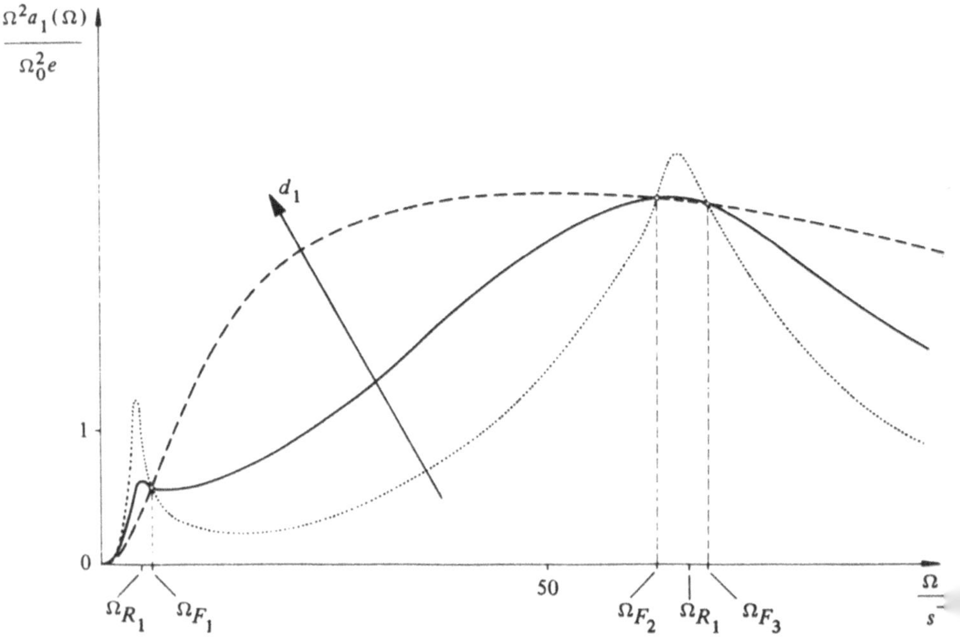

Fig. 8.9. Amplitude frequency response of the acceleration of the car body for various values of damping d_1

the passengers. The frequencies Ω_{F_i}, $i = 1(1)3$, of the fixpoints shown in Fig. 8.9 will be considered in Section 8.6.

8.4. Absorption

In multi-dof-vibrating systems the amplitudes of some state variables may vanish for certain excitation frequencies. This phenomenon when in spite of harmonic excitation individual elements of the vibrating system remain motionless, is called *absorption of vibrations.* This absorption phenomenon will be now illustrated by the example of a harmonically excited undamped motor car, Fig. 8.10. For a correct tune-up the mass m_1 vibrates with precisely that amplitude which causes the mutual cancellation of the forces acting on the mass m_2, i.e., of the excitation force acting through the second spring and the reaction force acting through the first spring. By (8.98), the amplitude frequency response of the displacement y_2 of the mass m_2 with vanishing damping is given

$$a_2(\Omega) = e\left|\frac{\kappa_{22}\left(\kappa_{11} - \Omega^2\right)}{\left(\kappa_{11} - \Omega^2\right)\left(\kappa_{22} + \kappa_{12} - \Omega^2\right) - \kappa_{11}\kappa_{12}}\right|. \tag{8.103}$$

For

$$\Omega = \Omega_A = \sqrt{\kappa_{11}} = \sqrt{\frac{k_1}{m_1}} \tag{8.104}$$

the mass m_2 remains motionless in spite of harmonic excitation. The mass m_1 can be regarded as a vibration absorber, connected through a spring with constant k_1 so that it counteracts the excitation force $k_2 e \cos \Omega_A t$ with an equal force $-k_2 e \cos \Omega_A t = k_1 a_1(\Omega_A) \cos(\Omega_A t - \pi)$. This is the case when

$$a_1(\Omega_A) = \frac{k_2}{k_1} e. \tag{8.105}$$

This example shows that under conditions of absorption a subsystem can be separated within the vibrating system which, if considered free, has a natural frequency coinciding with the excitation frequency and which cancels the excitating forces. This indicates the general *principle of vibration absorption*: A separated subsystem is a model for a harmonically excited vibration and it can partly cancel the effect of the exciting force by a motion in counterphase. The separated system is always marginally stable. Nevertheless, the whole system can be asymptotically stable. The *absorption condition* can be expressed mathematically by means of the frequency response vector (7.42):

$$g_i(\Omega_A) = 0 \quad \text{for at least one } i \tag{8.106}$$

under the assumption that the excitation is harmonic and nontrivial,

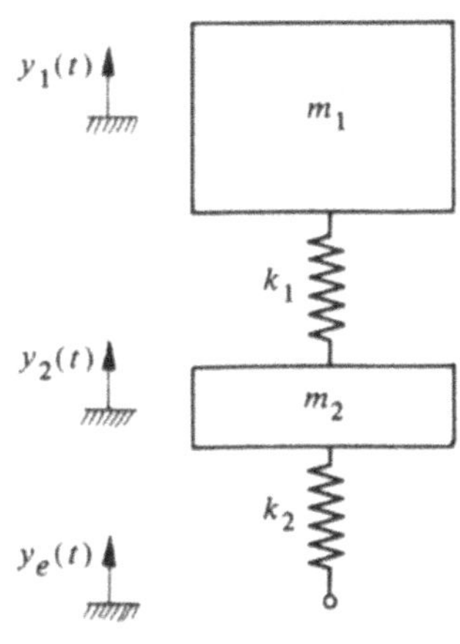

Fig. 8.10. Model for the vertical vibrations of an undamped motor car

$\boldsymbol{b}_{\mathrm{P}}(t)=\boldsymbol{b}e^{\mathrm{i}\Omega t}+\boldsymbol{b}e^{-\mathrm{i}\Omega t}$ with $\boldsymbol{b}\neq\boldsymbol{0}$, and as a rule, $\det(\mathrm{i}\Omega_A\boldsymbol{E}-\boldsymbol{A})\neq 0$. Condition (8.106) for absorption can also be stated as follows:

$$\boldsymbol{b}=(\mathrm{i}\Omega_A\boldsymbol{E}-\boldsymbol{A})c^{(i)} \tag{8.107}$$

with an $n\times 1$ vector $\boldsymbol{c}^{(i)}$ whose coordinates, except the i-th one, are arbitrarily chosen, and the i-th coordinate vanishes:

$$\boldsymbol{c}^{(i)}=[c_1 \quad c_2\ldots c_{i-1} \quad 0 \quad c_{i+1}\ldots c_n]^{\mathrm{T}}. \tag{8.108}$$

If the input vector $\boldsymbol{b}$ of the harmonic excitation is taken to be an arbitrary linear combination of the column vectors of the inverse frequency response matrix, excluding the i-th column, then absorption of vibrations is obtained in the i-th state coordinate. If absorption should take place not only in the i-th but also in the other coordinates, then more than one coordinate of the vector $\boldsymbol{c}^{(i)}$ must vanish. For absorption frequencies Ω_A which do not coincide with any of the resonance frequencies Ω_{R}, one gets from (8.107) the equation

$$\boldsymbol{g}(\Omega_A)=(\mathrm{i}\Omega_A\boldsymbol{E}-\boldsymbol{A})^{-1}\boldsymbol{b}=\boldsymbol{c}^{(i)}. \tag{8.109}$$

Using (13.14), one can also write the absorption condition in the form

$$\operatorname{adj}(\mathrm{i}\Omega_A\boldsymbol{E}-\boldsymbol{A})\boldsymbol{b}=k\boldsymbol{c}^{(i)} \tag{8.110}$$

where k is a constant.

Equations (8.107 to 110) imply that absorption in a vibrating system cannot occur simultaneously in all state coordinates. Indeed, if $\boldsymbol{c}^{(i)}=0$, then it follows from $\det(\mathrm{i}\Omega_A\boldsymbol{E}-\boldsymbol{A})\neq 0$ that also $\boldsymbol{b}=\boldsymbol{0}$, hence the system is not excited at all. In other words, if the excitation is nontrivial, $\boldsymbol{b}\neq\boldsymbol{0}$, then the system vibrates in at least one coordinate.

The discussion of vibration absorption in the case when $\det(\mathrm{i}\Omega_A\boldsymbol{E}-\boldsymbol{A})=0$, i.e., when an absorption frequency Ω_A coincides with a resonance frequency Ω_{R}, is of interest only if there is no strict resonance but when pseudoresonance takes place. In this case there is absorption of vibrations when at least one coordinate of the vectors (8.58) or (8.59) vanishes. Due to the pseudoresonance condition (8.54), the excitation vector $\boldsymbol{b}$ must have the same structure as in (8.107), while the vector $\boldsymbol{c}^{(i)}$ has to satisfy not (8.108, 8.109) but

$$g_i^{+}(\Omega_A)=\boldsymbol{e}_i^{\mathrm{T}}\boldsymbol{T}_2(\Omega_A)\boldsymbol{T}_4^{*}(\Omega_A)\boldsymbol{c}^{(i)}=0. \tag{8.111}$$

Example 8.4. Vibration absorption of the vertical vibrations of an automobile by the engine

The mechanical vibrations acting upon the passengers can be reduced by an elastically suspended absorption mass. In order to avoid at the same time an increasing weight of the vehicle, it has been suggested to apply the engine mass for the purpose of absorbing vibrations. Neglecting the masses of the wheels and axels, one then obtains the model shown on Fig. 8.11. By Problem 2.2, the corresponding equations of motion are given by

$$\boldsymbol{M}\ddot{\boldsymbol{y}}(t) + \boldsymbol{D}\dot{\boldsymbol{y}}(t) + \boldsymbol{K}\boldsymbol{y}(t) = \boldsymbol{h}(t) \tag{8.112}$$

with the matrices

$$\left.\begin{aligned} \boldsymbol{M} &= \begin{bmatrix} m_1 & 0 \\ 0 & m_2 \end{bmatrix}, & \boldsymbol{D} &= \begin{bmatrix} 0 & 0 \\ 0 & d_2 \end{bmatrix}, \\ \boldsymbol{K} &= \begin{bmatrix} k_1 & -k_1 \\ -k_1 & k_1 + k_2 \end{bmatrix}, & \boldsymbol{h}(t) &= \begin{bmatrix} 0 \\ k_2 y_e(t) + d_2 \dot{y}_e(t) \end{bmatrix}. \end{aligned}\right\} \tag{8.113}$$

For an undulating road profile, $y_e(t) = e \cos \Omega t$, the excitation function in complex notation is

$$\boldsymbol{h}(t) = \frac{e}{2}\begin{bmatrix} 0 \\ k_2 + \mathrm{i} d_2 \Omega \end{bmatrix} \mathrm{e}^{\mathrm{i}\Omega t} + \frac{e}{2}\begin{bmatrix} 0 \\ k_2 - \mathrm{i} d_2 \Omega \end{bmatrix} \mathrm{e}^{-\mathrm{i}\Omega t}. \tag{8.114}$$

Using the frequency response matrix of the mechanical system (8.112),

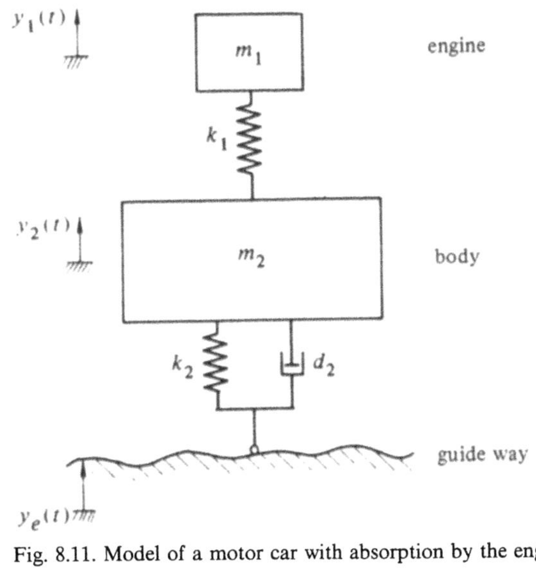

Fig. 8.11. Model of a motor car with absorption by the engine

$$-\Omega^2\boldsymbol{M}+\mathrm{i}\Omega\boldsymbol{D}+\boldsymbol{K}=\begin{bmatrix} k_1-m_1\Omega^2 & -k_1 \\ -k_1 & k_1+k_2-m_2\Omega^2+\mathrm{i}\Omega d_2 \end{bmatrix} \tag{8.115}$$

and the determinant

$$\det(-\Omega^2\boldsymbol{M}+\mathrm{i}\Omega\boldsymbol{D}+\boldsymbol{K})$$
$$=m_1m_2\Omega^4-im_1d_2\Omega^3-\left[m_1(k_1+k_2)+m_2k_1\right]\Omega^2+\mathrm{i}k_1d_2\Omega+k_1k_2 \tag{8.116}$$

one obtains the frequency response matrix (7.61):

$$\boldsymbol{F}_{\mathrm{M}}(\Omega)=\frac{1}{\det(-\Omega^2\boldsymbol{M}+\mathrm{i}\Omega\boldsymbol{D}+\boldsymbol{K})}$$
$$\times\begin{bmatrix} k_1+k_2-m_2\Omega^2+\mathrm{i}\Omega d_2 & k_1 \\ k_1 & k_1-m_1\Omega^2 \end{bmatrix}. \tag{8.117}$$

Employing the formula for the complex frequency response vector

$$\boldsymbol{q}=\boldsymbol{F}_{\mathrm{M}}\boldsymbol{h}=\frac{1}{2}\frac{e(k_2+\mathrm{i}d_2\Omega)}{\det(-\Omega^2\boldsymbol{M}+\mathrm{i}\Omega\boldsymbol{D}+\boldsymbol{K})}\begin{bmatrix} k_1 \\ k_1-m_1\Omega^2 \end{bmatrix} \tag{8.118}$$

and the abbreviations

$$\kappa_{ij}=\frac{k_i}{m_j},\quad i,\ j=1,2,\quad \vartheta_2=\frac{d_2}{m_2} \tag{8.119}$$

one gets the amplitude frequency responses

$$a_1^2(\Omega)=e^2\frac{\kappa_{11}^2\left(\kappa_{22}^2+\vartheta_2^2\Omega^2\right)}{\left[\left(\kappa_{11}-\Omega^2\right)\left(\kappa_{22}+\kappa_{12}-\Omega^2\right)-\kappa_{11}\kappa_{12}\right]^2+\vartheta_2^2\Omega^2\left(\kappa_{11}-\Omega^2\right)^2}, \tag{8.120}$$

$$a_2^2(\Omega)=e^2\frac{\left(\kappa_{11}-\Omega^2\right)^2\left(\kappa_{22}^2+\vartheta_2^2\Omega^2\right)}{\left[\left(\kappa_{11}-\Omega^2\right)\left(\kappa_{22}+\kappa_{12}-\Omega^2\right)-\kappa_{11}\kappa_{12}\right]^2+\vartheta_2^2\Omega^2\left(\kappa_{11}-\Omega^2\right)^2}. \tag{8.121}$$

Strict resonances are obtained only for vanishing damping, $\vartheta_2=0$, at the resonance frequencies $\Omega_{R_{1,2}}$ according to (8.100). If the damping is finite, $\vartheta_2 > 0$, the vibrating system (8.112) is asymptotically stable. The desired absorption phenomenon is obtained for the second coordinate, i.e., for the position of the car body, at the excitation frequency

$$\Omega_A^2 = \kappa_{11} = \frac{k_1}{m_1}: \qquad a_2(\Omega_A) = 0. \tag{8.122}$$

Moreover this absorption does not depend on the body spring k_1 nor on the shock absorber d_2.

A more detailed investigation of the amplitude frequency response (8.121) leads to the result shown in Fig. 8.12. For vanishing damping the system has two resonance frequencies and one absorption frequency. With increasing damping strict resonance turns over in a resonance phenomenon while the absorption remains unchanged. For very strong damping the resonance phenomena disappear while the absorption still remains. On the other hand, there is a steady increase in sensitivity towards frequency variations in the neighbourhood of the absorption frequency. More attention will be devoted in Section 8.6 to the interesting phenomenon that the fixpoints Ω_{F_i}, $i = 1, 3, 4$, are invariant with respect to the damping d_2. ◆

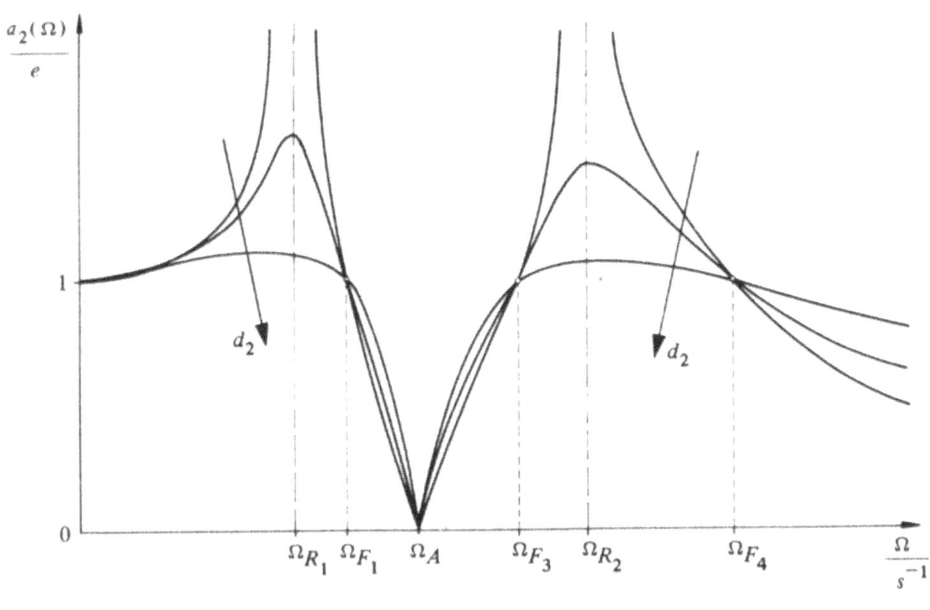

Fig. 8.12. Amplitude frequency response for various values of damping d_2

8.5. Pseudoresonance and absorption in mechanical systems

The essential condition for pseudoresonance and absorption also remain unchanged for mechanical vibrating systems (7.55) where the $f \times f$ frequency matrix $\boldsymbol{F}_{\mathrm{M}}$ is applied according to (7.61).

At a resonance frequency Ω_{R} with

$$\det\left[-\Omega_{\mathrm{R}}^2 \boldsymbol{M}+\mathrm{i} \Omega_{\mathrm{R}}(\boldsymbol{D}+\boldsymbol{G})+(\boldsymbol{K}+\boldsymbol{N})\right]=0 \tag{8.123}$$

the pseudoresonance conditions are, by (8.57) and (8.54),

$$\operatorname{adj}\left[-\Omega_{\mathrm{R}}^2 \boldsymbol{M}+\mathrm{i} \Omega_{\mathrm{R}}(\boldsymbol{D}+\boldsymbol{G})+(\boldsymbol{K}+\boldsymbol{N})\right] \boldsymbol{h}=\mathbf{0} \tag{8.124}$$

for simple eigenvalues $\lambda=\mathrm{i} \Omega_{\mathrm{R}}$, and

$$\boldsymbol{h}=\left[-\Omega_{\mathrm{R}}^2 \boldsymbol{M}+\mathrm{i} \Omega_{\mathrm{R}}(\boldsymbol{D}+\boldsymbol{G})+(\boldsymbol{K}+\boldsymbol{N})\right] \boldsymbol{d} \tag{8.125}$$

for eigenvalues $\lambda=\mathrm{i} \Omega_{\mathrm{R}}$ of arbitrary multiplicity, where $\boldsymbol{d}$ should is now an arbitrary $f \times 1$-vector. It follows from (7.64) that the pseudoresonance conditions for the position coordinates imply those for the velocity coordinates.

From (8.107) and (8.110) follow the absorption conditions for the j-th coordinate of the position vector and hence also for the j-th coordinate of the velocity vector:

$$\boldsymbol{h}=\left[-\Omega_A^2 \boldsymbol{M}+\mathrm{i} \Omega_A(\boldsymbol{D}+\boldsymbol{G})+(\boldsymbol{K}+\boldsymbol{N})\right] \boldsymbol{c}^{(j)} \tag{8.126}$$

or

$$\operatorname{adj}\left[-\Omega_A^2 \boldsymbol{M}+\mathrm{i} \Omega_A(\boldsymbol{D}+\boldsymbol{G})+(\boldsymbol{K}+\boldsymbol{N})\right] \boldsymbol{h}=k c^{(j)} \tag{8.127}$$

where $\boldsymbol{c}^{(j)}$ is now an arbitrary $f \times 1$-vector with $c_j^{(j)}=0$ and k is a constant. Here it has bèen assumed that

$$\det\left[-\Omega_A^2 \boldsymbol{M}+\mathrm{i} \Omega_A(\boldsymbol{D}+\boldsymbol{G})+(\boldsymbol{K}+\boldsymbol{N})\right] \neq 0. \tag{8.128}$$

For a single-dof-mechanical vibrating system

$$m \ddot{y}(t)+k y(t)=h \mathrm{e}^{\mathrm{i} \Omega t}+\bar{h} \mathrm{e}^{-\mathrm{i} \Omega t} \tag{8.129}$$

there can be neither pseudoresonance nor absorption. The inverse frequency response matrix is for $f=1$ a scalar

$$F_{\mathrm{M}}^{-1}(\Omega)=-m \Omega^2+k. \tag{8.130}$$

This implies, by (8.123), the condition for the resonance frequency

$$\det\left(-m\Omega_R^2 + k\right) = 0: \qquad \Omega_R^2 = \frac{k}{m} = \omega^2. \tag{8.131}$$

The condition for pseudoresonance (8.125) yields

$$h = \left[-m\omega^2 + k\right]d = 0 \cdot d = 0, \tag{8.132}$$

so that pseudoresonance occurs only in the trivial case $h = 0$. For a real excitation $h \neq 0$ and $\Omega = \Omega_R = \omega$ strict resonance always appears.

One finds from (8.126) that the absorption condition is

$$h = \left[-m\Omega_T^2 + k\right] \cdot 0 = 0, \tag{8.133}$$

since $c^{(1)} = 0$. This result says that under real excitation absorption cannot occur in a single-dof-mechanical vibrating system.

The phenomena of pseudoresonance and absorption appear only in multi-dof-vibrating systems.

8.6. Fixpoints

In Figures 8.9 and 8.12 amplitude frequency responses for various damping parameters are shown. It can be seen that there exist certain frequencies Ω_F such that the corresponding values of the amplitude frequency response do not depend on the damping. These invariant points of the amplitude frequency response are called *fixpoints.* They are characterized by the condition

$$|g_i(\Omega_F)| = a_i(\Omega_F) = \text{const} \tag{8.134}$$

where invariance is required with respect to the varying parameter. If this parameter is denoted by p, and under assumption of $\det(\mathrm{i}\Omega_F \boldsymbol{E} - \boldsymbol{A}) \neq 0$ for all values of p, then the frequencies Ω_F for the fixpoints in the i-th state coordinate can be determined from the equations

$$\frac{\mathrm{d}^k a_i(\Omega_F)}{\mathrm{d}p^k} = 0, \quad k = 1(1)\infty. \tag{8.135}$$

In practice, however, the frequencies Ω_F for fixpoints will be indicated by the necessary condition

$$a_i(\Omega_F)|_{p=0} = a_i(\Omega_F)|_{p\to\infty}. \tag{8.136}$$

The fixpoints are then found by substituting these possible values Ω_{F} and verifying that the amplitudes $a_i(\Omega_{\mathrm{F}})$ do not depend on p.

Example 8.5. Vertical vibrations of an automobile

In Example 8.3 the amplitude frequency response (8.97, 8.98) for the vertical vibrations of a motor car were calculated. In Fig. 8.9 one can clearly see fixpoints in the amplitude frequency response of the car body acceleration $\Omega^2 a_1(\Omega)$ relative to the varying damping parameter $d_1 = m_1\vartheta_1 = m_2\vartheta_2$. If one forms

$$\Omega_F^2 a_1(\Omega_F)|_{d_1=0} = \Omega_F^2 a_1(\Omega_F)|_{d_1\to\infty}, \tag{8.137}$$

according to (8.136), then one gets from (8.97)

$$\Omega_F^2 e\frac{\kappa_{11}\kappa_{22}}{\left|(\kappa_{11}-\Omega_F^2)(\kappa_{22}+\kappa_{12}-\Omega_F^2)-\kappa_{11}\kappa_{12}\right|}$$

$$= \Omega_F^2 e\frac{\Omega_F\kappa_{22}\cdot\dfrac{1}{m_1}}{\Omega_F\left|-\left(\dfrac{1}{m_1}+\dfrac{1}{m_2}\right)\Omega_F^2+\dfrac{1}{m_1}\kappa_{22}\right|}, \tag{8.138}$$

i.e.,

$$\frac{1}{\kappa_{11}}\left\{(\kappa_{11}-\Omega_F^2)(\kappa_{22}+\kappa_{12}-\Omega_F^2)-\kappa_{11}\kappa_{12}\right\} = \pm\left\{\kappa_{22}-\left(1+\frac{m_1}{m_2}\right)\Omega_F^2\right\}. \tag{8.139}$$

This quadratic equation for Ω_{F}^2 has the four solutions

$$\left.\begin{aligned} &\Omega_{F_0}^2 = 0, \quad \Omega_{F_2}^2 = \kappa_{22},\\ &\Omega_{F_{1,3}}^2 = \frac{1}{2}\left[\left(2+\frac{m_1}{m_2}\right)\kappa_{11}+\kappa_{22}+\kappa_{12}\right]\\ &\qquad \pm\sqrt{\frac{1}{4}\left[\left(2+\frac{m_1}{m_2}\right)\kappa_{11}+\kappa_{22}+\kappa_{12}\right]^2-2\kappa_{11}\kappa_{22}}\,. \end{aligned}\right\} \tag{8.140}$$

The corresponding amplitudes are given by

$$\Omega_{F_i}^2 a_1(\Omega_{F_i}) = e\Omega_{F_i}^2\frac{\kappa_{22}}{\left|\kappa_{22}-\left(1+\dfrac{m_1}{m_2}\right)\Omega_{F_i}^2\right|}, \qquad i = 0(1)3, \tag{8.141}$$

which yields for the trivial fixpoint

$$\Omega_{F_0}^2 a_1(\Omega_{F_0}) = 0 \tag{8.142}$$

and for the essential one

$$\Omega_{F_2}^2 a_1(\Omega_{F_2}) = \frac{k_2}{m_1} e. \tag{8.143}$$

Since (8.143) is almost the maximum value of the amplitude frequency response in Fig. 8.9, one obtains small car body accelerations when the tire stiffness is kept low. ◆

Example 8.6. Vibration absorption of the vertical vibrations of a motor car by the engine

According to (8.136), the fixpoints of the car body frequency response $a_2(\Omega)$ given in (8.121) and shown in Fig. 8.12 can be determined from

$$a_2(\Omega_F)|_{\vartheta_2=0} = a_2(\Omega_F)|_{\vartheta_2 \to \infty}, \tag{8.144}$$

i.e., from

$$\left.\begin{aligned} &0 = 0 \quad \text{for} \quad \kappa_{11} = \Omega_F^2, \\ &e\kappa_{22}\left|\frac{\kappa_{11} - \Omega_F^2}{(\kappa_{11} - \Omega_F^2)(\kappa_{22} + \kappa_{12} - \Omega_F^2) - \kappa_{11}\kappa_{12}}\right| = e \quad \text{for} \quad \kappa_{11} \neq \Omega_F^2. \end{aligned}\right\} \tag{8.145}$$

One obtains

$$\left.\begin{aligned} &\Omega_{F_0}^2 = 0, \quad \Omega_{F_2}^2 = \Omega_A^2 = \kappa_{11}, \quad \Omega_{F_3}^2 = \kappa_{11} + \kappa_{12}, \\ &\Omega_{F_{1,4}}^2 = \tfrac{1}{2}[\kappa_{11} + \kappa_{12} + 2\kappa_{22}] \pm \sqrt{\tfrac{1}{4}[\kappa_{11} + \kappa_{12} + 2\kappa_{22}]^2 - 2\kappa_{11}\kappa_{22}}, \end{aligned}\right\} \tag{8.146}$$

where the absorption fixpoint $\Omega_{F_2} = \Omega_A = \sqrt{\kappa_{11}}$ in (8.145) has to be considered separately. All the amplitudes, except of absorption, have equal values:

$$a_2(\Omega_{F_i}) = e, \quad i = 0, 1, 3, 4, \quad a_2(\Omega_{F_2}) = 0. \tag{8.147}$$

Since in this example, by (8.120, 8.121),

$$a_1(\Omega) = \frac{\kappa_{11}}{|\kappa_{11} - \Omega^2|} a_2(\Omega), \quad \Omega^2 \neq \kappa_{11}, \tag{8.148}$$

holds, one not only has fixpoints for the position coordinate y_2 of the car body, but also for the position coordinate y_1 of the motor. The values

$$a_1(\Omega_{F_i}) = \frac{\kappa_{11}}{\left|\kappa_{11} - \Omega_{F_i}^2\right|} e, \qquad i = 0, 1, 3, 4, \tag{8.149}$$

do not depend on the damping $d_2 = \vartheta_2 m_2$. ◆

8.7. Optimal frequency responses

The design of a vibrating system should usually ensure *good natural behavior* and *good disturbance behavior.* The optimization of natural vibrations was the subject of Section 6.2; the optimization of disturbance behavior is the purpose of this section. Disturbance behavior is measured by the frequency responses, especially the amplitude frequency responses of the state coordinates. Although the idea of a "good" frequency response is intuitively clear one also has to define a cost functional with respect to the frequency response, in analogy with the proceedings in Section 6.2. The aim is to diminish the amplitudes of the frequency response, i.e. to reduce the sensitivity of the vibrating system to disturbances. Here is where the first difficulties appear. For the centrifuge shown in Fig. 8.8, an increase of the damping factor D implies in the entire frequency range a reduction of the amplitudes. For the motor car shown in Fig. 8.9, the behavior of the amplitude frequency response is essentially more complicated. In certain frequency ranges the amplitude diminishes as the damping value d_1 is increasing, but in the remaining ranges the opposite behavior can be observed. Precisely this discerning behavior with respect to parameter changes can be frequently observed in engineering systems. It is a consequence of the so called *equilibrium theorem,* valid for a large class of vibrating systems, which states that under parameter variations the logarithmic mean over all frequencies of the amplitude frequency response remains unaltered. Consequently, an optimization of the frequency response is meaningful only within a limited frequency range.

A variety of cost functions can be used on any frequency range $\bar{\Omega} = [\Omega_1, \Omega_2]$, enabling one to make statements as to what is "acceptable".

As the first cost functional consider the *maxima* of the amplitude frequency response $a(\Omega)$ in $\bar{\Omega} = [\Omega_1, \Omega_2]$,

$$\max_{\Omega \in \bar{\Omega}} \{a(\Omega)\} \rightarrow \text{Minimum}, \tag{8.150}$$

which in a parameter optimization should be made as small as possible.

If an amplitude frequency response contains fixpoints with respect to some parameter, then the optimization can frequently be carried out in two steps: First, those parameters which influence the fixpoints are chosen so as to minimize the fixpoints as far as possible; second, the remaining parameters are varied so that the maxima of $a(\Omega)$ coincide precisely with the fixpoints. One should observe however, that this design procedure cannot always be successfully applied. For instance, by the result (8.147), the amplitudes of the nonvanishing fixpoints of the car vibration in Fig. 8.12 do not depend on the spring and damping constants, $a_2(\Omega_{\mathrm{F}_i})=e$, and the maxima of $a_2(\Omega)$ cannot coincide with these fixpoints because of $a_2(\Omega_{\mathrm{R}_1})>e$ and $a_2(\Omega_{\mathrm{R}_2})>e$. However, one can apply this method with greater success for optimizing the amplitude frequency response of a motor car shown in Fig. 8.9.

In multi-dof-vibrating systems usually more than one amplitude frequency response is of interest. If one has simultaneously to consider in a parameter optimization the amplitudes $a_1(\Omega),\ldots,a_k(\Omega)$, then the cost functional (8.150) should be replaced by the requirement

$$\max_{\Omega\in\bar{\Omega}}\{s_1a_1(\Omega),\ldots,s_ka_k(\Omega)\}\rightarrow \text{Minimum} \tag{8.151}$$

where the nonnegative weighting factors $s_i\geqslant 0,\ i=1(1)k$, still allow different weighting of the coordinates.

A second possibility of measuring the magnitude of amplitudes over a frequency range $\bar{\Omega}=[\Omega_1,\ \Omega_2]$ is rendered by the *linear amplitude frequency response integral*

$$\int_{\Omega_1}^{\Omega_2}a(\Omega)\mathrm{d}\Omega\rightarrow \text{Minimum.} \tag{8.152}$$

And if more amplitudes $a_i(\Omega),\ i=1(1)k$, have to be considered, then (8.152) can be generalized either to

$$\max_{i=1(1)k}\left\{s_i\int_{\Omega_1}^{\Omega_2}a_i(\Omega)\mathrm{d}\Omega\right\}\rightarrow \text{Minimum} \tag{8.153}$$

or to

$$\sum_{i=1}^{k}s_i\int_{\Omega_1}^{\Omega_2}a_i(\Omega)\mathrm{d}\Omega\rightarrow \text{Minimum} \tag{8.154}$$

where the quantities $s_i\geqslant 0,\ i=1(1)k$, are weighting factors to be chosen according to the engineering requirements.

As a third cost functional one can also use the *squared amplitude frequency response integral*

$$\int_{\Omega_1}^{\Omega_2} a^2(\Omega)\mathrm{d}\Omega \rightarrow \text{Minimum} \tag{8.155}$$

or

$$\sum_{i=1}^{k} s_i \int_{\Omega_1}^{\Omega_2} a_i^2(\Omega)\mathrm{d}\Omega \rightarrow \text{Minimum.} \tag{8.156}$$

Here the larger amplitudes are evaluated more strongly so that the high maxima of $a_i(\Omega)$ are more reduced by an optimization according to the squared frequency integral criterion. Since $s_i a_i^2(\Omega) = \bar{g}_i(\Omega) s_i g_i(\Omega)$, by (7.53), the criterion (8.156) can also be stated as

$$\int_{\Omega_1}^{\Omega_2} \boldsymbol{g}^*(\Omega)\boldsymbol{S}\boldsymbol{g}(\Omega)\mathrm{d}\Omega \rightarrow \text{Minimum} \tag{8.157}$$

where $\boldsymbol{g}(\Omega)$ is the complex frequency response vector (7.42), $\boldsymbol{g}^*$ is the conjugate transpose vector to $\boldsymbol{g}$, and

$$\boldsymbol{S} = \mathrm{diag}\{s_1, s_2, \ldots, s_n\} \tag{8.158}$$

is a diagonal positive semidefinite weighting matrix. A generalization of (8.158) can be attained by a symmetric positive semidefinite matrix $\boldsymbol{S}$ which also evaluates the linear combinations of the $a_i(\Omega)$.

The calculation of the cost functionals (8.150 to 157) is in general quite laborious. One proceeds by evaluating the functionals directly. However, to get some idea about the size of the squared amplitude frequency integral (8.157), in the special case when the frequency range is unbounded, $\Omega_1 = -\infty$, $\Omega_2 = +\infty$, one can evaluate the integral (8.157) by complex path integrals or with the help of the Fourier transform:

$$\begin{aligned}\int_{-\infty}^{+\infty} \boldsymbol{g}^*(\Omega)\boldsymbol{S}\boldsymbol{g}(\Omega)\mathrm{d}\Omega &= \boldsymbol{b}^* \int_{-\infty}^{+\infty} (-\mathrm{i}\Omega\boldsymbol{E} - \boldsymbol{A}^{\mathrm{T}})^{-1} \boldsymbol{S} (\mathrm{i}\Omega\boldsymbol{E} - \boldsymbol{A})^{-1} \mathrm{d}\Omega\, \boldsymbol{b} \\ &= 2\pi \boldsymbol{b}^* \int_0^{\infty} e^{\boldsymbol{A}^{\mathrm{T}}\tau} \boldsymbol{S} e^{\boldsymbol{A}\tau} \mathrm{d}\tau\, \boldsymbol{b}.\end{aligned} \tag{8.159}$$

Here it is assumed that the excitation vector $\boldsymbol{b}$ does not depend on Ω, and $\boldsymbol{A}$ is the asymptotically stable system matrix of the vibrating system (8.1). The integral (8.159) can be solved by means of the Ljapunov matrix equation (13.43), according to (13.51),

$$\int_{-\infty}^{+\infty} \boldsymbol{g}^*(\Omega)\boldsymbol{S}\boldsymbol{g}(\Omega)\mathrm{d}\Omega = 2\pi \boldsymbol{b}^* \boldsymbol{R} \boldsymbol{b} \tag{8.160}$$

where $\boldsymbol{R}$ satisfies the equation

$$\boldsymbol{A}^{\mathrm{T}}\boldsymbol{R} + \boldsymbol{R}\boldsymbol{A} = -\boldsymbol{S}. \tag{8.161}$$

Thus the optimization according to the squared amplitude frequency response integral for an infinite frequency range is equivalent to an optimization according to the integral of squared error (6.94) if one substitutes $\boldsymbol{x}_0 = \sqrt{2\pi}\,\boldsymbol{b}$ in (6.104). Hence a connection is now established between the determination of optimal natural vibrations in Section 6.2 and optimal frequency responses in this section. Under the given conditions, integrals of squared error (6.94) and squared amplitude frequency integrals (8.157) lead to the same optimal parameters.

Another functional which deals simultaneously with natural behavior and disturbance behavior is the *damping factor* (6.92),

$$D_0 = -\max_{i=1(1)n}\left\{\frac{\operatorname{Re}\lambda_i}{|\lambda_i|}\right\}, \tag{8.162}$$

which is determined by the eigenvalues λ_i, $i = 1(1)n$, of the system matrix $\boldsymbol{A}$ (of course one assumes asymptotical stability, i.e., $\operatorname{Re}\lambda_i < 0$, $i = 1(1)n$). As can be seen in Fig. 8.5, the magnification functions (8.25) of the second order elementary frequency responses are small when the damping factor is large. Thus, under the transformation (8.7) one also obtains small values for the frequency response $g_i(\Omega)$ of the individual state coordinates.

To obtain by means of the above cost functionals the optimal parameters for an acceptable disturbance behavior of a vibrating system, one may use the methods of Section 6.2.3.

A further possibility of attaining an acceptable frequency response consists of applying an additional *vibration absorber*. The *absorber tuning* can be explained with the help of Fig. 8.12. In the neighborhood of the absorption frequency $\Omega_A^2 = \kappa_{11} = k_1/m_1$ the amplitude frequency response (8.121) of the car body is satisfactory. The motor mass m_1 acts as a vibration absorber for the car body mass m_2. Although the excitation acts directly on the mass m_2 (Fig. 8.11), this mass does not vibrate if the excitation frequency Ω coincides with Ω_A. The use of vibration absorbers is always interesting when there is only a small frequency range $\overline{\Omega}$ for the optimal frequency response. In engineering practice for instance, this is the case with machine foundations. Vibration absorption will be further illustrated in the following example.

Example 8.7. Vibration absorber for the vertical vibrations of a motor car

Consider the amplitude frequency response of the car body acceleration during the vertical vibrations of a motor car, as discussed in

Example 8.3, Fig. 8.9. The curve fully drawn in Fig. 8.9 represents an acceptable amplitude frequency response during the variation of the damping parameters d_1. To improve this curve even further, one can apply an additional vibration absorber. A model of car suspension with absorber mass is shown in Fig. 8.13. The methods explained in Chapter 2 yield for this system with three degrees of freedom the following matrix differential equation:

$$\begin{bmatrix} m_1 & 0 & 0 \\ 0 & m_2 & 0 \\ 0 & 0 & m_3 \end{bmatrix} + \begin{bmatrix} d_1 & -d_1 & 0 \\ -d_1 & d_1+d_3 & -d_3 \\ 0 & -d_3 & d_3 \end{bmatrix} \begin{bmatrix} \dot{y}_1 \\ \dot{y}_2 \\ \dot{y}_3 \end{bmatrix} +$$

$$+ \begin{bmatrix} k_1 & -k_1 & 0 \\ -k_1 & k_1+k_2+k_3 & -k_3 \\ 0 & -k_3 & k_3 \end{bmatrix} \begin{bmatrix} y_1 \\ y_2 \\ y_3 \end{bmatrix} = \begin{bmatrix} 0 \\ k_2 y_e(t) \\ 0 \end{bmatrix}. \tag{8.163}$$

The curves in Fig. 8.14 show the various amplitude frequency responses obtained for the parameter data as in Example 4.3, the additional absorber data $m_3 = 20$ kg, $k_3 = 600$ N/cm, and the different damping values $d_3 = 0$, 300, 600 and 1000 Ns/m.

A total vibration absorption takes place for vanishing absorption damping $d_3 = 0$ and $\Omega_A = \sqrt{\kappa_3} = \sqrt{\kappa_3/m_3}$. But in other frequency ranges of interest the amplitude becomes larger. With growing damping d_3 one actually loses the effect of total absorption, but this is compensated by the diminishing of amplitudes in other frequency ranges. Thus the

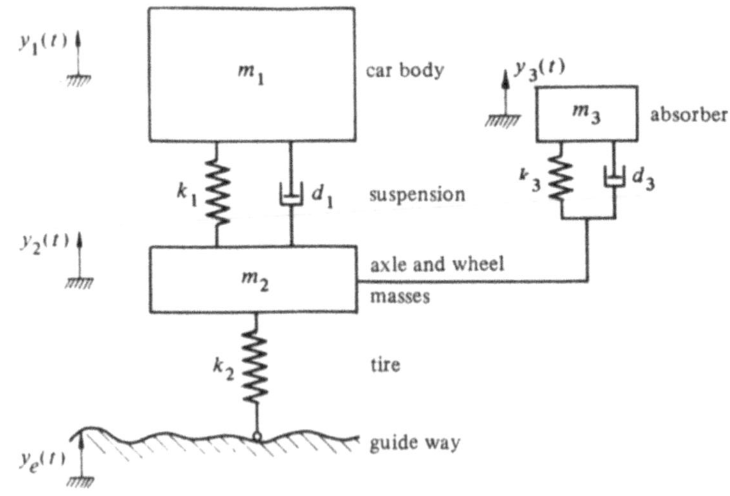

Fig. 8.13. Motor car suspension with vibration absorber

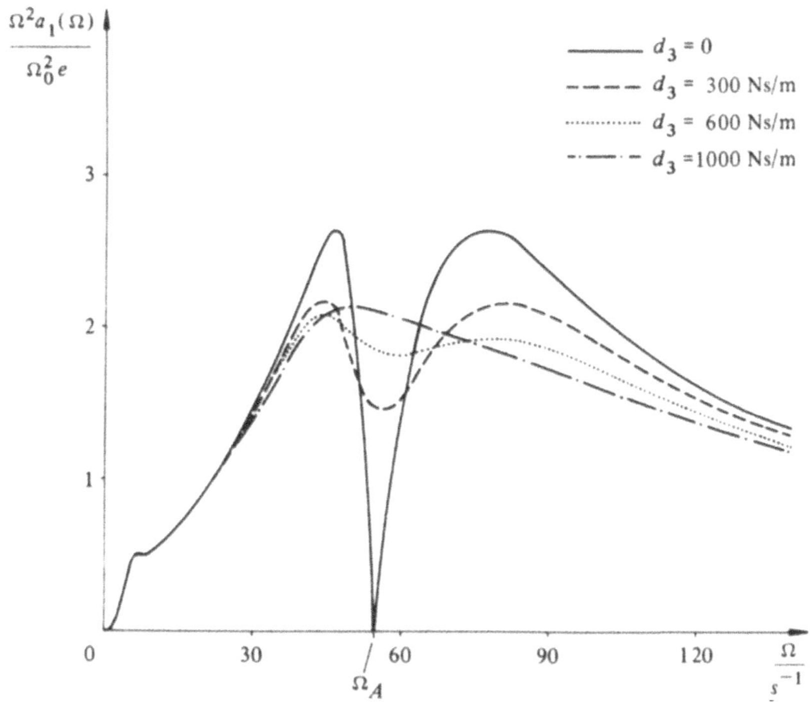

Fig. 8.14. Amplitude frequency response of the body acceleration of a motor car with vibration absorber

damped vibration absorber with $d_3 \approx 1000$ Ns/m represents an acceptable compromise for the considered frequency range; it reduces the amplitude maxima by about 20%. ◆

8.8. Parameter identification

The preceding sections were devoted to the analysis of the behavior of a vibrating system with fixed or free system parameters under harmonic disturbance. The typical mathematical tool used for this purpose was the frequency response calculated from the system parameters. Often the parameters are not exactly known and the question arises if one can determine, i.e., identify the values of the parameters by measuring the frequency response. It is always possible to determine the frequency response of a vibrating system without knowing the system parameters by repeated measurement of resulting harmonic system responses for known harmonic excitation signals. The question of *parameter identification* is a problem which is dealt with especially in control system theory. Hence only a parameter identification procedure will be indicated here in principal. A broad treatment of such problems, with special consideration of parameter identification for mechanical vibrating systems, can be found in the publications of Zeller and Schwarz.

The following procedure can be given for the *identification* of the system matrix $\boldsymbol{A}$ of an asymptotically stable dynamic system (8.1) from a sequence of measured polar plots $g_i(\Omega)$, $i = 1(1)n$: The measured vectors $\boldsymbol{g}(\Omega_j)$ are known for n distinct excitation frequencies Ω_j, $j = 1(1)n$. One then has, by (7.42, 7.43),

$$(\mathrm{i}\Omega_j \boldsymbol{E} - \boldsymbol{A})\boldsymbol{g}(\Omega_j) = \boldsymbol{b}(\Omega_j), \quad j = 1(1)n, \tag{8.164}$$

or, in a different notation,

$$\left[(\mathrm{i}\Omega_1 \boldsymbol{E} - \boldsymbol{A})\boldsymbol{g}(\Omega_1) \vdots (\mathrm{i}\Omega_2 \boldsymbol{E} - \boldsymbol{A})\boldsymbol{g}(\Omega_2) \vdots \dots \vdots (\mathrm{i}\Omega_n \boldsymbol{E} - \boldsymbol{A})\boldsymbol{g}(\Omega_n)\right]$$
$$= \left[\boldsymbol{b}(\Omega_1) \vdots \boldsymbol{b}(\Omega_2) \vdots \dots \vdots \boldsymbol{b}(\Omega_n)\right]. \tag{8.165}$$

Solving (8.165) with respect to $\boldsymbol{A}$ yields

$$\boldsymbol{A} = \left\{\mathrm{i}\left[\Omega_1 \boldsymbol{g}(\Omega_1) \vdots \dots \vdots \Omega_n \boldsymbol{g}(\Omega_n)\right]\right.$$
$$\left. - \left[\boldsymbol{b}(\Omega_1) \vdots \dots \vdots \boldsymbol{b}(\Omega_n)\right]\right\}\left[\boldsymbol{g}(\Omega_1) \vdots \dots \vdots \boldsymbol{g}(\Omega_n)\right]^{-1} \tag{8.166}$$

If the matrix of the n measured vectors $[g(\Omega_1) \vdots \dots \vdots g(\Omega_n)]$ is regular and the input vectors $\boldsymbol{b}(\Omega_i)$, $i = 1(1)n$, and the measured frequencies Ω_j, $j = 1(1)n$, are known, then the system matrix $\boldsymbol{A}$ can be determined according to (8.166). This accomplishes the parameter identification.

The assumption of regularity for the matrix $[g(\Omega_1) \vdots \dots \vdots g(\Omega_n)]$ can be verified for a constant excitation vector $\boldsymbol{b}(\Omega_j) = \boldsymbol{b}$, $j = 1(1)n$. Due to the theorem of Cayley and Hamilton (4.130), there exist certain functions $\gamma_k(\Omega)$ such that $g(\Omega)$ can be written as

$$\boldsymbol{g}(\Omega) = (\mathrm{i}\Omega \boldsymbol{E} - \boldsymbol{A})^{-1}\boldsymbol{b} = \sum_{k=0}^{n-1} \gamma_k(\Omega)\boldsymbol{A}^k \boldsymbol{b}. \tag{8.167}$$

Thus the n vectors $\boldsymbol{g}(\Omega_j)$, $j = 1(1)n$, are for different excitation frequencies Ω_j precisely linearly independent when the controllability matrix

$$\left[\boldsymbol{b} \vdots \boldsymbol{A}\boldsymbol{b} \vdots \boldsymbol{A}^2\boldsymbol{b} \vdots \dots \vdots \boldsymbol{A}^{n-1}\boldsymbol{b}\right] \tag{8.168}$$

is regular (see Chapter 12). However, if the excitation vector is not chosen in this way, then only the part of the system matrix $\boldsymbol{A}$ which is controllable by $\boldsymbol{b}$ can be identified.

The parameters identification according to (8.166) represents only one of the basic possibilities of solving this kind of problem. In particular, it is assumed that the polar plot measurements are exact and not burdened with measurement errors. Applicable identification procedures always have to take into account the errors of measurements.

Problem 8.1. Centrifuge

According to Example 8.2, the unbalanced undamped centrifuge in Fig. 3.4 has, by (8.80), the equations of motion

$$\ddot{y}(t) + g\Omega \boldsymbol{S}\dot{y}(t) + \kappa y(t) = e\Omega^2 \begin{bmatrix} \cos \Omega t \\ \sin \Omega t \end{bmatrix}. \tag{8.169}$$

The excitation function in complex representation is

$$\boldsymbol{h}(t) = \boldsymbol{h}e^{\mathrm{i}\Omega t} + \bar{\boldsymbol{h}}e^{-\mathrm{i}\Omega t} \quad \text{with} \quad \boldsymbol{h} = \frac{e\Omega^2}{2}\begin{bmatrix} 1 \\ -\mathrm{i} \end{bmatrix}. \tag{8.170}$$

a) What is the inverse frequency response matrix $[-\Omega^2 \boldsymbol{M} + \mathrm{i}\Omega \boldsymbol{G} + \boldsymbol{K}]$?

- ☐ $\begin{bmatrix} -\Omega^2 + \kappa & \mathrm{i}\Omega g \\ -\mathrm{i}\Omega g & -\Omega^2 + \kappa \end{bmatrix}$,
- ☐ $\begin{bmatrix} -\Omega^2 + \kappa & \mathrm{i}\Omega^2 g \\ -\mathrm{i}\Omega^2 g & -\Omega^2 + \kappa \end{bmatrix}$,
- ☐ $\begin{bmatrix} -\Omega^2(1-g) + \kappa & \mathrm{i}\Omega^2 g \\ -\mathrm{i}\Omega^2 g & -\Omega^2(1-g) + \kappa \end{bmatrix}$,
- ☐ $\begin{bmatrix} -\Omega^2(1-g) + \kappa & \mathrm{i}\Omega g \\ -\mathrm{i}\Omega g & -\Omega^2(1-g) + \kappa \end{bmatrix}$.

b) What is the determinant $\det[-\Omega^2 \boldsymbol{M} + \mathrm{i}\Omega \boldsymbol{G} + \boldsymbol{K}]$?

- ☐ $[\kappa - (1+g)\Omega^2][\kappa - (1-g)\Omega^2]$,
- ☐ $(1+g)(1-g)\Omega^4 - 2\kappa\Omega^2 + \kappa^2$,
- ☐ $\Omega^4 - (2\kappa + g^2\Omega^2)\Omega^2 + \kappa^2$,
- ☐ $(1-g^2)\Omega^4 - 2\kappa\Omega^2 + \kappa^2$.

c) What are the necessary conditions for strict resonance?

- ☐ $\Omega_{R_1}^2 = \kappa, \qquad \Omega_{R_2}^2 = \kappa;$
- ☐ $\Omega_{R_1}^2 = \kappa(1+g), \qquad \Omega_{R_2}^2 = \kappa(1-g);$
- ☐ $\Omega_{R_1}^2 = \dfrac{\kappa}{1+g}, \qquad \Omega_{R_2}^2 = \dfrac{\kappa}{1-g};$
- ☐ $\Omega_{R_1}^2 = \dfrac{\kappa}{g}, \qquad \Omega_{R_2}^2 = \dfrac{\kappa}{g}.$

d) What is the inverse frequency response matrix $[-\Omega^2 \boldsymbol{M} + \mathrm{i}\Omega \boldsymbol{G} + \boldsymbol{K}]$ at the first resonance frequency $\Omega = \Omega_{R_1}$?

□ $\kappa g \begin{bmatrix} 0 & \mathrm{i} \\ -\mathrm{i} & 0 \end{bmatrix}$, □ $\kappa g \begin{bmatrix} 1 & \mathrm{i} \\ -\mathrm{i} & 1 \end{bmatrix}$,

□ $\frac{\kappa g}{1+g} \begin{bmatrix} 0 & \mathrm{i} \\ -\mathrm{i} & 0 \end{bmatrix}$, □ $\frac{\kappa g}{1+g} \begin{bmatrix} 1 & \mathrm{i} \\ -\mathrm{i} & 1 \end{bmatrix}$.

e) What is the adjoint matrix $\mathrm{adj}[-\Omega^2 \boldsymbol{M} + \mathrm{i}\Omega \boldsymbol{G} + \boldsymbol{K}]$ at the first resonance frequency $\Omega = \Omega_{R_1}$?

□ $\kappa g \begin{bmatrix} 0 & -\mathrm{i} \\ \mathrm{i} & 0 \end{bmatrix}$, □ $\kappa g \begin{bmatrix} 1 & -\mathrm{i} \\ \mathrm{i} & 1 \end{bmatrix}$,

□ $\frac{\kappa g}{1+g} \begin{bmatrix} 0 & -\mathrm{i} \\ \mathrm{i} & 0 \end{bmatrix}$, □ $\frac{\kappa g}{1+g} \begin{bmatrix} 1 & -\mathrm{i} \\ \mathrm{i} & 1 \end{bmatrix}$.

f) Evaluate the pseudoresonance condition:

$$\mathrm{adj}\left[-\Omega_{R_1}^2 \boldsymbol{M} + \mathrm{i}\Omega_{R_1} \boldsymbol{G} + \boldsymbol{K}\right] \boldsymbol{h} = \begin{bmatrix} \quad & \quad \\ \quad & \quad \end{bmatrix} \begin{bmatrix} \quad \\ \quad \end{bmatrix} = \begin{bmatrix} \quad \\ \quad \end{bmatrix}.$$

At the first resonance frequency there is
□ strict resonance,
□ pseudoresonance.

g) What is the inverse frequency response matrix $[-\Omega^2 \boldsymbol{M} + \mathrm{i}\Omega \boldsymbol{G} + \boldsymbol{K}]$ at the second resonance frequency $\Omega = \Omega_{R_2}$?

□ $\frac{\kappa g}{1-g} \begin{bmatrix} 0 & \mathrm{i} \\ -\mathrm{i} & 0 \end{bmatrix}$, □ $\frac{\kappa g}{1-g} \begin{bmatrix} -1 & \mathrm{i} \\ -\mathrm{i} & -1 \end{bmatrix}$,

□ $\frac{\kappa g}{1-g} \begin{bmatrix} 0 & -\mathrm{i} \\ \mathrm{i} & 0 \end{bmatrix}$, □ $\frac{\kappa g}{1-g} \begin{bmatrix} 1 & -\mathrm{i} \\ \mathrm{i} & 1 \end{bmatrix}$.

h) What is the adjoint matrix $\mathrm{adj}[-\Omega^2 \boldsymbol{M} + \mathrm{i}\Omega \boldsymbol{G} + \boldsymbol{K}]$ at the second resonance frequency $\Omega = \Omega_{R_2}$?

□ $\frac{\kappa g}{1-g} \begin{bmatrix} 0 & -\mathrm{i} \\ \mathrm{i} & 0 \end{bmatrix}$, □ $\frac{\kappa g}{1-g} \begin{bmatrix} 1 & \mathrm{i} \\ -\mathrm{i} & 1 \end{bmatrix}$,

□ $\frac{\kappa g}{1-g} \begin{bmatrix} 0 & \mathrm{i} \\ -\mathrm{i} & 0 \end{bmatrix}$, □ $\frac{\kappa g}{1-g} \begin{bmatrix} -1 & -\mathrm{i} \\ \mathrm{i} & -1 \end{bmatrix}$.

i) Evaluate the pseudoresonance condition:

$$\mathrm{adj}\left[-\Omega_{R_2}^2 \boldsymbol{M} + \mathrm{i}\Omega_{R_2} \boldsymbol{G} + \boldsymbol{K}\right] \boldsymbol{h} = \begin{bmatrix} \quad & \quad \\ \quad & \quad \end{bmatrix} \begin{bmatrix} \quad \\ \quad \end{bmatrix} = \begin{bmatrix} \quad \\ \quad \end{bmatrix}.$$

At the second resonance frequency there is
□ strict resonance,
□ pseudoresonance.

j) Which excitation vector generates pseudoresonance at the second resonance frequency?

□ $h = \frac{e\Omega^2}{2}\begin{bmatrix} 1 \\ -\mathrm{i} \end{bmatrix}$, □ $h = \frac{e\Omega^2}{2}\begin{bmatrix} 1 \\ \mathrm{i} \end{bmatrix}$,

□ $h = \frac{e\Omega^2}{2}\begin{bmatrix} \mathrm{i} \\ -1 \end{bmatrix}$, □ $h = \frac{e\Omega^2}{2}\begin{bmatrix} \mathrm{i} \\ 1 \end{bmatrix}$. ◆

Problem 8.2. Double pendulum

Assume that the double pendulum in Fig. 3.3 is excited by harmonic forces. The equations of motion are

$$\ddot{y}(t) + \frac{g}{L}\begin{bmatrix} 3 & -1 \\ -1 & 1 \end{bmatrix} y(t) = \boldsymbol{h}\,\mathrm{e}^{\mathrm{i}\Omega t} + \bar{\boldsymbol{h}}\,\mathrm{e}^{-\mathrm{i}\Omega t}. \tag{8.171}$$

a) What is the inverse frequency response matrix $[-\Omega^2 \boldsymbol{M} + \boldsymbol{K}]$?

□ $\begin{bmatrix} \Omega^2 + 3\frac{g}{L} & -\frac{g}{L} \\ -\frac{g}{L} & \Omega^2 + \frac{g}{L} \end{bmatrix}$, □ $\begin{bmatrix} \Omega^2 - 3\frac{g}{L} & \frac{g}{L} \\ \frac{g}{L} & \Omega^2 - \frac{g}{L} \end{bmatrix}$,

□ $\begin{bmatrix} -\Omega^2 - 3\frac{g}{L} & \frac{g}{L} \\ \frac{g}{L} & -\Omega^2 - \frac{g}{L} \end{bmatrix}$, □ $\begin{bmatrix} -\Omega^2 + 3\frac{g}{L} & -\frac{g}{L} \\ -\frac{g}{L} & -\Omega^2 + \frac{g}{L} \end{bmatrix}$.

b) Which excitation vector $\boldsymbol{h}$ causes at the frequency $\Omega_A = \sqrt{g/L}$ amplitude absorption of the first particle?

□ $\boldsymbol{h} = \begin{bmatrix} 1 \\ 1 \end{bmatrix}$, □ $\boldsymbol{h} = \begin{bmatrix} 2 \\ -1 \end{bmatrix}$, □ $\boldsymbol{h} = \begin{bmatrix} 1 \\ 0 \end{bmatrix}$, □ $\boldsymbol{h} = \begin{bmatrix} 1 \\ -1 \end{bmatrix}$.

c) Which excitation vector $\boldsymbol{h}$ causes at the frequency $\Omega_A = \sqrt{g/L}$ amplitude absorption of the second particle?

□ $\boldsymbol{h} = \begin{bmatrix} 1 \\ 1 \end{bmatrix}$, □ $\boldsymbol{h} = \begin{bmatrix} 2 \\ -1 \end{bmatrix}$, □ $\boldsymbol{h} = \begin{bmatrix} 1 \\ 0 \end{bmatrix}$, □ $\boldsymbol{h} = \begin{bmatrix} 1 \\ -1 \end{bmatrix}$.

d) Which excitation vector $\boldsymbol{h}$ causes at the frequency $\Omega_A = \sqrt{2g/L}$ amplitude absorption of the first particle?

$\square\ \boldsymbol{h} = \begin{bmatrix} 1 \\ 1 \end{bmatrix}$, $\square\ \boldsymbol{h} = \begin{bmatrix} 2 \\ -1 \end{bmatrix}$, $\square\ \boldsymbol{h} = \begin{bmatrix} 1 \\ 0 \end{bmatrix}$, $\square\ \boldsymbol{h} = \begin{bmatrix} 1 \\ -1 \end{bmatrix}$.

e) Which excitation vector $\boldsymbol{h}$ causes at the frequency $\Omega_A = \sqrt{2g/L}$ amplitude absorption of the second particle?

$\square\ \boldsymbol{h} = \begin{bmatrix} 1 \\ 1 \end{bmatrix}$, $\square\ \boldsymbol{h} = \begin{bmatrix} 2 \\ -1 \end{bmatrix}$, $\square\ \boldsymbol{h} = \begin{bmatrix} 1 \\ 0 \end{bmatrix}$, $\square\ \boldsymbol{h} = \begin{bmatrix} 1 \\ -1 \end{bmatrix}$.

f) Which excitation vector $\boldsymbol{h}$ causes at the frequency $\Omega_A = \sqrt{2g/L}$ absorption at both particles?

$\square\ \boldsymbol{h} = \begin{bmatrix} 1 \\ 1 \end{bmatrix}$, $\square\ \boldsymbol{h} = \begin{bmatrix} 2 \\ -1 \end{bmatrix}$, $\square\ \boldsymbol{h} = \begin{bmatrix} 1 \\ 0 \end{bmatrix}$, $\square\ \boldsymbol{h} = \begin{bmatrix} 1 \\ -1 \end{bmatrix}$.

Problem 8.3. Single-dof-vibrating system

For a harmonically excited single dof-vibrating system

$$m\ddot{y}(t) + d\dot{y}(t) + ky(t) = h^{(1)} \cos \Omega t \tag{8.172}$$

the phase lags and amplitudes are known from frequency response measurements for various excitation frequencies:

$$\begin{aligned}
&\Omega_1 = 0 &&: \psi = 0 &&, \quad a = 4h^{(1)};\\
&\Omega_2 = 5(\sqrt{2}-1) &&: \psi = 45° &&, \quad a = (2+\sqrt{2})h^{(1)}\\
&\Omega_3 = 5 &&: \psi = 90° &&, \quad a = 2h^{(1)};\\
&\Omega = \infty &&: \psi = 180° &&, \quad a = 0.
\end{aligned}$$

a) What are the values of the mass m, damping d, and spring constant k of the system (8.172)?

□ $m = 1, \quad d = \frac{1}{10}, \quad k = \frac{1}{4}$;
□ $m = \frac{1}{100}, \quad d = \frac{1}{10}, \quad k = \frac{1}{4}$;
□ $m = \frac{1}{100}, \quad d = 1, \quad k = \frac{1}{4}$;
□ $m = 1, \quad d = 2, \quad k = 1$;
□ $m = 1, \quad d = \frac{1}{10}, \quad k = 1$.

b) What are the eigenvalues of the system (8.172)?

☐ $\lambda_{1,2} = -\frac{1}{20} \pm \mathrm{i}$;

☐ $\lambda_{1,2} = -1$;

☐ $\lambda_1 = 0, \qquad \lambda_2 = -100$;

☐ $\lambda_{1,2} = -5$;

☐ $\lambda_{1,2} = -\frac{1}{20} \pm \frac{1}{2}i$.

c) What is the equation of the amplitude frequency response $a(\Omega)$?

☐ $a = h^{(1)}\left[\left(\frac{1}{4} - \frac{\Omega^2}{100}\right)^2 + \frac{\Omega^2}{100}\right]^{-1/2}$;

☐ $a = h^{(1)}\left[(1 - \Omega^2)^2 + \frac{\Omega^2}{100}\right]^{-1/2}$;

☐ $a = h^{(1)}\left[\left(\frac{1}{4} - \Omega^2\right)^2 + \frac{\Omega^2}{100}\right]^{-1/2}$;

☐ $a = h^{(1)}\left[(1 - \Omega^2)^2 + 4\Omega^2\right]^{-1/2}$;

☐ $a = h^{(1)}\left[\left(\frac{1}{4} - \frac{\Omega^2}{100}\right)^2 + \Omega^2\right]^{-1/2}$

d) Which relation is satisfied by the phase lag $\psi(\Omega)$?

☐ $\tan \psi = \dfrac{\Omega}{10 - 10\Omega^2}$;

☐ $\tan \psi = \dfrac{2\Omega}{5 - 20\Omega^2}$;

☐ $\tan \psi = \dfrac{2\Omega}{1 - \Omega^2}$;

☐ $\tan \psi = \dfrac{100\Omega}{25 - \Omega^2}$;

☐ $\tan \psi = \dfrac{10\Omega}{25 - \Omega^2}$. ◆

9. Random vibrations

Mechanical vibrating systems can also be excited by random forces. For instance, a motor car travelling along a rough road is subject to random excitations. A rough road is characterized by a rough surface without conspicuous potholes or road surface undulations. Random vibrations can also be observed on centrifuges with erratic charge. Random excitations also play an important role in the vibrations of buildings since wind forces display a very irregular behavior. Random forces cannot be represented by functions of time, their description requires the notion of a *stochastic process.* Consequently the response, i.e., the motion of the vibrating system, becomes a stochastic process. Thus this chapter will start with a survey on properties of stochastic processes. Next a discussion of random vibrations excited by white noise will follow. The analysis of these vibrations is carried out with the aid of the methods of spectral density analysis and covariance analysis. In conclusion random vibrations excited by colored noise processes will be considered.

9.1. Stochastic vector processes

A scalar *stochastic process* $v(t)$ can be viewed as a family of functions of time. Each individual function of time is then called a realization $v^{(1)}(t)$, $v^{(2)}(t), \ldots$ of the process, Fig. 9.1.

For any given instant, e.g. $t = t_1$, the stochastic process is characterized by a random variable $v(t_1)$. This random variable is described by its probability distribution

$$\mathscr{P}\{v(t_1) \leqslant v_1\} = \mathscr{P}(v_1). \tag{9.1}$$

The probability distribution is a measure of the probability that the value attained by $v(t_1)$ does not exceed v_1. The boundary values are given by

$$\mathscr{P}(-\infty) = 0, \quad \mathscr{P}(+\infty) = \int_{-\infty}^{+\infty} \mathrm{d}\,\mathscr{P} = 1. \tag{9.2}$$

If $v_i(t)$, $i=1(1)n$, are n scalar stochastic processes, then

$$\boldsymbol{v}(t)=\left[v_1(t)\ v_2(t)\dots v_n(t)\right]^{\mathrm{T}} \tag{9.3}$$

is an $n\times 1$ stochastic *vector process*. This vector process can be described by its joint probability distribution

$$\mathscr{P}\left\{\boldsymbol{v}(t):\ \boldsymbol{v}(t_j)\leqslant \boldsymbol{v}_j,\qquad j=1(1)m\right\}, \tag{9.4}$$

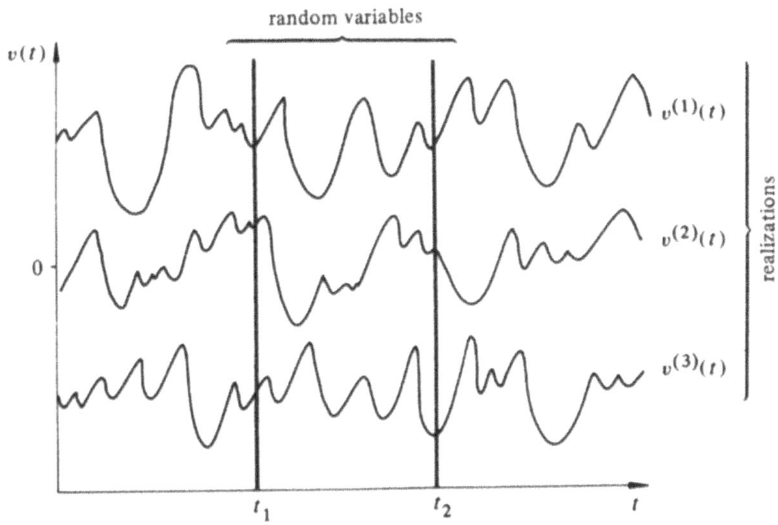

Fig. 9.1. Scalar stochastic process

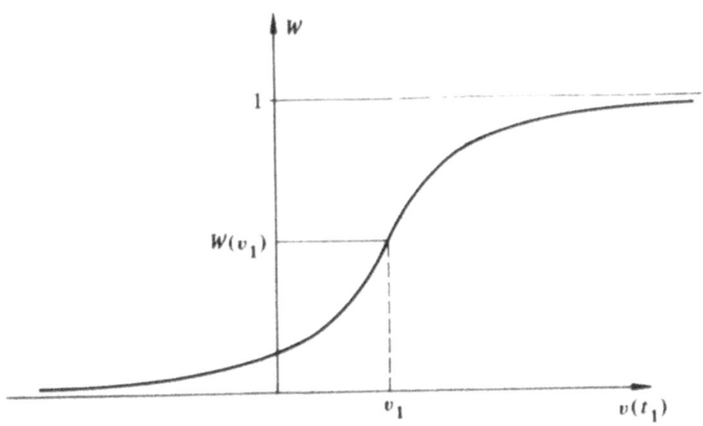

Fig. 9.2. Example of a probability distribution

for all v_j, all $t_j \geqslant t_0$, and every number m. Apart from the probability distribution (9.4), the first and second order moments, i.e., the mean vector and the correlation matrix are of special importance for the description of a stochastic process $v(t)$. The $n \times 1$ *mean vector*

$$\boldsymbol{m}_v(t) = E\{\boldsymbol{v}(t)\} = \int_{-\infty}^{\infty} \boldsymbol{v}(t) \mathrm{d}\mathscr{P} \tag{9.5}$$

is the linear expectation of the random variable $v(t)$. It is equal to the first moment of the distribution density and indicates the vector around which the values of the random variables appear. The $n \times n$ *correlation matrix*

$$\boldsymbol{C}_v(t, \tau) = E\{\boldsymbol{v}(t)\boldsymbol{v}^{\mathrm{T}}(\tau)\} = \int_{-\infty}^{\infty} \boldsymbol{v}(t)\boldsymbol{v}^{\mathrm{T}}(\tau) \mathrm{d}\mathscr{P} \tag{9.6}$$

and the central $n \times n$ correlation matrix

$$\boldsymbol{N}_v(t, \tau) = E\{[\boldsymbol{v}(t) - \boldsymbol{m}_v(t)][\boldsymbol{v}(\tau) - \boldsymbol{m}_v(\tau)]^{\mathrm{T}}\} \tag{9.7}$$

are defined by the quadratic expectations and they describe the second moments of the distribution density at instants t and τ. Thus the correlation matrices represent a measure of the coupling of the process at various time moments. If the mean value of the considered process vanishes, $\boldsymbol{m}_v(t) \equiv \boldsymbol{0}$, then the correlation matrix (9.6) is equal to the central correlation matrix (9.7). Moreover, it is

$$\left.\begin{aligned} \boldsymbol{C}_v(t, \tau) &= \boldsymbol{C}_v^{\mathrm{T}}(\tau, t), \\ \boldsymbol{N}_v(t, \tau) &= \boldsymbol{N}_v^{T}(\tau, t), \\ \boldsymbol{C}_v(t, \tau) &= \boldsymbol{N}_v(t, \tau) + \boldsymbol{m}_v(t)\boldsymbol{m}_v^{\mathrm{T}}(\tau) \end{aligned}\right\} \tag{9.8}$$

for all t, τ. The central correlation matrix yields for $t = \tau$ the symmetric $n \times n$ *covariance matrix*

$$\boldsymbol{P}_v(t) = \boldsymbol{N}_v(t, t) \geqslant \boldsymbol{0}, \tag{9.9}$$

which is nonnegative definite for all t. The covariance matrix describes the mutual coupling of the n scalar stochastic processes $v_i(t)$, $i = 1(1)n$, of the considered vector process at the instant t. In particular, $P_{vii}(t)$ is the variance of the i-th scalar stochastic process $y_i(t)$, $i = 1(1)n$, often also termed dispersion or quadratic mean. Moreover

$$\sigma_{vi}(t) = \sqrt{P_{vii}(t)}, \qquad i = 1(1)n, \tag{9.10}$$

is the corresponding *standard deviation* or variance of the i-th scalar process $v_i(t)$.

Some special stochastic processes will now be considered; these will have to be assumed to permit a closed mathematical treatment of random vibrations.

A stochastic process is called *stationary* if the probability distribution is time-invariant:

$$\mathscr{P}\{\boldsymbol{v}(t):\ \boldsymbol{v}(t_j)\leqslant\boldsymbol{v}_j\}=\mathscr{P}\{\boldsymbol{v}(t+T):\ \boldsymbol{v}(t_j+T)\leqslant\boldsymbol{v}_j\},\quad \text{for all } T. \tag{9.11}$$

This implies that the characteristic properties of a stationary process are likewise time-invariant:

$$\left.\begin{aligned}&\boldsymbol{m}_v(t)=\text{const},\\&\boldsymbol{N}_v(t,\tau)=\boldsymbol{N}_v(t-\tau),\\&\boldsymbol{P}_v(t)=\boldsymbol{P}_v(0)=\text{const.}\end{aligned}\right\} \tag{9.12}$$

For a stationary stochastic process one can introduce the *power spectral density matrix*

$$\boldsymbol{S}_v(\Omega)=\int_{-\infty}^{\infty}\mathrm{e}^{-\mathrm{i}\Omega s}\boldsymbol{N}_v(s)\mathrm{d}s, \tag{9.13}$$

i.e., the matrix which is the Fourier transform of $\boldsymbol{N}_v(t-\tau)$ with $s=t-\tau$. The complex power spectral density matrix satisfies

$$\left.\begin{aligned}&\boldsymbol{S}_v^T(\Omega)=\boldsymbol{S}_v(-\Omega),\\&\boldsymbol{S}_v^*(\Omega)=\boldsymbol{S}_v(\Omega),\\&\boldsymbol{S}_v(\Omega)\geqslant\boldsymbol{0}.\end{aligned}\right\} \tag{9.14}$$

where $\boldsymbol{S}_v^*(\Omega)$ is the conjugate transpose matrix of $\boldsymbol{S}_v(\Omega)$. Moreover, the inverse Fourier transform

$$\boldsymbol{N}_v(s)=\frac{1}{2\pi}\int_{-\infty}^{\infty}\boldsymbol{S}_v(\Omega)\,\mathrm{e}^{\mathrm{i}\Omega s}\mathrm{d}\Omega \tag{9.15}$$

again yields the correlation matrix. The power spectral density is a useful means of processing the experimental data of stationary stochastic processes.

A stochastic process $\boldsymbol{v}(t)$ is a Gauss process if, for any collection of instants t_j, $j=1(1)m$, the $n\times 1$ random vector $\boldsymbol{v}(t_j)$ has a Gaussian probability distribution. The Gauss or normal distribution always occurs when the random variables are composed of infinitely many independent quantities, such that each of them has a vanishing influence upon the random variables. The Gauss distribution is totally described by the first and second order moments. Hence a Gaussian stochastic process $\boldsymbol{v}(t)$ is uniquely described by the mean vector $\boldsymbol{m}_v(t)$ and the central correlation matrix $\boldsymbol{N}_v(t, \tau)$.

A stochastic process $\boldsymbol{v}(t)$ is a Markov process if

$$\mathscr{P}\{\boldsymbol{v}(t_l) \leqslant \boldsymbol{v}_l | \boldsymbol{v}(t_1), \boldsymbol{v}(t_2), \ldots, \boldsymbol{v}(t_{l-1})\} = \mathscr{P}\{\boldsymbol{v}(t_l) \leqslant \boldsymbol{v}_l | \boldsymbol{v}(t_{l-1})\}, \tag{9.16}$$

is satisfied for all l, all $t_1, t_2, \ldots, t_l$, arranged in $t_l \geqslant t_{l-1} \geqslant \ldots \geqslant t_1$, and all v_l. The symbol $\mathscr{P}\{A|B\}$ stands for conditional probability, i.e., the probability of A if it is known that the event B already took place. Thus the Markov process does not depend on events in the past.

One says that a stochastic process $v(t)$, $t \geqslant t_0$, has *independent increments* if

$$\left.\begin{aligned} &\boldsymbol{v}(t_0) = \boldsymbol{0}, \\ &E\{\boldsymbol{v}(t_2) - \boldsymbol{v}(t_1)\} = E\{\boldsymbol{v}(t_4) - \boldsymbol{v}(t_3)\} = \boldsymbol{0}, \\ &E\{[\boldsymbol{v}(t_2) - \boldsymbol{v}(t_1)][\boldsymbol{v}(t_4) - \boldsymbol{v}(t_3)]^{\mathrm{T}}\} = \boldsymbol{0} \end{aligned}\right\} \tag{9.17}$$

for every sequence of instants t_1, t_2, t_3, t_4 satisfying $t_0 \leqslant t_1 \leqslant t_2 \leqslant t_3 \leqslant t_4$. A stochastic process $\boldsymbol{v}(t)$ with independent increments is called a Wiener process if each of the increments $\boldsymbol{v}(t_2) - \boldsymbol{v}(t_1)$ is a Gaussian random vector with vanishing mean value and a covariance matrix $\boldsymbol{P} = \boldsymbol{Q}(t_2 - t_1)$, where $\boldsymbol{Q}$ is a constant nonnegative definite intensity matrix.

A stochastic process $\boldsymbol{v}(t)$ is called a stationary *white noise* if it is an idealized Gauss-Markov process with independent increments for $(t_2 - t_1) \to 0$. Such a process has the properties

$$\left.\begin{aligned} &\boldsymbol{m}_v(t) = \boldsymbol{0}, \\ &\boldsymbol{N}_v(t, \tau) = \boldsymbol{Q}_v \delta(t - \tau), \quad \boldsymbol{Q}_v \geqslant \boldsymbol{0}, \end{aligned}\right\} \tag{9.18}$$

where $\boldsymbol{Q}_v$ is the constant nonnegative definite *intensity matrix* of the white noise, and $\delta(t - \tau)$ denotes the Dirac function (7.2). One obtains the power spectral density matrix of the white noise from (9.13) and (9.18):

$$\boldsymbol{S}_v(\Omega) = \boldsymbol{Q}_v. \tag{9.19}$$

Thus for white noise the intensity matrix $\boldsymbol{Q}_v$ is equal to the power spectral density matrix. Due to the very small correlation between any two neighboring random vectors $\boldsymbol{v}(t_1)$ and $\boldsymbol{v}(t_2)$ of the white noise, the process is very irregular and it contains signals of very high frequencies. Consequently the covariance matrix of the white noise has an infinite value,

$$\boldsymbol{P}_v(t) = \boldsymbol{Q}_v\delta(0) \to \infty. \tag{9.20}$$

This shows that white noise is not a real process. But when white noise has been subject to integration, it becomes a Wiener process, and thus becomes physically meaningful. In the next section the response of vibrating systems to stochastic excitations will be considered. But this involves at least one integration, and therefore white noise can be used with success as an excitation process.

A comprehensive discussion of the theory of stochastic processes can be found for instance in the books of Papoulis, Jazwinski, Kwakernaak and Sivan.

9.2. Stochastic excitation

Consider a vibrating system under the action of random forces which are represented by stationary white noise $\boldsymbol{b}(t)$,

$$\left.\begin{aligned} \boldsymbol{m}_b(t) &= \boldsymbol{0}, \\ \boldsymbol{N}_b(t, \tau) &= \boldsymbol{Q}\delta(t-\tau), \end{aligned}\right\} \tag{9.21}$$

where $\boldsymbol{Q}$ is the $n \times n$ intensity matrix of excitation. Because of the stochastic excitation, the motion of the system is now described by a stochastic differential equation. In the special case of a vibrating system excited by white noise, it is possible to formally retain the deterministic differential equation

$$\dot{\boldsymbol{x}}(t) = \boldsymbol{A}\boldsymbol{x}(t) + \boldsymbol{b}(t), \quad \boldsymbol{x}(0) = \boldsymbol{x}_0. \tag{9.22}$$

Nevertheless, the initial state $\boldsymbol{x}_0$ must now be treated as a Gaussian random vector

$$\boldsymbol{x}_0 \sim (\boldsymbol{m}_0, \boldsymbol{P}_0) \tag{9.23}$$

which does not depend on the excitation (9.21). Thus the initial state (9.23) is described by the $n \times 1$ mean vector $\boldsymbol{m}_0$ and the $n \times n$ covariance matrix $\boldsymbol{P}_0$.

The *stochastic response* of the system (9.22) is formally given by

$$x(t) = \Phi(t)x_0 + \int_0^t \Phi(t-\tau)b(\tau)\mathrm{d}\tau, \tag{9.24}$$

which also contains, besides the fundamental matrix $\Phi(t) = \mathrm{e}^{At}$ and the initial state x_0, a stochastic integral. It can now be shown that the stochastic response $x(t)$ is a nonstationary Gauss-Markov process and such a process can be uniquely described by its mean vector $m_x(t)$ and the correlation matrix $N_x(t_1, t_2)$. To determine these quantities from (9.24) one needs the following two properties of the stochastic integral, stated below without proof:

$$E\left\{\int_{t_0}^t \Phi(t-\tau)b(\tau)\mathrm{d}\tau\right\} = \mathbf{0}, \tag{9.25}$$

$$E\left\{\left[\int_{t_0}^{t_1} \Phi(t_1-\tau_1)b(\tau_1)\mathrm{d}\tau_1\right]\left[\int_{t_0}^{t_2} \Phi(t_2-\tau_2)b(\tau_2)\mathrm{d}\tau_2\right]^{\mathrm{T}}\right\}$$

$$= \int_{t_0}^{t_1}\int_{t_0}^{t_2} \Phi(t_1-\tau_1)Q\delta(\tau_2-\tau_1)\Phi^T(t_2-\tau_2)\mathrm{d}\tau_1\mathrm{d}\tau_2$$

$$= \int_{t_0}^{\min(t_1,t_2)} \Phi(t_1-\tau)Q\Phi^{\mathrm{T}}(t_2-\tau)\mathrm{d}\tau. \tag{9.26}$$

This and (9.24) imply for the $n \times 1$ mean vector

$$m_x(t) = \Phi(t)m_0 \tag{9.27}$$

and for the $n \times n$ correlation matrix

$$N_x(t_1, t_2) = \Phi(t_1)P_0\Phi^{\mathrm{T}}(t_2) + \int_{t_0}^{\min(t_1,t_2)} \Phi(t_1-\tau)Q\Phi^{\mathrm{T}}(t_2-\tau)\mathrm{d}\tau. \tag{9.28}$$

Further, for $t_0 = 0$ and $t_1 = t_2 = t$ the symmetric $n \times n$ covariance matrix

$$P_x(t) = \Phi(t)P_0\Phi^{\mathrm{T}}(t) + \int_0^t \Phi(t-\tau)Q\Phi^{\mathrm{T}}(t-\tau)\mathrm{d}\tau \tag{9.29}$$

is obtained. If the fundamental matrix $\Phi(t) = \mathrm{e}^{At}$ is substituted into the integral in (9.29), this integral then becomes

$$\int_0^t \mathrm{e}^{A(t-\tau)}Q\,\mathrm{e}^{A^{\mathrm{T}}(t-\tau)}\mathrm{d}\tau = \mathrm{e}^{At}\int_0^t \mathrm{e}^{-A\tau}Q\,\mathrm{e}^{-A^{\mathrm{T}}\tau}\mathrm{d}\tau\,\mathrm{e}^{A^{\mathrm{T}}t}. \tag{9.30}$$

The remaining integral has the solution

$$\int_0^t e^{-A\tau} Q \, e^{-A^{\mathrm{T}}\tau} d\tau = e^{-A t} P \, e^{-A^{\mathrm{T}} t} - P, \tag{9.31}$$

where the constant $n \times n$ matrix P satisfies the Ljapunov matrix equation

$$AP + PA^{\mathrm{T}} + Q = 0. \tag{9.32}$$

One can verify the validity of the solution (9.31) either by differentiation or prove it directly by means of the series expansion (4.14). The existence and uniqueness of the solution (9.31) is always guaranteed, provided the system is asymptotically stable.

The characteristic properties of the first and second order of a stochastic response $x(t)$ can now be summarized as follows:

$$\left. \begin{aligned} m_x(t) &= \Phi(t) m_0, \\ P_x(t) &= \Phi(t)(P_0 - P)\Phi^T(t) + P. \end{aligned} \right\} \tag{9.33}$$

In particular, one obtains the variance $\sigma_{xi}(t) = \sqrt{P_{xii}(t)}$ of the i-th state variable $x_i(t)$ from the i-th diagonal element of the matrix $P_x(t)$. For asymptotically stable systems the stationary stochastic response is

$$m_x(t \to \infty) = 0, \tag{9.34}$$

$$P_x(t \to \infty) = P = \int_0^{\infty} e^{A\tau} Q \, e^{A^{\mathrm{T}}\tau} d\tau = \text{const.} \tag{9.35}$$

This shows that the covariance matrix P carries all the information about the stationary response. The matrix P can be found either by spectral density analysis or by covariance analysis.

9.3. Spectral density analysis

The spectral density analysis is based on the simple connection between the power spectral density of the excitation process and the power spectral density of the solution process. An additional inverse Fourier transform yields the covariance matrix.

The correlation matrix of a stationary stochastic process, $x(t \to \infty)$, of an asymptotically stable system is obtained from (9.26) and (9.27):

$$N_x(s) = \int_{-\infty}^{\infty} \int_{-\infty}^{\infty} \Phi(z_1) Q \delta(s + z_2 - z_1) \Phi^{\mathrm{T}}(z_2) dz_1 dz_2 \tag{9.36}$$

where $s = t_1 - t_2$, $z_1 = t_1 - \tau_1$, $z_2 = t_2 - \tau_2$. The Fourier transform (9.13) now yields the power spectral density of the solution process

$$\boldsymbol{S}_x(\Omega) = \boldsymbol{F}(-\Omega)\boldsymbol{Q}\boldsymbol{F}^{\mathrm{T}}(\Omega), \tag{9.37}$$

where $\boldsymbol{F}(\Omega)$ is the frequency response matrix (7.43). Often, the importance of the power spectral density matrix $\boldsymbol{S}_x(\Omega)$ is limited; in the engineering view of the solution process one needs the covariance matrix $\boldsymbol{P}$. Relation (9.15) implies for $s = 0$ the result

$$\boldsymbol{P} = \frac{1}{2\pi}\int_{-\infty}^{\infty} \boldsymbol{S}_x(\Omega)\mathrm{d}\Omega. \tag{9.38}$$

Thereby one has to solve an improper integral for every element of the power spectral density matrix. The solution of these improper integrals for the diagonal elements P_{ii}, respectively for the standard deviations σ_i^2, was given by James, Nichols and Philips:

$$\sigma_i^2 = \frac{1}{2\pi}\int_{-\infty}^{\infty} S_{xii}(\Omega)\mathrm{d}\Omega = \frac{1}{2\pi}\int_{-\infty}^{\infty} \frac{g_n(\Omega)}{h_n(\mathrm{i}\Omega)h_n(-\mathrm{i}\Omega)}\mathrm{d}\Omega = I_n. \tag{9.39}$$

Here $g_n(\Omega)$ and $h_n(\mathrm{i}\Omega)$ are polynomials of the form

$$\left.\begin{aligned} g_n(\Omega) &= b_0\Omega^{2n-2} + b_1\Omega^{2n-4} + \ldots + b_{n-1}, \\ h_n(\mathrm{i}\Omega) &= a_0(\mathrm{i}\Omega)^n + a_1(\mathrm{i}\Omega)^{n-1} = \ldots + a_n = \det(\mathrm{i}\Omega\boldsymbol{E} - \boldsymbol{A}). \end{aligned}\right\} \tag{9.40}$$

For $n \leqslant 7$ a complete evaluation of the integrals (9.39) can be achieved with a reasonable amount of effort. The following are the results for $n = 1(1)4$:

$$I_1 = \frac{b_0}{2a_0a_1}, \tag{9.41}$$

$$I_2 = \frac{a_0b_1 + a_2b_0}{2a_0a_1a_2}, \tag{9.42}$$

$$I_3 = \frac{-a_0a_3b_1 - a_0a_1b_2 - a_2a_3b_0}{2a_0a_3(a_0a_3 - a_1a_2)}, \tag{9.43}$$

$$I_4 = \frac{-a_0a_1a_4b_2 + a_0b_3(a_oa_3 - a_1a_2) - a_4b_0(a_2a_3 - a_1a_4) - a_0a_3a_4b_1}{2a_0a_4(a_0a_3^2 + a_1^2a_4 - a_1a_2a_3)}. \tag{9.44}$$

It is easily seen that the size of these formulae increases quickly with increasing order.

Example 9.1. Random vibrations of a motor car – spectral density analysis

A motor car proceeds with constant velocity on an rough road, Fig. 9.3. The model consists of the car body mass m with the body spring k and the shock-absorber d, while the masses of the wheels and axles are neglected. In terms of the relative displacement $y(t)$ the equation of motion for the vertical vibration is

$$m\ddot{y}(t) + d\dot{y}(t) + ky(t) = h(t) \tag{9.45}$$

where

$$h(t) = -m\ddot{y}_e(t) \tag{9.46}$$

is the scalar white noise excitation process of intensity m^2q. In this simplest approximation, the car is excited by a stochastic process with constant acceleration intensity q. As a rule, the unevenness with a large wavelength has a large amplitude and for a small wavelength a very small amplitude is observed. This implies an approximately constant acceleration for all frequencies, i.e., white noise (9.46). More accurate models of the excitation processes caused by rough roads lead to white velocity noise or colored noise, see Müller, Popp and Schiehlen. The motor car in Fig. 9.3 is an asymptotically stable system since $k > 0$ and $d > 0$ are assumed for the spring and shock absorber. The equations (9.45) and

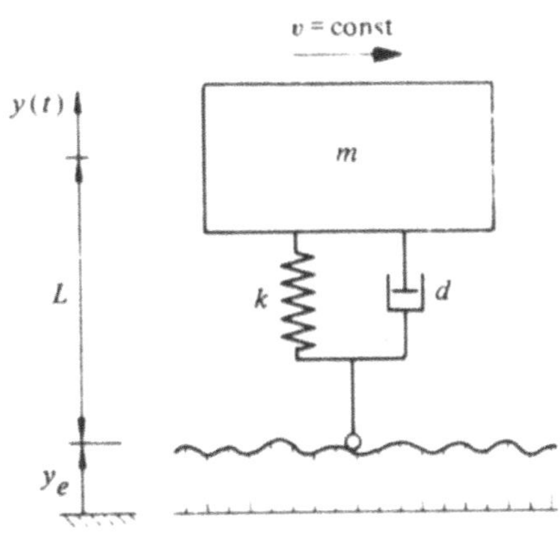

Fig. 9.3. Motor car on rough road

(9.46) imply the state equation with the abbreviations $\vartheta = d/m$, $\kappa = k/m$:

$$\underbrace{\begin{bmatrix} \dot{y} \\ \ddot{y} \end{bmatrix}}_{\dot{\boldsymbol{x}}(t)} = \underbrace{\begin{bmatrix} 0 & 1 \\ -\kappa & -\vartheta \end{bmatrix}}_{\boldsymbol{A}} \underbrace{\begin{bmatrix} y \\ \dot{y} \end{bmatrix}}_{\boldsymbol{x}(t)} + \underbrace{\begin{bmatrix} 0 \\ -\ddot{y}_e \end{bmatrix}}_{\boldsymbol{b}(t)}, \tag{9.47}$$

where the excitation process $\boldsymbol{b}(t)$ is determined by the intensity matrix

$$\boldsymbol{Q} = \begin{bmatrix} 0 & 0 \\ 0 & q \end{bmatrix}. \tag{9.48}$$

To begin with, for the spectral density analysis one needs the frequency response matrix (7.43) of the system (9.47):

$$\boldsymbol{F}(\Omega) = \frac{1}{\kappa - \Omega^2 + \mathrm{i}\Omega\vartheta} \begin{bmatrix} \mathrm{i}\Omega + \vartheta & 1 \\ -\kappa & \mathrm{i}\Omega \end{bmatrix}. \tag{9.49}$$

This leads to, by (9.37), the power spectral density matrix of the solution process

$$\boldsymbol{S}_x(\Omega) = \frac{1}{(\kappa - \Omega^2)^2 + (\Omega\vartheta)^2} \begin{bmatrix} \mathrm{i}\Omega + \vartheta & 1 \\ -\kappa & \mathrm{i}\Omega \end{bmatrix} \begin{bmatrix} 0 & 0 \\ 0 & q \end{bmatrix} \begin{bmatrix} -\mathrm{i}\Omega + \vartheta & -\kappa \\ 1 & -\mathrm{i}\Omega \end{bmatrix}$$

$$= \frac{9}{(\kappa - \Omega^2)^2 + (\Omega\vartheta)^2} \begin{bmatrix} 1 & -\mathrm{i}\Omega \\ \mathrm{i}\Omega & \Omega^2 \end{bmatrix}. \tag{9.50}$$

The covariance matrix (9.38) of the solution process is

$$\boldsymbol{P} = \frac{q}{2\pi} \int_{-\infty}^{\infty} \begin{bmatrix} 1 & -\mathrm{i}\Omega \\ \mathrm{i}\Omega & \Omega^2 \end{bmatrix} \frac{\mathrm{d}\Omega}{(\kappa - \Omega^2)^2 + (\Omega\vartheta^2)}. \tag{9.51}$$

The standard deviations of the displacement y and the velocity $\dot{y}$ are given by integrals of the form (9.39):

$$\sigma_y^2 = P_{11} = \frac{q}{2\pi} \int_{-\infty}^{\infty} \frac{1}{(-\Omega^2 + \mathrm{i}\Omega\vartheta + \kappa)(-\Omega^2 - \mathrm{i}\Omega\vartheta + \kappa)} \mathrm{d}\Omega, \tag{9.52}$$

$$\sigma_{\dot{y}}^2 = P_{22} = \frac{q}{2\pi} \int_{-\infty}^{\infty} \frac{\Omega^2}{(-\Omega^2 + \mathrm{i}\Omega\vartheta + \kappa)(-\Omega^2 - \mathrm{i}\Omega\vartheta + \kappa)} \mathrm{d}\Omega. \tag{9.53}$$

By comparing coefficients in (9.52) (or in (9.53)) with those in (9.42) one finds that $a_0 = 1$, $a_1 = \vartheta$, $a_2 = \kappa$ and $b_0 = 0$, $b_1 = 1$ (or $b_0 = 1$, $b_1 = 0$). Therefore

$$\sigma_y^2 = \frac{q}{2\vartheta\kappa} = \frac{m^2 q}{2dk}, \qquad \sigma_{\dot{y}}^2 = \frac{q}{2\vartheta} = \frac{mq}{2d}. \tag{9.54}$$

Moreover one obtains from (9.51), by a somewhat laborious integral evaluation,

$$\sigma_{y\dot{y}} = P_{12} = 0. \tag{9.55}$$

First of all one finds the trivial result that for $d \to \infty$ and $k \to \infty$ and thus for a blocked wheel suspension, the relative motion $y(t)$ vanishes. But for an engineering specification of the riding comfort the relevant quantity is the absolute acceleration $\ddot{y}_a$ of the car body which satiesfies, by (9.45) and (9.46), the equation

$$\ddot{y}_a = \ddot{y} + \ddot{y}_e = -\vartheta\dot{y} - \kappa y \tag{9.56}$$

or, due to (9.47),

$$\ddot{y}_a = [-\kappa \quad -\vartheta]\begin{bmatrix} y \\ \dot{y} \end{bmatrix} = \boldsymbol{Tx}, \tag{9.57}$$

introducing the 1×2 transformation matrix $\boldsymbol{T}$. If one now recalls the transformation law for Gaussian random vectors

$$P_{\ddot{y}_a} = \boldsymbol{TPT}^{\mathrm{T}}, \tag{9.58}$$

then (9.52) through (9.55) and (9.57) imply that the standard deviation of the absolute acceleration is given by

$$\sigma_{\ddot{y}_a}^2 = P_{\ddot{y}_a} = \kappa^2\sigma_y^2 + \vartheta^2\sigma_{\dot{y}}^2. \tag{9.59}$$

Thus, by (9.59) the quadratic means of the stochastic processes $y(t)$ and $\dot{y}(t)$, which are uncoupled according to (9.55), are added. Hence (9.54) implies

$$\sigma_{\ddot{y}_a} = \sqrt{\frac{q}{2}}\sqrt{\frac{k}{d} + \frac{d}{m}}. \tag{9.60}$$

The optimal riding comfort, $\sigma_{\ddot{y}_a} \to 0$, is obtained for a soft wheel suspension $d \to 0$ and $k/d \to 0$. However for static reasons, the spring constant

k is bounded from below. For finite k one finds from (9.60) the optimal damping

$$d_{\text{opt}}^2 = km \tag{9.61}$$

which corresponds to an optimal damping factor

$$D_{\text{opt}} = \tfrac{1}{2}. \tag{9.62}$$

This confirms the known rule of thumb in motor car design: Very soft body spring, and a shock absorber calibrated at $D_{\text{opt}} = 0.5$. ◆

9.4. Covariance analysis

The *covariance analysis* reduces the calculation of the covariance matrix to finding a solution of the Ljapunov matrix equation (9.32). In Section 13.2 various methods of solving Ljapunov's equation are presented. Here the implicit solution (13.50) and the explicit solution (13.55) will be applied to random vibrations.

The *implicit* or *direct* solution (13.50) leads to a linear system of algebraic equations

$$\hat{\boldsymbol{A}}\boldsymbol{p} = \boldsymbol{q}, \tag{9.63}$$

where the $n(n+1)/2$ vectors

$$\left.\begin{aligned} \boldsymbol{p} &= [P_{11} \quad P_{12} \dots P_{1n} \quad P_{22} \dots P_{2n} \dots P_{nn}]^{\mathrm{T}}, \\ \boldsymbol{q} &= [Q_{11} \quad Q_{12} \dots Q_{1n} \quad Q_{22} \dots Q_{2n} \dots Q_{nn}]^{T} \end{aligned}\right\} \tag{9.64}$$

contain the essential elements of the symmetric $n \times n$ matrices $\boldsymbol{P}$ and $\boldsymbol{Q}$. Further, $\hat{\boldsymbol{A}}$ is an auxiliary $n(n+1)/2 \times n(n+1)/2$ matrix. However for ordinary mechanical systems the order of the system of equations (9.63) can be reduced. For this purpose one makes use of the special structures of the system matrix $\boldsymbol{A}$ and the intensity matrix $\boldsymbol{Q}$, as indicated by (3.38) and (3.39). Accordingly,

$$\boldsymbol{A} = \left[\begin{array}{c|c} \boldsymbol{0} & \boldsymbol{E} \\ \hline -\boldsymbol{M}^{-1}(\boldsymbol{K}+\boldsymbol{N}) & -\boldsymbol{M}^{-1}(\boldsymbol{D}+\boldsymbol{G}) \end{array}\right], \quad \boldsymbol{Q} = \left[\begin{array}{c|c} \boldsymbol{0} & \boldsymbol{0} \\ \hline \boldsymbol{0} & \boldsymbol{M}^{-1}\boldsymbol{V}\boldsymbol{M}^{-1} \end{array}\right], \tag{9.65}$$

hold where $\boldsymbol{V}$ is the $f \times f$ intensity matrix of the stochastic excitation process $\boldsymbol{h}(t)$. Furthermore, one can subdivide the symmetric $n \times n$ covariance matrix $\boldsymbol{P}$ in $f \times f$ submatrices $\boldsymbol{P}_{\mathrm{I}}$, $\boldsymbol{P}_{\mathrm{II}}$, $\boldsymbol{P}_{\mathrm{III}}$:

$$\boldsymbol{P} = \left[\begin{array}{c|c} \boldsymbol{P}_{\mathrm{I}} & \boldsymbol{P}_{\mathrm{II}} \\ \hline \boldsymbol{P}_{\mathrm{II}}^{\mathrm{T}} & \boldsymbol{P}_{\mathrm{III}} \end{array}\right]. \tag{9.66}$$

If one now substitutes (9.65) and (9.66) into the Ljapunov equation (9.32), one gets

$$\left.\begin{array}{l} \boldsymbol{P}_{\mathrm{II}} + \boldsymbol{P}_{\mathrm{II}}^{\mathrm{T}} = \boldsymbol{0}, \\ (\boldsymbol{K}+\boldsymbol{N})\boldsymbol{P}_{I} + (\boldsymbol{D}+\boldsymbol{G})\boldsymbol{P}_{II}^{\mathrm{T}} - \boldsymbol{M}\boldsymbol{P}_{III} = \boldsymbol{0}, \\ (\boldsymbol{K}+\boldsymbol{N})\boldsymbol{P}_{II}\boldsymbol{M} + (\boldsymbol{D}+\boldsymbol{G})\boldsymbol{P}_{III}\boldsymbol{M} + \boldsymbol{M}\boldsymbol{P}_{II}^{\mathrm{T}}(\boldsymbol{K}-\boldsymbol{N}) \\ +\boldsymbol{M}\boldsymbol{P}_{III}(\boldsymbol{D}-\boldsymbol{G}) - \boldsymbol{V} = \boldsymbol{0}. \end{array}\right\} \tag{9.67}$$

It can be seen that the submatrix $\boldsymbol{P}_{\mathrm{II}}$ is skewsymmetric and thus the number of essential elements of the covariance matrix is reduced by $n(n/2+1)/4$. Thus there remains in Ljapunov's equation only $n(3n/2 + 1)/4$ essential elements, and one obtains the system of linear algebraic equations

$$\hat{\boldsymbol{B}}\boldsymbol{r} = \boldsymbol{s}, \tag{9.68}$$

where

$$\left.\begin{array}{l} \boldsymbol{r} = [P_{11} \;\; P_{12} \;\; \ldots \;\; P_{1f} \;\; P_{22} \;\; \ldots \;\; P_{2,f+1} \;\; \ldots \,\vdots\, P_{f+1,f+1} \;\; \ldots \;\; P_{nn}]^{\mathrm{T}}, \\ \boldsymbol{s} = [0 \;\; 0 \;\; \ldots \;\; 0 \;\; 0 \;\; \ldots \;\; 0 \;\; \ldots \,\vdots\, V_{11} \;\; \ldots \;\; V_{ff}]^{\mathrm{T}} \end{array}\right\} \tag{9.69}$$

are $n(3n/2+1)/4 \times 1$ vectors and $\hat{\boldsymbol{B}}$ is an auxiliary matrix of the same dimension. As the implicit solution increases the system order, it is in principle applicable when an efficient digital computer is used. But also in any other method of solution, the submatrix $\boldsymbol{P}_{\mathrm{II}}$ remains skewsymmetric. This offers a good possibility to check the correctness of the result.

Example 9.2. Random vibrations of a motor car - covariance analysis

The implicit method of solution will be applied in order to determine the standard deviations of the displacement $y(t)$ and the velocity $\dot{y}(t)$ for the motor car considered in Example 9.1. According to (9.45), (9.46), (9.66), the details in this case are: $M = m$, $D = d$, $G = 0$, $K = k$, $N = 0$,

$V = m^2 q,\ P_{\mathrm{I}} = \sigma_y^2,\ P_{\mathrm{II}} = \sigma_{y\dot{y}},\ P_{\mathrm{III}} = \sigma_{\dot{y}}^2$. Hence it follows directly from (9.67) that

$$\left.\begin{aligned} &2\sigma_{y\dot{y}} = 0, \\ &k\sigma_y^2 + d\sigma_{y\dot{y}} - m\sigma_{\dot{y}}^2 = 0, \\ &2km\sigma_{y\dot{y}} - 2dm\sigma_{\dot{y}}^2 - m^2 q = 0 \end{aligned}\right\} \tag{9.70}$$

or

$$\sigma_y^2 = \frac{m^2 q}{2dk}, \quad \sigma_{\dot{y}}^2 = \frac{mq}{2d}, \quad \sigma_{y\dot{y}} = 0. \tag{9.71}$$

This was a much more direct way of finding (9.54). ◆

The *explicit solution* can be written as the following closed formula for the covariance matrix:

$$\boldsymbol{P} = \frac{1}{2a_0 \det \boldsymbol{H}} \sum_{k=0}^{n-1} H_{k+1,1} \sum_{m=0}^{2k} (-1)^m \boldsymbol{A}_m \boldsymbol{Q} \boldsymbol{A}_{2k-m}^{\mathrm{T}}, \tag{9.72}$$

where

$$\boldsymbol{A}_m = \boldsymbol{A}\boldsymbol{A}_{m-1} + a_m \boldsymbol{E}, \qquad m = 0(1)2n-2, \tag{9.73}$$

are auxiliary $n \times n$ matrices, a_m is the m-th characteristic coefficient, H is the $n \times n$ Hurwitz matrix (5.39), and $H_{k+1,1}$ is the cofactor of the element $(k+1,\ 1)$ of the Hurwitz matrix. The explicit solution (9.72) is well suited for analytic investigation of low order systems. For systems of higher order the numerical evaluation of the Hurwitz determinant det $\boldsymbol{H}$ can result in difficulties. Thus for systems of higher order, one of the numerical procedures outlined in Section 14.2 is recommended.

Example 9.3. Centrifuge under random loading

A rotating centrifuge is charged with a liquid which stochastically flows out of two pipes, Fig. 9.4. Let the centrifuge be a symmetric rigid body rotating with constant angular velocity Ω and suspended by symmetric springs and dampers. Then, according to Problem 3.1, the state equation is

$$\underbrace{\begin{bmatrix} \dot{\alpha} \\ \dot{\beta} \\ \hline \ddot{\alpha} \\ \ddot{\beta} \end{bmatrix}}_{\dot{\boldsymbol{x}}(t)} = \underbrace{\left[\begin{array}{cc|cc} 0 & 0 & 1 & 0 \\ 0 & 0 & 0 & 1 \\ \hline -\kappa & 0 & -\vartheta & -w \\ 0 & -\kappa & w & -\vartheta \end{array}\right]}_{\boldsymbol{A}} \underbrace{\begin{bmatrix} \alpha \\ \beta \\ \hline \dot{\alpha} \\ \dot{\beta} \end{bmatrix}}_{\boldsymbol{x}(t)} + \underbrace{\begin{bmatrix} 0 \\ 0 \\ \hline v(t) \\ 0 \end{bmatrix}}_{\boldsymbol{b}(t)}, \tag{9.74}$$

where κ, ϑ, $w = g\Omega$ are the parameter introduced in Problem 3.1 and $v(t)$ is the stochastic process of white noise with intensity q. The random forces act only in the y-direction, Fig. 9.4, thus in (9.74) only the moment with respect to the x-axis appears. Therefore the intensity matrix of the system is

$$\boldsymbol{Q} = \left[\begin{array}{cc|cc} 0 & 0 & 0 & 0 \\ 0 & 0 & 0 & 0 \\ \hline 0 & 0 & q & 0 \\ 0 & 0 & 0 & 0 \end{array}\right]. \tag{9.75}$$

The system is asymptotically stable, as was shown in Example 5.8.

In this case the implicit solution leads to a system of equations of order 7. Thus the explicit solution will now be applied;

$$\boldsymbol{P} = \frac{1}{2a_0 \det H} \sum_{k=0}^{3} H_{k+1,1} \sum_{m=0}^{2k} (-1)^m \boldsymbol{A}_m \boldsymbol{Q} \boldsymbol{A}_{2m-k}^{\mathrm{T}}. \tag{9.76}$$

According to (9.73), the auxiliary matrices $\boldsymbol{A}_m$ are

$$\boldsymbol{A}_0 = \boldsymbol{E}, \tag{9.77}$$

$$\boldsymbol{A}_1 = \begin{bmatrix} 2\vartheta & 0 & 1 & 0 \\ 0 & 2\vartheta & 0 & 1 \\ -\kappa & 0 & \vartheta & -w \\ 0 & -\kappa & w & \vartheta \end{bmatrix}, \tag{9.78}$$

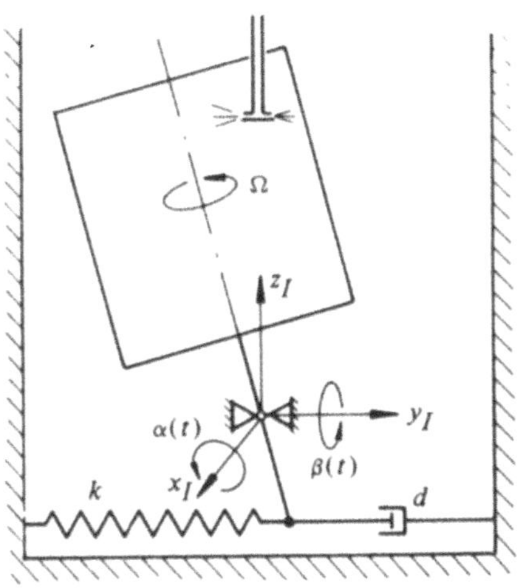

Fig. 9.4. Centrifuge with random load distribution

$$A_2 = \begin{bmatrix} \kappa + \vartheta^2 + w^2 & 0 & \vartheta & -w \\ 0 & \kappa + \vartheta^2 + w^2 & w & \vartheta \\ -\kappa\vartheta & \kappa w & \kappa & 0 \\ -\kappa w & -\kappa\vartheta & 0 & \kappa \end{bmatrix}, \tag{9.79}$$

$$A_3 = \begin{bmatrix} \kappa\vartheta & \kappa w & \kappa & 0 \\ -\kappa w & k\vartheta & 0 & \kappa \\ -\kappa^2 & 0 & 0 & 0 \\ 0 & -\kappa^2 & 0 & 0 \end{bmatrix}, \tag{9.80}$$

$$A_4 = A_5 = A_6 = \mathbf{0}. \tag{9.81}$$

The condition $A_4 = \mathbf{0}$ offers good check of calculation. Further, the coefficients a_m of the characteristic equation are given by

$$\left.\begin{array}{llll} a_0 = 1, & a_1 = 2\vartheta, & a_2 = 2\kappa + \vartheta^2 + w^2, & \\ a_3 = 2\vartheta\kappa, & a_4 = \kappa^2, & a_5 = 0, & a_6 = 0 \end{array}\right\} \tag{9.82}$$

and the Hurwitz matrix (5.39) is

$$H = \begin{bmatrix} a_1 & 1 & 0 & 0 \\ a_3 & a_2 & a_1 & 1 \\ 0 & a_4 & a_3 & a_2 \\ 0 & 0 & 0 & a_4 \end{bmatrix}. \tag{9.83}$$

This permits the determination of the cofactors:

$$\left.\begin{aligned} H_{11} &= 2\vartheta\kappa^3(\kappa + \vartheta^2 + w^2), \\ H_{21} &= -2\vartheta\kappa^3, \\ H_{31} &= 2\vartheta\kappa^2, \\ H_{41} &= -2\vartheta(\kappa + \vartheta^2 + w^2) \end{aligned}\right\} \tag{9.84}$$

and consequently the Hurwitz determinant is

$$\det H = a_1 H_{11} + a_3 H_{21} = 4\vartheta^2\kappa^3(\vartheta^2 + w^2). \tag{9.85}$$

If one now substitutes (9.77) through (9.85) into (9.76), one gets as a result the covariance matrix

$$P = \frac{q}{4\vartheta\kappa(\vartheta^2 + w^2)} \left[\begin{array}{cc|cc} 2\vartheta^2 + w^2 & \vartheta w & 0 & 0 \\ \vartheta w & w^2 & 0 & 0 \\ \hline 0 & 0 & (2\vartheta^2 + w^2)\kappa & \vartheta\kappa w \\ 0 & 0 & \vartheta\kappa w & \kappa w^2 \end{array}\right]. \tag{9.86}$$

The submatrices $\boldsymbol{P}_{\mathrm{I}}$ and $\boldsymbol{P}_{\mathrm{III}}$ are symmetric while the submatrix $\boldsymbol{P}_{\mathrm{II}}$ vanishes, which does not contradict its skewsymmetry stated in (9.67). The standard deviations σ_α, σ_β, $\sigma_{\dot\alpha}$, $\sigma_{\dot\beta}$ are the square roots of the variances P_{11}, P_{22}, P_{33}, P_{44}. Thus they will not be written out individually. ◆

A comparison of the covariance and spectral analyses leads to the conclusion that covariance analysis is better suited for the investigation of vibrating systems. Nevertheless, power spectral density investigations are much more common. The justification for this is mainly historical. About 1950, the theory of random vibrations was developed in the frequency domain, based on Wiener's filter theory. The further development of filter theory in the time domain by Kalman since 1960 found little application in the study of vibrations. Only in recent years a certain break-through of the covariance methods in the theory of random vibrations takes place.

9.5. Colored noise excitation processes

The process of white noise has proved to be a good auxiliary device for overcoming mathematical difficulties encountered with stochastic differential equations. Nevertheless white noise is not a real process and therefore the intensity matrix (9.19) cannot be directly measured. This problem is overcome by the process of *colored noise* which is defined as a stationary Gauss-Markov process which can be viewed as the response of a dynamic shape filter excited by white noise. The $n \times 1$ process $\boldsymbol{b}(t)$ of colored noise is given by

$$\left.\begin{aligned} \boldsymbol{b}(t) &= \boldsymbol{B}\boldsymbol{\xi}(t) \\ \dot{\boldsymbol{\xi}}(t) &= \boldsymbol{C}\boldsymbol{\xi}(t) + \boldsymbol{D}\boldsymbol{v}(t), \end{aligned}\right\} \tag{9.87}$$

where $\boldsymbol{v}(t)$ is an $n \times 1$ process of white noise, $\boldsymbol{\xi}(t)$ is an $s \times 1$ process of colored noise, and $\boldsymbol{B}$, $\boldsymbol{C}$, $\boldsymbol{D}$ are constant matrices of suitable dimensions. If the excitation process (9.87) is acting on a vibrating system (9.22), then one obtains the following total system:

$$\underbrace{\begin{bmatrix} \dot{\boldsymbol{x}} \\ \hline \dot{\boldsymbol{\xi}} \end{bmatrix}}_{\dot{\bar{\boldsymbol{x}}}(t)} = \underbrace{\left[\begin{array}{c|c} \boldsymbol{A} & \boldsymbol{B} \\ \hline \boldsymbol{0} & \boldsymbol{C} \end{array}\right]}_{\bar{\boldsymbol{A}}} \underbrace{\begin{bmatrix} \boldsymbol{x} \\ \hline \boldsymbol{\xi} \end{bmatrix}}_{\bar{\boldsymbol{x}}(t)} + \underbrace{\begin{bmatrix} \boldsymbol{0} \\ \hline \boldsymbol{D} \end{bmatrix} \boldsymbol{v}.}_{\bar{\boldsymbol{b}}(t)} \tag{9.88}$$

Thus there again remains a linear system of the form (9.22) which, by (9.88), has the increased order $(n+s)$. For the solution of (9.88) and the related determination of $\boldsymbol{x}(t)$ one can again apply any of the methods explained in Sections 9.2 to 9.4.

However, in many cases it is not necessary to introduce the form filter (9.87); instead the intensity of the white noise can be approximately determined from the power spectral density of the colored noise. This will be illustrated for the example of a scalar process. In Fig. 9.5 a typical diagram of the power spectral density $S_b(\Omega)$ of a real process $b(t)$ is shown obtained from measurements. For determining the intensity it is convenient to distinguish between the following three cases:

1. A slow system. If the eigenvalues λ_i, $i = 1(1)n$, of the system matrix $\boldsymbol{A}$ are relatively small compared to the characteristic frequency Ω_0 of the process $b(t)$, $\max_i |\lambda_i| \ll \Omega_0$, then one has $Q_b = 2S_0$, where Ω_0 and S_0 are defined by Fig. 9.5.
2. An intermediate system. If the eigenvalues λ_i, $i = 1(1)n$, of the system matrix $\boldsymbol{A}$ have the same order of magnitude as the characteristic frequency Ω_0 of the process $b(t)$, $\max_i |\lambda_i| \approx \Omega_0$, then one has $Q_b = S_0$ or $Q_b = \sigma_b^2/2\Omega_0$ where Ω_0 and S_0 are defined by Fig. 9.5 and σ_b is the standard deviation of $b(t)$.
3. A fast system. If the eigenvalues λ_i, $i = 1(1)n$, of the system matrix $\boldsymbol{A}$ are large in comparison to the characteristic frequency of the exciting process, then the form filter (9.37) has to be used. One than has

$$\dot{b}(t) = cb(t) + dv(t), \quad v(t) \sim [0, q_v], \tag{9.89}$$

where the constants c and d are to be determined according to (9.37) so that the best possible approximation of the measured power spectral density $S_b(\Omega)$ is obtained.

If the excitation of a vibrating system occurs not by a scalar but rather by a vector process, then the above approximation theorems can be applied to each element of the power spectral density matrix.

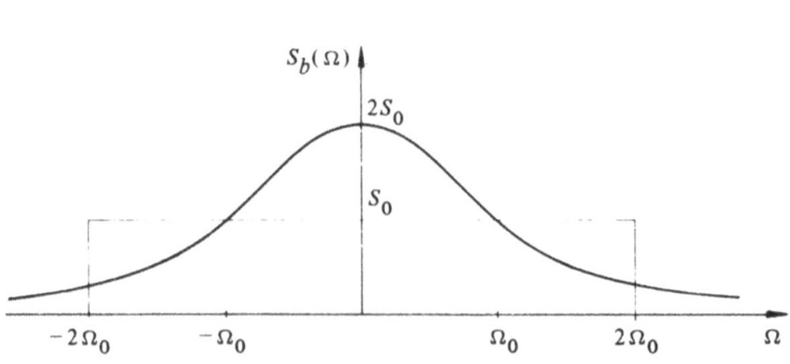

Fig. 9.5. Typical diagram of the measured spectral density of a real process

Example 9.4. Random vibrations of a motor car – excitation by colored noise

For the investigation of the riding comfort of a motor car depending on various road surfaces, the excitation of a motor car by white noise, as in (9.46), will now be replaced by colored noise excitation $h(t)$:

$$\dot{h}(t) = -\gamma h(t) + v(t), \quad v(t) \sim \left[0, \gamma^2 m^2 q\right]. \tag{9.90}$$

This leads, according to (9.37), to the power spectral density of the colored noise

$$S_h(\Omega) = \frac{\gamma^2 m^2 q}{\gamma^2 + \Omega^2}, \tag{9.91}$$

shown in Fig. 9.6 for various values of γ. One sees that for large values $\gamma \to \infty$ the process is that of white noise. To determine the standard deviations of the displacement $y(t)$ and the velocity $\dot{y}(t)$, one needs the extended system (9.88):

$$\underbrace{\begin{bmatrix} \dot{y} \\ \ddot{y} \\ \dot{h} \end{bmatrix}}_{\dot{\bar{x}}(t)} = \underbrace{\begin{bmatrix} 0 & 1 & 0 \\ -\kappa & -\vartheta & \frac{1}{m} \\ 0 & 0 & -\gamma \end{bmatrix}}_{\bar{A}} \underbrace{\begin{bmatrix} y \\ \dot{y} \\ h \end{bmatrix}}_{\bar{x}(t)} + \underbrace{\begin{bmatrix} 0 \\ 0 \\ v \end{bmatrix}}_{\bar{b}(t)} \tag{9.92}$$

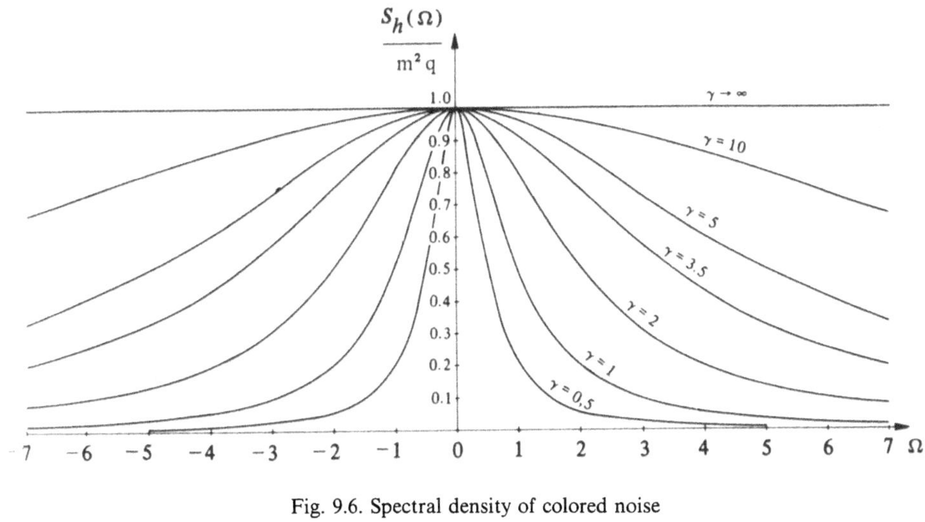

Fig. 9.6. Spectral density of colored noise

with the extended covariance matrix $\bar{\boldsymbol{P}}$ and the extended intensity matrix $\bar{\boldsymbol{Q}}$,

$$\bar{\boldsymbol{P}} = \begin{bmatrix} P_{11} & P_{12} & P_{13} \\ P_{12} & P_{22} & P_{23} \\ P_{13} & P_{23} & P_{33} \end{bmatrix}, \qquad \bar{\boldsymbol{Q}} = \begin{bmatrix} 0 & 0 & 0 \\ 0 & 0 & 0 \\ 0 & 0 & \gamma^2 m^2 q \end{bmatrix}. \tag{9.93}$$

A covariance analysis by means of the implicit solution yields the result

$$P_{11} = \sigma_y^2 = \frac{(\gamma + \vartheta)\gamma q}{2\vartheta\kappa(\gamma^2 + \vartheta\gamma + \kappa)}, \tag{9.94}$$

$$P_{22} = \sigma_{\dot{y}}^2 = \frac{\gamma^2 q}{2\vartheta(\gamma^2 + \vartheta\gamma + \kappa)}. \tag{9.95}$$

For ϑ, $\kappa \ll \gamma$, i.e., if the system is slow, (9.94) and (9.95) move to (9.54). This shows that white noise is a limit case of real, colored noise processes. ◆

Problem 9.1. Random vibrations of a magnetically levitated vehicle

In example 3.5 the state equation of a magnetically levitated vehicle as shown in Fig. 3.2 was established. Using the control coefficients (4.120) and the normalized parameters $c = 1$, $d = 1$, $m = 1$, $T = 1$ one obtains the state equation

$$\underbrace{\begin{bmatrix} \dot{y} \\ \ddot{y} \\ \dot{i} \end{bmatrix}}_{\dot{\boldsymbol{x}}(t)} \; = \; \underbrace{\begin{bmatrix} 0 & 1 & 0 \\ 1 & 0 & 1 \\ -\frac{33}{32} & -\frac{21}{16} & -1 \end{bmatrix}}_{\boldsymbol{A}} \underbrace{\begin{bmatrix} y \\ \dot{y} \\ i \end{bmatrix}}_{\boldsymbol{x}(t)} + \underbrace{\begin{bmatrix} 0 \\ -\ddot{y}_e \\ 0 \end{bmatrix}}_{\boldsymbol{b}(t)}. \tag{9.96}$$

According to Example 4.5, the system (9.96) is asymptotically stable. If one now assumes that the acceleration $\ddot{y}_e(t)$ caused by the rail roughness is an excitation process of white noise of intensity q, then the covariance matrix is

$$\boldsymbol{Q} = \begin{bmatrix} 0 & 0 & 0 \\ 0 & q & 0 \\ 0 & 0 & 0 \end{bmatrix}. \tag{9.97}$$

The random vibrations will be investigated by means of the covariance analysis described in Section 9.4.

1. What is the order of the system of linear algebraic equations (9.63) needed for the implicit solution of the Ljapunov matrix equation?

☐ 1, ☐ 3, ☐ 6, ☐ 10.

2. What is the vector $\boldsymbol{p}$ of the essential elements of the covariance matrix?

☐ $p = P_{11}$,

☐ $\boldsymbol{p} = [P_{11} \quad P_{22} \quad P_{33}]^{\mathrm{T}}$,

☐ $\boldsymbol{p} = [P_{11} \quad P_{12} \quad P_{13} \quad P_{22} \quad P_{23} \quad P_{33}]^{T}$.

3. What is the vector $\boldsymbol{q}$ of the essential elements of the intensity matrix?

☐ $\boldsymbol{q} = [0 \quad q \quad 0]^{\mathrm{T}}$,

☐ $\boldsymbol{q} = [0 \quad 0 \quad 0 \quad q \quad 0 \quad 0]^{\mathrm{T}}$,

☐ $\boldsymbol{q} = [q \quad 0 \quad 0 \quad 0 \quad 0 \quad 0]^{\mathrm{T}}$.

4. What is the form of the matrix $\hat{\boldsymbol{A}}$ in (9.63)?

☐ $$\hat{\boldsymbol{A}} = \begin{bmatrix} 0 & 2 & 0 \\ 0 & 0 & 2 \\ -\frac{33}{16} & -\frac{21}{8} & -2 \end{bmatrix},$$

☐ $$\hat{\boldsymbol{A}} = \begin{bmatrix} 0 & 2 & 0 & 0 & 0 & 0 \\ 0 & 0 & 1 & 1 & 0 & 0 \\ -\frac{1}{32} & -\frac{5}{16} & -1 & 0 & 1 & 0 \\ 0 & 0 & 0 & 0 & 2 & 0 \\ 0 & -\frac{1}{32} & 0 & -\frac{5}{16} & -1 & 1 \\ 0 & 0 & -\frac{1}{16} & 0 & -\frac{5}{8} & -2 \end{bmatrix},$$

☐ $$\hat{\boldsymbol{A}} = \begin{bmatrix} 0 & 2 & 0 & 0 & 0 & 0 \\ 1 & 0 & 1 & 1 & 0 & 0 \\ -\frac{33}{32} & -\frac{21}{16} & -1 & 0 & 1 & 0 \\ 0 & 2 & 0 & 0 & 2 & 0 \\ 0 & -\frac{33}{32} & 1 & -\frac{21}{16} & -1 & 1 \\ 0 & 0 & -\frac{33}{16} & 0 & -\frac{21}{8} & -2 \end{bmatrix}.$$

5. What are the numerical components of the solution vector p?

☐ $p = \left[\frac{176}{3} \quad \frac{7}{3} \quad \frac{1017}{16}\right]^{\mathrm{T}} q,$

☐ $p = \left[\frac{176}{3} \quad 0 \quad -\frac{7}{3} \quad \frac{7}{3} \quad -\frac{1}{2} \quad \frac{11}{48}\right]^{\mathrm{T}} q,$

☐ $p = \left[\frac{176}{3} \quad 0 \quad -61 \quad \frac{7}{3} \quad -\frac{1}{2} \quad \frac{1017}{16}\right]^{\mathrm{T}} q.$

The riding comfort of the magnetically levitated vehicle is determined by the absolute acceleration $\ddot{y}_a$. By (9.96) it follows

$$\ddot{y}_a = \ddot{y} + \ddot{y}_e = y + i = \boldsymbol{T}\boldsymbol{x} \tag{9.98}$$

with the 1×3 transformation matrix

$$\boldsymbol{T} = [1 \quad 0 \quad 1]. \tag{9.99}$$

This and (9.58) imply the following formula for the variance $P_{\ddot{y}_a}$ of the absolute acceleration:

$$P_{\ddot{y}_a} = \boldsymbol{T}\boldsymbol{P}\boldsymbol{T}^{\mathrm{T}} = P_{11} + 2P_{13} + P_{33}. \tag{9.100}$$

6. What is the standard deviation $\sigma_{\ddot{y}_a}$ of the absolute acceleration?

☐ $\sigma_{\ddot{y}_a} = \sqrt{\frac{11}{48} q}\,,$

☐ $\sigma_{\ddot{y}_a} = \frac{11}{48} q,$

☐ $\sigma_{\ddot{y}_a} = \sqrt{\frac{1017}{16} q}\,.$ ◆

PART III

Time-variant vibrating systems

Surveying the equations of motion found in Part I of this book, one will notice that time-variant vibrating systems were associated with rheonomic constraints and time-dependent forces. The corresponding linearized state equations contain time-variant coefficients. The presence of time-variant coefficients makes the solving of such systems more difficult. In particular, the methods for investigating time-invariant systems described in Part II cannot be applied to time-variant systems. The methods of solving time-variant vibrating systems will be considered in the present, third part of the book. Here a predominant role will play the periodically time-variant systems. First of all, the periodicity of the parameters permits the application of a whole series of mathematically justified solution procedures, and secondly, one often finds in engineering applications periodic processes causing parametrically excited vibrations. Among the examples are: vibrations in periodically moved vehicles or in periodic magnetic fields, vibrations of rotors with asymmetries in the moments of inertia or in the stiffness and damping forces; linearized vibrations in the neighborhood of a periodic solution of a nonlinear system. To begin with in Chapter 10 the general solution of general time-variant and periodically time-variant vibrating systems will be given. It will follow the stability investigation for periodically time-variant systems. In Chapter 11 the parametrically excited vibrations will be discussed as well as the forced parametrically excited vibrations for various typical excitation functions. The results will be illustrated by the examples of a centrifuge with unsymmetric rotor and of a pendulum with a periodically moving point of suspension.

10. General solution and stability

Just as in the case of time-invariant systems, the general solution of a time-variant system is determined by its *fundamental matrix*. Although for general time-variant systems the fundamental matrix cannot be explicitly calculated, for periodically time-variant systems the precise structure of the fundamental matrix is given by the Floquet theory. The criteria for stability and boundedness can also be found by means of the Floquet theory. This section concludes with an analysis of the particular properties of mechanical systems.

10.1. General time-variant systems

The state equation of a linear time-variant vibrating system is

$$\dot{\boldsymbol{x}}(t)=\boldsymbol{A}(t)\boldsymbol{x}(t)+\boldsymbol{b}(t), \qquad \boldsymbol{x}(t_0)=\boldsymbol{x}_0, \tag{10.1}$$

where for ordinary mechanical systems the $n\times n$ system matrix $\boldsymbol{A}(t)$ and the $n\times 1$ excitation vector $\boldsymbol{b}(t)$ have, by (3.38), (3.39), the special structure

$$\boldsymbol{A}(t)=\left[\begin{array}{c|c}\boldsymbol{0} & \boldsymbol{E}\\ \hline -\boldsymbol{M}^{-1}(t)\boldsymbol{Q}(t) & -\boldsymbol{M}^{-1}(t)\boldsymbol{P}(t)\end{array}\right],\ \boldsymbol{b}(t)=\left[\begin{array}{c}\boldsymbol{0}\\ \hline \boldsymbol{M}^{-1}(t)\boldsymbol{h}(t)\end{array}\right]. \tag{10.2}$$

The existence and uniqueness of a solution $\boldsymbol{x}(t)$ of (10.1) is always assured, provided the functions $\boldsymbol{A}(t)$, and $\boldsymbol{b}(t)$, or $\boldsymbol{M}(t)$, $\boldsymbol{P}(t)$, $\boldsymbol{Q}(t)$ and $\boldsymbol{h}(t)$ depend continuously on time for $t>t_0$.

The general solution of the homogeneous system

$$\dot{\boldsymbol{x}}(t)=\boldsymbol{A}(t)\boldsymbol{x}(t), \qquad \boldsymbol{x}(t_0)=\boldsymbol{x}_0 \tag{10.3}$$

is

$$\boldsymbol{x}(t)=\boldsymbol{\Phi}(t,t_0)\boldsymbol{x}_0 \tag{10.4}$$

where $\boldsymbol{\Phi}(t, t_0)$ is the regular $n \times n$ fundamental matrix. Contrary to what has been the case for time-invariant systems, the solution is now influenced by the *initial time instant* t_0. Hence the fundamental matrix in (10.4) will now be written with the double argument (t, t_0). The fundamental matrix now possesses the following properties, generalizing those mentioned in Chapter 4:

1. The fundamental matrix satisfies the matrix differential equation

$$\dot{\boldsymbol{\Phi}}(t, t_0) = \boldsymbol{A}(t)\boldsymbol{\Phi}(t, t_0), \qquad \boldsymbol{\Phi}(t_0, t_0) = \boldsymbol{E}, \tag{10.5}$$

with the $n \times n$ identity matrix $\boldsymbol{E}$ as initial condition.

2. The fundamental matrix possesses the transition property

$$\boldsymbol{\Phi}(t_2, t_0) = \boldsymbol{\Phi}(t_2, t_1)\boldsymbol{\Phi}(t_1, t_0). \tag{10.6}$$

3. The inverse fundamental matrix is obtained by interchanging the arguments,

$$\boldsymbol{\Phi}^{-1}(t, t_0) = \boldsymbol{\Phi}(t_0, t). \tag{10.7}$$

4. The determinant of the fundamental matrix is given by the exponential function of the integrated trace of the system matrix $\boldsymbol{A}(t)$,

$$\det \boldsymbol{\Phi}(t, t_0) = \exp \int_{t_0}^{t} \mathrm{Tr}\, \boldsymbol{A}(\tau)\mathrm{d}\tau. \tag{10.8}$$

For time-invariant systems the fundamental matrix can be represented as an exponential function of the system matrix, by (4.13). This is however not the case for general time-variant systems. The implications of this fact are profound, for neither the Cayley-Hamilton theorem (4.130) nor the eigenvalue problem (4.53) can be applied for the calculation of the fundamental matrix. The only possibility for definite calculation consists in integrating (10.5) numerically. According to the discussion on Chapter 14, the integration can be performed on an analog computer and equally well on a digital computer by means of an integration procedure. Approximately one can also replace the time-variant system by a piecewise time-invariant system, and then apply any of the calculation methods mentioned in Chapter 4. To be sure, the problem of the numerical accuracy must then be investigated in detail.

To obtain the general solution of the inhomogeneous system (10.1), one also applies here the superposition principle, according to which the general solution (10.4) of the homogeneous system and a particular integral of (10.1) yield together the general solution of (10.1). The result is (cf. (4.42))

$$x(t)=\Phi(t,\ t_0)x_0+\int_{t_0}^{t}\Phi(t,\ \tau)b(\tau)\mathrm{d}\tau. \tag{10.9}$$

For time-invariant systems it was possible, at least for some special excitation functions, to solve the integral (10.9) in a closed form. But in the case of general time-variant systems a closed solution of the integral can be obtained only in the case of impulsive excitations. In all other cases only numerical integration is available for investigating forced vibrations.

10.2. Periodic time-variant systems

A periodic time-variant system is given if the system matrix $A(t)$ in (10.1) has not generally time-variant but periodically time-variant coefficients,

$$A(t)=A(t+T) \tag{10.10}$$

where T is the period. Starting from the Floquet theory, one can obtain from (10.10) far-reaching conclusions concerning the general solution and the fundamental matrix. The periodic time-variant systems are also of greater importance in engineering than the general time-variant systems.

By the theory of Floquet, the fundamental matrix $\Phi(t,\ t_0)$ satisfies the equation

$$\Phi(t+T,\ t_0)=\Phi(t,\ t_0)C, \tag{10.11}$$

where C is a constant nonsingular $n\times n$ matrix. This implies that $\Phi(t+T,\ t_0)$ is, just as $\Phi(t,\ t_0)$, a fundamental matrix of the system (10.3), (10.10). One can show this directly by differentiating (10.11) and taking into account (10.5) and (10.10),

$$\begin{aligned}\dot{\Phi}(t+T,\ t_0)&=\dot{\Phi}(t,\ t_0)C=A(t)\Phi(t,\ t_0)C\\&=A(t+T)\Phi(t+T,\ t_0),\end{aligned} \tag{10.12}$$

If one now introduces the natural $n\times n$ matrix logarithm R,

$$R=\frac{1}{T}\ln C,\quad C=\mathrm{e}^{RT}=\Phi(t_0+T,\ t_0), \tag{10.13}$$

then one can also write the fundamental matrix as

$$\boldsymbol{\Phi}(t, t_0) = \boldsymbol{Z}(t)\, \mathrm{e}^{\boldsymbol{R}(t-t_0)}, \quad \boldsymbol{Z}(t_0) = \boldsymbol{E}. \tag{10.14}$$

Here $\boldsymbol{Z}(t) = \boldsymbol{Z}(t+T)$ is a nonsingular periodic $n \times n$ matrix, as can be easily seen from (10.11), (10.13) and (10.14):

$$\boldsymbol{Z}(t+T)\, \mathrm{e}^{\boldsymbol{R}(t+T-t_0)} = \boldsymbol{Z}(t)\, \mathrm{e}^{\boldsymbol{R}(t+T-t_0)}. \tag{10.15}$$

Thus the fundamental matrix (10.14) can be written as a product of a periodic matrix and an exponential matrix.

By Ljapunov's reducibility theorem, one can reduce by means of the matrix $\boldsymbol{Z}(t)$ every system (10.1) with periodic coefficients (10.10) to a time-invariant system. Under the transformation

$$\boldsymbol{x}(t) = \boldsymbol{Z}(t)\boldsymbol{z}(t) \tag{10.16}$$

(10.1), (10.10) is carried over into

$$\dot{\boldsymbol{z}}(t) = \boldsymbol{R}\boldsymbol{z}(t) + \bar{\boldsymbol{b}}(t) \tag{10.17}$$

with the constant $n \times n$ matrix

$$\boldsymbol{R} = \boldsymbol{Z}^{-1}(t)\boldsymbol{A}(t)\boldsymbol{Z}(t) - \boldsymbol{Z}^{-1}(t)\dot{\boldsymbol{Z}}(t) \tag{10.18}$$

and the $n \times 1$ excitation vector

$$\bar{\boldsymbol{b}}(t) = \boldsymbol{Z}^{-1}(t)\boldsymbol{b}(t). \tag{10.19}$$

Ljapunov's reducibility theorem is mainly of theoretical interest since, as a rule, the transformation matrix $\boldsymbol{Z}(t)$ is not known. Certainly, it is possible to determine the transformation matrix $\boldsymbol{Z}(t)$ by integration over a period from (10.13) and (10.14), but when this information is available, then one also knows the fundamental matrix as will be shown below. Thus (10.17) is of only small practical importance.

Example 10.1. Asymmetric centrifuge

Consider the centrifuge in Problem 3.1 and assume now that its rotor is asymmetric due to an asymmetric mass distribution relative to a bodyfixed system of principal axes of inertia, Fig. 10.1. Then the principal moments of inertia with respect to the x-axis and the y-axis are different: $I_x \neq I_y$. Consequently periodic coefficients appear in the tensor of inertia, and hence also in the inertia matrix $\boldsymbol{M}(t)$ and the velocity matrix $\boldsymbol{P}(t)$,

$$\boldsymbol{M}(t)=\begin{bmatrix} I+U\cos 2\Omega t & U\sin 2\Omega t \\ U\sin 2\Omega t & I-U\cos 2\Omega t \end{bmatrix}, \tag{10.20}$$

$$\boldsymbol{P}(t)=\begin{bmatrix} d-2U\Omega\sin 2\Omega t & 2U\Omega\cos 2\Omega t+I_z\Omega \\ 2U\Omega\cos 2\Omega t-I_z\Omega & d+2U\Omega\sin 2\Omega t \end{bmatrix}, \tag{10.21}$$

while the position matrix

$$\boldsymbol{Q}=\begin{bmatrix} k-mgL & 0 \\ 0 & k-mgL \end{bmatrix}. \tag{10.22}$$

remains unchanged. Here the following abbreviations are introduced:

$$I=\tfrac{1}{2}(I_x+I_y)+mL^2, \quad U=\tfrac{1}{2}(I_x-I_y). \tag{10.23}$$

If the asymmetry U vanishes, then (10.20), (10.21) move to the matrices known from Problem 3.1. The homogeneous state equation (10.3), (10.10) for this example is

$$\underbrace{\begin{bmatrix} \dot\alpha \\ \dot\beta \\ \hline \ddot\alpha \\ \ddot\beta \end{bmatrix}}_{\dot{\boldsymbol{x}}} = \underbrace{\left[\begin{array}{c|c} \boldsymbol{0} & \boldsymbol{E} \\ \hline -\boldsymbol{M}^{-1}(t)\boldsymbol{Q} & -\boldsymbol{M}^{-1}(t)\boldsymbol{P}(t) \end{array}\right]}_{\boldsymbol{A}(t)} \underbrace{\begin{bmatrix} \alpha \\ \beta \\ \hline \dot\alpha \\ \dot\beta \end{bmatrix}}_{\boldsymbol{x}}, \tag{10.24}$$

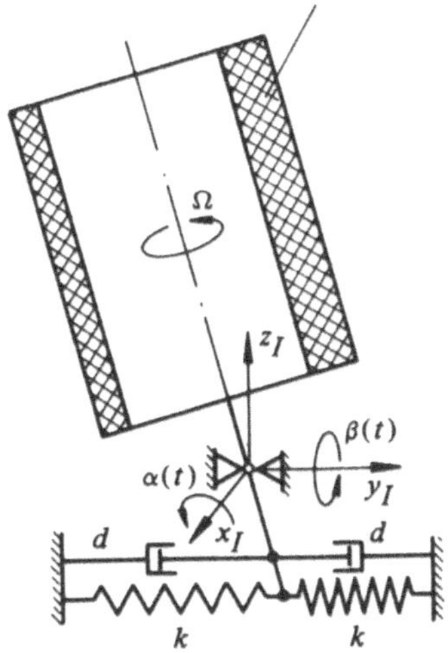

Fig. 10.1. Asymmetric centrifuge

with the periodic system matrix $A(t)=A(t+T)$, $T=2\pi/\Omega$. On the basis of purely kinematic considerations one can now find a periodic transformation matrix $\boldsymbol{Z}(t)$ which satisfies condition (10.18):

$$\boldsymbol{Z}(t)=\boldsymbol{Z}(t+T)=\left[\begin{array}{c|c} \boldsymbol{S}(t) & \mathbf{0} \\ \hline \dot{\boldsymbol{S}}(t) & \boldsymbol{S}(t) \end{array}\right] \tag{10.25}$$

with

$$\boldsymbol{S}(t)=\boldsymbol{S}(t+T)=\begin{bmatrix} \cos\Omega t & -\sin\Omega t \\ \sin\Omega t & \cos\Omega t \end{bmatrix}. \tag{10.26}$$

By (10.26), the transformation (10.16) yields

$$\left.\begin{aligned} \alpha &= z_1\cos\Omega t - z_2\sin\Omega t, \\ \beta &= z_1\sin\Omega t + z_2\cos\Omega t, \\ \dot\alpha &= -z_1\,\Omega\sin\Omega t - z_2\Omega\cos\Omega t + z_3\cos\Omega t - z_4\sin\Omega t, \\ \dot\beta &= z_1\,\Omega\cos\Omega t - z_2\Omega\sin\Omega t + z_3\sin\Omega t + z_4\cos\Omega t, \end{aligned}\right\} \tag{10.27}$$

which expresses the transformation from an inertial system of coordinates to one that rotates with the centrifuge. In particular, the structure of an ordinary mechanical system remains preserved, since (10.27) implies the conditions $\dot z_1 = z_3$ and $\dot z_2 = z_4$.

The relations (10.17) and (10.18) imply the final result

$$\begin{bmatrix} \dot z_1 \\ \dot z_2 \\ \hline \ddot z_1 \\ \ddot z_2 \end{bmatrix} = \left[\begin{array}{cc|cc} 0 & 0 & 1 & 0 \\ 0 & 0 & 0 & 1 \\ \hline -\kappa_x & \Omega\vartheta_x & -\vartheta_x & -w_x \\ -\Omega\vartheta_y & -\kappa_y & w_y & -\vartheta_y \end{array}\right] \begin{bmatrix} z_1 \\ z_2 \\ \hline z_3 \\ z_4 \end{bmatrix}, \tag{10.28}$$

with the abbreviations

$$\left.\begin{aligned} \kappa_x &= \frac{k - mgL + \left(I_z - I_y - mL^2\right)\Omega^2}{I_x + mL^2}, \\ \kappa_y &= \frac{k - mgL + \left(I_z - I_x - mL^2\right)\Omega^2}{I_y + mL^2}, \\ \vartheta_x &= \frac{d}{I_x + mL^2}, \qquad \vartheta_y = \frac{d}{I_y + mL^2}, \\ w_x &= \frac{I_z - I_x - I_y - 2mL^2}{I_x + mL^2}, \qquad w_y = \frac{I_z - I_x - I_y - 2mL^2}{I_y + mL^2}. \end{aligned}\right\} \tag{10.29}$$

Thus in the rotating coordinate system the asymmetric centrifuge has a state equation with time-invariant coefficients. One can explain this by the symmetric suspension, i.e. the rotation-symmetric spring and damping arrangement.

This shows that the Ljapunov reducibility theorem can possess great advantages, provided one can find a suitable transformation matrix $\boldsymbol{Z}(t)$ without solving the equation for the fundamental matrix. ◆

As a rule, the fundamental matrix $\boldsymbol{\Phi}(t, t_0)$ can only be found by integrating (10.5). However for periodically time-variant systems the integration can be restricted to the first period

$$t_0 \leqslant t = t_0 + t^* \leqslant t_0 + T. \tag{10.30}$$

If one describes the running time by the number of periods ν and the time t^* which is bounded according to (10.30),

$$t = t_0 + t^* + \nu T \quad \text{with} \quad \nu \leqslant \frac{t - t_0}{T} < (\nu + 1),\ \nu \text{ integer}, \tag{10.31}$$

then (10.11) and (10.13) imply the following equation for the fundamental matrix:

$$\boldsymbol{\Phi}(t, t_0) = \boldsymbol{\Phi}(t_0 + t^* + \nu T, t_0) = \boldsymbol{\Phi}(t_0 + t^*, t_0)\boldsymbol{\Phi}^{\nu}(t_0 + T, t_0). \tag{10.32}$$

Due to (10.32), the calculation of the fundamental matrix is reduced to an *integration* of (10.5) *over one period* and *taking* powers of the constant matrix $\boldsymbol{\Phi}(t_0 + T, t_0)$. This certainly implies a large numerical advantage. The integration can be carried out by an analog computer as well as by a digital computer. But the matrix powers can be calculated only by a digital computer. Thus for the evaluation of (10.32) one needs either a digital computer or a hybrid computer allowing the combination of analog and digital calculations.

The matrix powers in (10.32) can be determined either by consecutive *matrix multiplication* or by solving the *eigenvalue problem*

$$\boldsymbol{x}(t + T) = \mu \boldsymbol{x}(t), \qquad \boldsymbol{x}(t) = \boldsymbol{\Phi}(t, t_0)\boldsymbol{x}_0. \tag{10.33}$$

The eigenvalue problem (10.33) describes how the state vector changes during one period. If one substitutes (10.32) with $\nu = 1$ in (10.33), one gets

$$[\mu \boldsymbol{E} - \boldsymbol{\Phi}(t_0 + T, t_0)]\tilde{\boldsymbol{x}}_0 = \boldsymbol{0}. \tag{10.34}$$

As was described at length in Chapter 4, one obtains as a result of solving the eigenvalue problem (10.34) the eigenvalues μ_i, $i = 1(1)n$, and the eigen and principal vectors $\tilde{\boldsymbol{x}}_{0i} = \tilde{\boldsymbol{x}}_0(\mu_i)$. The eigenvectors can be arranged into an $n \times n$ modal matrix

$$\boldsymbol{X} = [\tilde{\boldsymbol{x}}_{01} \vdots \tilde{\boldsymbol{x}}_{02} \vdots \dots \vdots \tilde{\boldsymbol{x}}_{0n}]. \tag{10.35}$$

A similarity transformation $x(t) = Xx(t)$ with the modal matrix (10.35) leads to the Jordan canonical form of the matrix $\boldsymbol{\Phi}(t_0 + T, t_0)$:

$$\boldsymbol{J} = \boldsymbol{X}^{-1}\boldsymbol{\Phi}(t_0 + T, t_0)\boldsymbol{X}. \tag{10.36}$$

The Jordan matrix is of a digital form,

$$\boldsymbol{J} = \text{diag}\{\mu_i\}, \qquad i = 1(1)n, \tag{10.37}$$

if for every eigenvalue μ_i with multiplicity v_i the nullity d_i of the matrix $[\mu_i\boldsymbol{E} - \boldsymbol{\Phi}(t_0 + T, t_0)]$ is also v_i. On the other hand, if the nullity satisfies $1 \leqslant d_i \leqslant v_i$, then the Jordan matrix is built up of the $r_{ik} \times r_{ik}$ submatrices

$$\boldsymbol{J}_{ik} = \begin{bmatrix} \mu_i & 1 & 0 & \dots & 0 \\ 0 & \mu_i & 1 & \dots & 0 \\ 0 & 0 & \mu_i & \dots & 0 \\ \vdots & \vdots & \vdots & & \vdots \\ 0 & 0 & 0 & \dots & \mu_i \end{bmatrix} \tag{10.38}$$

as was explained in detail in Section 4.3. By (10.36), the matrix powers in (10.32) satisfy

$$\boldsymbol{\Phi}^{\nu}(t_0 + T, t_0) = \boldsymbol{X}\boldsymbol{J}^{\nu}\boldsymbol{X}^{-1}, \tag{10.39}$$

i.e., no powers of the modal matrix need be calculated. It follows from (10.37) that the powers of the diagonal matrix are given by

$$\boldsymbol{J}^{\nu} = \text{diag}\{\mu_i^{\nu}\}, \qquad i = 1(1)n, \tag{10.40}$$

and the powers of the $r_{ik} \times r_{ik}$ submatrix of the Jordan matrix are

$$\boldsymbol{J}_{ik}^{\nu} = \begin{bmatrix} \mu_i^{\nu} & \dfrac{\nu}{1!}\mu_i^{\nu-1} & \dfrac{\nu(\nu-1)}{2!}\mu_i^{\nu-2} & \dots & \dfrac{\prod\limits_{m=1}^{r_{ik}-1}(\nu - m + 1)}{(r_{ik}-1)!}\mu^{(\nu - r_{ik} + 1)} \\ 0 & \mu_i^{\nu} & \dfrac{\nu}{1!}\mu_i^{\nu-1} & \dots & \dfrac{\prod\limits_{m=1}^{r_{ik}-2}(\nu - m + 1)}{(r_{ik}-2)!}\mu^{(\nu - r_{ik} + 2)} \\ 0 & 0 & \mu_i^{\nu} & \dots & \dfrac{\prod\limits_{m=1}^{r_{ik}-3}(\nu - m + 1)}{(r_{ik}-3)!}\mu^{(\nu - r_{ik} + 3)} \\ \vdots & \vdots & \vdots & & \vdots \\ 0 & 0 & 0 & \dots & \mu_i^{\nu} \end{bmatrix} \tag{10.41}$$

The calculation of matrix powers according to (10.39), as compared to matrix multiplication, is especially interesting for higher numbers of period.

10.3. Stability of periodic time-variant systems

The stability definitions of Section 5.1 also apply to the present case of time-variant systems. The time-invariance of the system was not among the essential assumptions in the stability definitions. Accordingly, the stability of the homogeneous periodic time-variant vibrating system (10.3), (10.10) is determined by the behaviour of the general solution

$$\boldsymbol{x}(t_0 + t^* + \nu T) = \boldsymbol{\Phi}(t_0 + t^*, t_0)\boldsymbol{X}\boldsymbol{J}^\nu \boldsymbol{X}^{-1}\boldsymbol{x}_0 \tag{10.42}$$

for arbitrary initial conditions and for all times $t \geqslant t_0$, i.e., $\nu \geqslant 0$. Since the fundamental matrix is bounded and regular for $t_0 \leqslant t^* \leqslant t_0 + T$ and the modal matrix $\boldsymbol{X}$ is a constant regular matrix, the stability can depend only on the powers of the Jordan matrix $\boldsymbol{J}^\nu$. By Section 10.2, $\boldsymbol{J}^\nu$ is either a diagonal matrix ($d_i = v_i$ for all eigenvalues μ_i) or a Jordan matrix ($d_i < v_i$ for at least one eigenvalue) with at least one submatrix of the form (10.41). Taking now $t \to \infty$ or $\nu \to \infty$ one obtains the following stability conditions in terms of the eigenvalues μ_i of the matrix $\boldsymbol{\Phi}(t_0 + T, t_0)$:

The periodic system (10.3), (10.10) is precisely

asymptotically stable when all eigenvalues of $\boldsymbol{\Phi}(t_0 + T, t_0)$ have magnitudes smaller than one

$$|\mu_i| < 1, \qquad i = 1(1)n;$$

marginally stable when all eigenvalues of $\boldsymbol{\Phi}(t_0 + T, t_0)$ have magnitudes not exceeding one, at least one eigenvalue has magnitude one and for each of the eigenvalues μ_j with magnitude one the multiplicity v_j equals the nullity d_j of the corresponding characteristic matrix:

$$|\mu_i| \leqslant 1,\ i = 1(1)n \text{ and } |\mu_j| = 1 \text{ with } d_j = v_j;$$

unstable when at least one eigenvalue of $\boldsymbol{\Phi}(t_0 + T, t_0)$ has magnitude greater than one, or there is at least one multiple eigenvalue with magnitude one whose multiplicity v_j exceeds the defect d_j of the corresponding characteristic matrix:

$$|\mu_i| > 1 \quad \text{or} \quad |\mu_j| = 1 \quad \text{with} \quad d_j < v_j.$$

The above conditions simplify for single eigenvalues $v_j = d_j = 1$, since in this case the nullity of the characteristic matrix has no effect.

These stability conditions are closely related to the stability conditions for time-invariant systems given in Section 5.2.1. The connection is provided by (10.13) since the eigenvalues also satisfy the equations corresponding to the matrrix relations:

$$\lambda_i = \frac{1}{T}\ln \mu_i, \quad \mu_i = e^{\lambda_i T}, \quad i = 1(1)n, \tag{10.43}$$

where λ_i are the eigenvalues of the matrix $\boldsymbol{R}$. As in the time-invariant case, the stability analysis can also be carried out with the aid of matrix equations. For periodic time-variant systems Stein's matrix equation takes over the fundamental role which was played by Ljapunov's equation in Section 5.2.3. However Stein's equation does not overcome the need for integrating the fundamental matrix.

In contrast to the theory for time-invariant systems in Section 5.2.2, it is not possible to give for time-variant systems exact stability criteria that would not require integration and the calculation of eigenvalues. Only for weak periodic systems

$$\dot{\boldsymbol{x}}(t) = (\boldsymbol{A}_0 + \epsilon \boldsymbol{A}_1(t))\boldsymbol{x}(t), \quad \boldsymbol{A}_1(t) = \boldsymbol{A}_1(t+T), \tag{10.44}$$

one can apply the following theorem due to Ljapunov:

If $\boldsymbol{A}_0$ is a constant $n \times n$ matrix such that all its eigenvalues have negative real parts, Re $\lambda_i < 0$, $i = 1(1)n$, and $\boldsymbol{A}_1(t)$ is a continuous periodic $n \times n$ matrix, then there exists a positive real constant ϵ_0 such that $0 < \epsilon < \epsilon_0$ implies the asymptotic stability of all solutions of (10.44).

Further, numerous approximate stability criteria have been developed but they all suffer the disadvantage of limited applicability and, frequently, are valid only for single-dof-systems. Hence they will be no presented here.

Example 10.2. Pendulum with periodically moving point of suspension

For small displacements $y(t)$ the motion of the periodically accelerated pendulum, Fig. 10.2, is described by the equation

$$m\ddot{y}(t) + d\dot{y}(t) + \left(\frac{mg}{L} + \frac{ma\Omega^2}{L}\cos \Omega t\right) y(t) = 0, \tag{10.45}$$

where d is the coefficient of the viscous friction in the bearing. Hence the state equation is

$$\underbrace{\begin{bmatrix} \dot{y} \\ \ddot{y} \end{bmatrix}}_{\dot{\boldsymbol{x}}(t)} = \underbrace{\begin{bmatrix} 0 & 1 \\ -\dfrac{g}{L} - \dfrac{a\Omega^2}{L}\cos \Omega t & -\dfrac{d}{m} \end{bmatrix}}_{\boldsymbol{A}(t)} \underbrace{\begin{bmatrix} y \\ \dot{y} \end{bmatrix}}_{\boldsymbol{x}(t)}, \tag{10.46}$$

the system matrix being periodic, $A(t) = A(t) = A(t + 2\pi/\Omega)$. To determine the fundamental matrix for $\Omega = 1$; $g/L = 0{,}5$; $a\Omega^2/L = 0{,}5$; $d/m = 0{,}5$ at $t_0 = 0$ an integration over the period $T = 2\pi/\Omega$ was carried out by an computer yielding

$$\boldsymbol{\Phi}(T, 0) = \begin{bmatrix} -0.233 & 0.037 \\ 0.425 & -0.253 \end{bmatrix}. \tag{10.47}$$

For the stability analysis the eigenvalues of the matrix (10.47) need to be known. The characteristic equation

$$\det[\mu \boldsymbol{E} - \boldsymbol{\Phi}(T, 0)] = \mu^2 + 0.486\mu + 0.043 = 0 \tag{10.48}$$

leads to the eigenvalues $\mu_1 = -0.369$ and $\mu_2 = -0.117$ which are both smaller than one in absolute value. Hence the equilibrium position of the pendulum is, for the given numerical values, asymptotically stable.

The periodically accelerated pendulum can also be viewed as a weak periodic system (10.44). Then (10.46) implies

$$\boldsymbol{A}_0 = \begin{bmatrix} 0 & 1 \\ -\dfrac{g}{L} & -\dfrac{d}{m} \end{bmatrix}, \quad \boldsymbol{A}_1(t) = \begin{bmatrix} 0 & 0 \\ -\dfrac{\Omega^2}{L}\cos \Omega t & 0 \end{bmatrix}, \quad \epsilon = a. \tag{10.49}$$

For $g/L = 0.5$ and $d/m = 0.5$ the constant matrix $\boldsymbol{A}_0$ has the eigenvalues $\lambda_{1,2} = -0.25 \pm 0.66\mathrm{i}$ which have both negative real parts. Hence, according to the stability theorem for weak periodic systems, the periodically accelerated pendulum is also asymptotically stable for all excitation amplitudes $a < L/2$ at the given excitation frequency $\Omega = 1$. ◆

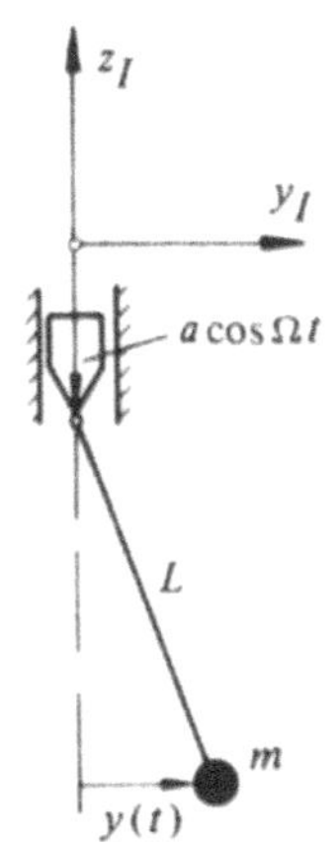

Fig. 10.2. Periodically accelerated pendulum

In Section 5.3 a connection was found between general boundedness and asymptotic stability of time-invariant systems. This connection remains valid for time-variant systems. The following theorem holds:

The inhomogeneous system (10.1, 10.10) is generally bounded if and only if the system (10.3, 10.10) is asymptotically stable.

For the connection between boundedness and stability the following theorem remains valid:

The inhomogeneous system (10.1) is bounded relative to decreasing functions $\int_0^\infty \|\boldsymbol{b}(t)\| \mathrm{d}t < \infty$ if the homogeneous system (10.3) is marginally stable.

No general statements about boundedness can be made relative to other functions, e.g., periodic excitation functions $\boldsymbol{b}(t) = \boldsymbol{b}(t+T)$.

10.4. Mechanical systems

One obtains a good insight into the stability behavior of mechanical vibrating systems by starting from Mathieu's differential equation for the periodically accelerated pendulum and considering single- and multi-dof-vibrations. One can then recogize that instabilities are often associated with certain relations between the frequency of periodic parameters and the natural frequency. Thus one also speaks of combination frequencies leading to parameterresonance.

10.4.1. Mathieu's differential equation

Equations of single-dof-vibrating systems with periodic coefficients, e.g., as in the case of a periodically accelerated pendulum, can often be reduced to Mathieu's differential equation

$$\ddot{y}(t) + 2D\dot{y}(t) + (\beta^2 + \gamma \cos t)y(t) = 0, \tag{10.50}$$

where D is the constant damping factor, β is a (real or imaginary) frequency ratio, and γ is the intensity of the periodic parameter.

The stability behavior of Mathieu's equation has been extensively investigated in the literature. The results of Ince, Strutt, Klotter and Kotowski have been collected in the stability chart in Fig. 10.3. The stability regions are plainly visible. For $D = 0$ one has either stability or instability, while for $D > 0$ there is asymptotic stability or instability. Further, for weak periodic undamped systems the stability boundaries can be written explicitly:

$$\beta_1^2 = 0, \quad \beta_{2,3}^2 = \tfrac{1}{4} \mp \tfrac{1}{2}\gamma, \quad \beta_{4,5}^2 = 1, \quad \ldots \quad \beta_{2k,2k+1}^2\left(\frac{k}{2}\right)^2, \quad k = 2, 3, 4, \ldots. \tag{10.51}$$

By (10.41), the frequency condition for instability or parameter resonance is

$$2\beta = m, \quad m = 0, 1, 2, \ldots. \tag{10.52}$$

However one can see from Fig. 10.3 that for small periodic parameters merely the case $m = 1$ or $\beta = 1/2$ is critical, i.e., (10.52) is a necessary condition for *parameter resonance*.

For statically stable systems, $\beta^2 > 0$, the periodic parameters diminish the region of stability. For statically unstable systems, $\beta^2 < 0$, periodic parameters can cause stabilization of the vibrations, Fig. 10.3. It should be observed here that for $\beta^2 < 0$, $D > 0$ one gets even asymptotic stability.

Example 10.3. Periodically accelerated pendulum with normalized equation of motion

By a suitable normalization of amplitude and time the equation of motion of the periodically accelerated pendulum can be transformed in Mathieu's differential equation (10.50). For this purpose one introduces the dimensionless time $\tau = \Omega t$, and one divides (10.45) by $m\Omega^2$. One then obtains

$$D = \frac{d}{2m\Omega}, \quad \beta^2 = \frac{g}{L\Omega^2}, \quad \gamma = \frac{a}{L} \tag{10.53}$$

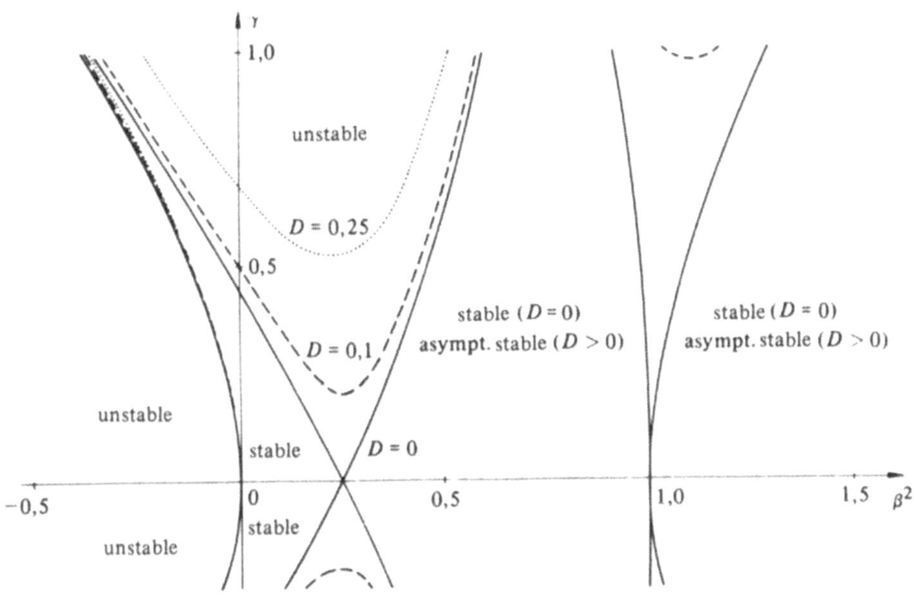

Fig. 10.3. Stability chart for Mathieu's differential equation

or, with the numerical data from Example 10.2, $D = 0.25$; $\beta^2 = 0.5$; $\gamma = 0.5$. Looking now at the stability chart in Fig. 10.3, one can again see that for these values the system is asymptotically stable.

However for the values $D = 0.25$; $\beta^2 = 0.3$; $\gamma = 1.0$ the statically stable position of equilibrium becomes unstable, while for $D = 0.25$; $\beta^2 = -0.3$; $\gamma = 1.0$ the statically unstable position of equilibrium becomes asymptotically stable, Fig. 10.4. These interesting properties can not only be checked by calculation but can also be easily verified by experiments. ◆

10.4.2. Single-degree-of-freedom mechanical systems

A single-dof-mechanical vibrating system with periodic parameters, normalized with mass $m(t) = m(t + T)$, has the equation of motion

$$\ddot{y}(t) + p(t)\dot{y}(t) + q(t)y(t) = 0 \tag{10.54}$$

with the coefficients

$$p(t) = p(t + T), \quad q(t) = q(t + T) \tag{10.55}$$

and period T. The stability of the position of equilibrium of the system (10.54), (10.55) is determined by the matrix

$$\boldsymbol{\Phi}(T, 0) = \begin{bmatrix} y_1(T) & y_2(T) \\ \dot{y}_1(T) & \dot{y}_2(T) \end{bmatrix} = \boldsymbol{C} \tag{10.56}$$

which is obtained by twice integrating (10.54) with the initial conditions

$$\left.\begin{aligned} y_1(0) = 1, \quad y_2(0) = 0, \\ \dot{y}_1(0) = 0, \quad \dot{y}_2(0) = 1 \end{aligned}\right\} \tag{10.57}$$

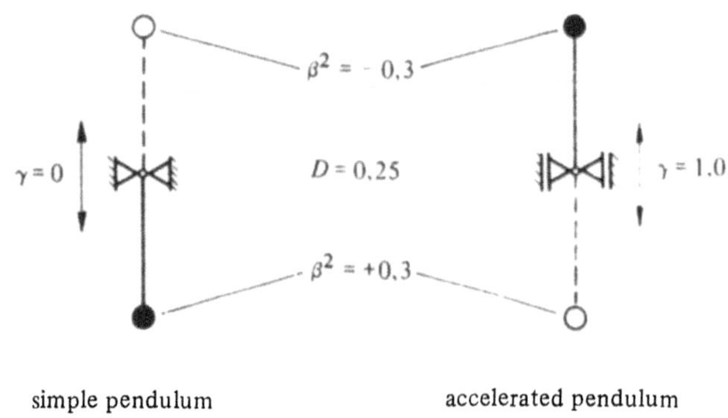

Fig. 10.4. Stability of the positions of equilibrium: ● stable, ○ unstable

and the initial time $t_0 = 0$. The characteristic equation is

$$\det[\mu \boldsymbol{E} - \boldsymbol{\Phi}(T, 0)] = \mu^2 + 2B\mu + \det \boldsymbol{\Phi}(T, 0) = 0, \tag{10.58}$$

where, due to (10.8), the abbreviations

$$2B = y_1(T) + \dot{y}_2(T) = \mathrm{Tr}\ \Phi(T, 0), \tag{10.59}$$

$$\det \boldsymbol{\Phi}(T, 0) = y_1(T)\dot{y}_2(T) - \dot{y}_1(T)y_2(T) = \exp \int_0^T (-p(\tau))\mathrm{d}\tau \tag{10.60}$$

were used. The eigenvalues of (10.58) are located either on the circle in the complex plaine with radius $\sqrt{\det \boldsymbol{\Phi}(T, 0)}$ and center 0, or one the real axis, Fig. 10.5. An analysis of (10.58) yields the following values for the maximal eigenvalues, decisive for stability:

$$\left.\begin{aligned} |\mu|_{\max} &= \sqrt{\det \boldsymbol{\Phi}(T, 0)} && \text{for} \quad B^2 \leqslant \det \boldsymbol{\Phi}(T, 0), \\ |\mu|_{\max} &= |B| + \sqrt{B^2 - \det \boldsymbol{\Phi}(T, 0)} > \det \boldsymbol{\Phi}(T, 0) \\ & && \text{for} \quad B^2 > \det \boldsymbol{\Phi}(T, 0). \end{aligned}\right\} \tag{10.61}$$

This leads to the stability conditions given in Table 10.1.

One can see that asymptotic stability is possible only when there is positive damping in the mean. In the undamped case, marginally stable vibrations can occur, while damping that is negative in the mean always

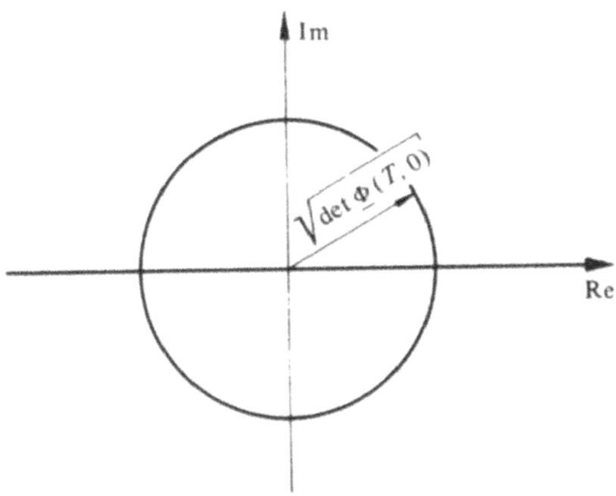

Fig. 10.5. Location of the eigenvalues of the matrix $\boldsymbol{\Phi}(T, 0)$ in the complex plane

yields unstable behavior. Thus Table 10.1 offer a good survey of the vibrational behavior of a single-dof-mechanical system with periodic parameters.

10.4.3. Multi-degree-of-freedom mechanical systems

It is not possible any more to make general statements concerning the vibrational phenomena for multi-dof-mechanical systems with periodic parameters. However, under restrictive assumptions a theorem of Cesari is known which permits certain sufficient conditions for instability or parameter resonance. This theorem is as follows:

A mechanical system of the form

$$\boldsymbol{M}\ddot{\boldsymbol{y}}(t)+\boldsymbol{K}_0\boldsymbol{y}(t)+\epsilon\boldsymbol{K}_1(t)\boldsymbol{y}(t)=\boldsymbol{0} \tag{10.62}$$

with a constant $f\times f$ inertia matrix $\boldsymbol{M}$, a constant $f\times f$ matrix $\boldsymbol{K}_0$, and an even periodic $f\times f$ excitation matrix

$$\boldsymbol{K}_1(t)=\boldsymbol{K}_1(-t)=\boldsymbol{K}_1(t+T) \tag{10.63}$$

has for sufficiently small ϵ marginally stable solutions if

$$\left.\begin{array}{l} m\omega\neq 2\omega_i,\quad m\omega\neq\omega_i\pm\omega_j,\\ i\neq j,\quad i,\ j=1(1)f,\quad m=0,1,2,\ldots \end{array}\right\} \tag{10.64}$$

holds. Here ω_i, $i=1(1)f$, are the natural frequencies of the system

$$\boldsymbol{M}\ddot{\boldsymbol{y}}(t)+\boldsymbol{K}_0\boldsymbol{y}(t)=\boldsymbol{0} \tag{10.65}$$

and $\omega=2\pi/T$ is the frequency of the periodic excitation.

In addition to the first necessary condition for *parameter resonance*

$$m\omega=2\omega_i,\quad m=0,1,2,\ldots, \tag{10.66}$$

Table 10.1. Stability conditions for a single-dof-mechanical vibrating system

damping	asymptotic stable	stable	unstable
$\int_0^T p(\tau)\mathrm{d}\tau>0$	$\lvert B\rvert<\frac{1}{2}(1+\det \boldsymbol{C})$	$\lvert B\rvert=\frac{1}{2}(1+\det \boldsymbol{C})$	$\lvert \boldsymbol{B}\rvert>\frac{1}{2}(1+\det \boldsymbol{C})$
$\int_0^T p(\tau)\mathrm{d}\tau=0$	–	$\lvert B\rvert<1$	$\lvert B\rvert\geqslant 1$
$\int_0^T p(\tau)\mathrm{d}\tau<0$	–	–	B arbitrary

which already appeared in (10.52) for the single-dof-vibrating system, one finds from (10.64) a second necessary frequency condition

$$m\omega = \omega_i \pm \omega_j, \quad m = 0, 1, 2, \ldots, \tag{10.67}$$

describing the combination frequencies. One can only verify by a more detailed investigation, according to Section 10.3, whether the combination frequencies are indeed critical frequencies, leading to instabilities. Nevertheless, for systems with periodic parameters one will attempt to satisfy the frequency conditions (10.64), i.e., to avoid (10.66) and (10.67).

Problem 10.1. State vectors of the periodically accelerated pendulum

The fundamental matrix of the periodically accelerated pendulum, Example 10.2, was determined by a computer for the time $t = 2\pi$ and the initial time $t_0 = 0$ as follows:

$$\boldsymbol{\Phi}(2\pi, 0) = \begin{bmatrix} -0.232 & 0.037 \\ 0.425 & -0.251 \end{bmatrix}. \tag{10.68}$$

What are the state vectors $\boldsymbol{x}(2\pi)$ for the following initial conditions $\boldsymbol{x}(0)$?

1.

$$\boldsymbol{x}(0) = \begin{bmatrix} 1 \\ 0.2 \end{bmatrix}:$$

□ $\boldsymbol{x}(2\pi) = \begin{bmatrix} -0.224 \\ 0.088 \end{bmatrix}$, □ $\boldsymbol{x}(2\pi) = \begin{bmatrix} -0.224 \\ 0.375 \end{bmatrix}$, □ $\boldsymbol{x}(2\pi) = \begin{bmatrix} -0.098 \\ 0.375 \end{bmatrix}$.

2.

$$\boldsymbol{x}(0) = \begin{bmatrix} 0.5 \\ 0.5 \end{bmatrix}:$$

□ $\boldsymbol{x}(2\pi) = \begin{bmatrix} -0.224 \\ 0.425 \end{bmatrix}$, □ $\boldsymbol{x}(2\pi) = \begin{bmatrix} -0.100 \\ -0.255 \end{bmatrix}$, □ $\boldsymbol{x}(2\pi) = \begin{bmatrix} -0.098 \\ 0.088 \end{bmatrix}$.

3.

$$\boldsymbol{x}(0) = \begin{bmatrix} -0.3 \\ 0.7 \end{bmatrix}:$$

□ $\boldsymbol{x}(2\pi) = \begin{bmatrix} 0.094 \\ -0.303 \end{bmatrix}$, □ $\boldsymbol{x}(2\pi) = \begin{bmatrix} 0.944 \\ -0.303 \end{bmatrix}$, □ $\boldsymbol{x}(2\pi) = \begin{bmatrix} 0.944 \\ 0.303 \end{bmatrix}$.

4.

$$x(0)=\begin{bmatrix}-0.4\\-0.9\end{bmatrix}:$$

$\square\; x(2\pi)=\begin{bmatrix}0.604\\0.303\end{bmatrix}$, $\square\; x(2\pi)=\begin{bmatrix}0.944\\-0.055\end{bmatrix}$, $\square\; x(2\pi)=\begin{bmatrix}0.060\\0.055\end{bmatrix}$. ◆

Problem 10.2. Damping constant of the periodically accelerated pendulum

According to Example 10.3, the state equation in normalized form of the periodically accelerated pendulum is

$$\begin{bmatrix}\dot{x}_1\\\dot{x}_2\end{bmatrix}=\begin{bmatrix}0 & 1\\-\beta^2-\gamma\cos t & -2D\end{bmatrix}\begin{bmatrix}x_1\\x_2\end{bmatrix}. \tag{10.69}$$

The fundamental matrix was integrated for $\beta^2=\gamma=0.5$ by a computer. The result for $t=T=2\pi$ is

$$\Phi(T,0)=\begin{bmatrix}-0.232 & 0.037\\0.425 & -0.253\end{bmatrix}. \tag{10.70}$$

1. What is the numerical evaluation of $\Phi(2T, 0)$?

$\square\; \Phi(2T,0)=\begin{bmatrix}-0.453 & 0.565\\-0.255 & -0.106\end{bmatrix}$,

$\square\; \Phi(2T,0)=\begin{bmatrix}-0.025 & 0.007\\0.081 & -0.028\end{bmatrix}$,

$\square\; \Phi(2T,0)=\begin{bmatrix}0.070 & -0.018\\-0.207 & 0.080\end{bmatrix}$.

2. Which formula determines the parameter D of the system (10.69)?

$\square\; D=\dfrac{-\ln\det\Phi(T,0)}{2\pi}$,

$\square\; D=\dfrac{-\ln\det\Phi(2T,0)}{2\pi}$,

$\square\; D=\dfrac{-\ln\det\Phi(T,0)}{4\pi}$.

Hint: Relation (10.8) should be applied.

3. Which numerical value D was used in the calculation of the fundamental matrix $\boldsymbol{\Phi}(T, 0)$?

☐ $D = 1.000$,

☐ $D = 0.500$,

☐ $D = 0.250$,

☐ $D = 0.125$,

☐ $D = 0$.

Hint:

$$\ln 0.0019 = -2\pi, \quad \ln 0.2079 = -\frac{\pi}{2},$$

$$\ln 0.0432 = -\pi, \quad \ln 1{,}000 = 0.$$ ◆

Problem 10.3. Stability of a periodic time-variant system

1. Are the linear systems with periodic parameters and the following eigenvalues of the matrix $\boldsymbol{\Phi}(T, 0)$

	asympt. stable,	stable,	un-stable
a) $\mu_{1,2} = 1;\ \mu_{3,4} = 0.5 \pm 0.7\mathrm{i};\ d_{1,2} = 1$	☐	☐	☐
b) $\mu_1 = -10;\ \mu_2 = -1;\ \mu_3 = -0.5;\ \mu_4 = -0.3$	☐	☐	☐
c) $\mu_{1,2} = -1;\ \mu_{3,4} = -0.7 \pm 0.8\mathrm{i};\ d_{1,2} = 2$	☐	☐	☐
d) $\mu_{1,2,3,4} = 0.5;\ d_{1,2,3,4} = 2$	☐	☐	☐
e) $\mu_{1,2} = 0.9 \pm 0.1\mathrm{i};\ \mu_{3,4} = -0.2 \pm 0.8\mathrm{i}$	☐	☐	☐
f) $\mu_{1,2} = 0.8 \pm 0.6\mathrm{i};\ \mu_3 = 0.9;\ \mu_4 = 0.1$	☐	☐	☐
g) $\mu_{1,2} = \pm 0.9;\ \mu_{3,4} = \pm 0.4$	☐	☐	☐
h) $\mu_{1,2,3,4} = 1;\ d_{1,2,3,4} = 4$	☐	☐	☐
i) $\mu_{1,2} = -0.5 \pm 0.8\mathrm{i};\ \mu_{3,4} = 0.7 \pm 0.9;$	☐	☐	☐
k) $\mu_{1,2} = 1;\ \mu_3 = 0.5;\ \mu_4 = 0.1;\ \mathrm{d}_{1,2} = 2$	☐	☐	☐

2. The linear system with periodic coefficients

$$\dot{\boldsymbol{x}}(t) = (\boldsymbol{A}_0 + \epsilon \boldsymbol{A}_1(t))\boldsymbol{x}(t), \tag{10.71}$$

is asymptotically stable for

$$\boldsymbol{A}_0 = \begin{bmatrix} 0 & 1 \\ -0{,}5 & -0{,}5 \end{bmatrix}, \quad \boldsymbol{A}_1 = \begin{bmatrix} 0 & 0 \\ -0{,}5\cos t & 0 \end{bmatrix}, \quad \epsilon = 1{,}4. \tag{10.72}$$

Which of the following systems are definitely asymptotically stable?

□ $A_0 = \begin{bmatrix} 0 & 1 \\ -0,5 & -1 \end{bmatrix}$, $A_1 = \begin{bmatrix} 0 & 0 \\ -0,5 \cos t & 0 \end{bmatrix}$, $\epsilon = 1,0$;

□ $A_0 = \begin{bmatrix} 0 & 1 \\ -0,5 & -0,5 \end{bmatrix}$, $A_1 = \begin{bmatrix} 0 & 0 \\ -0,5 \cos t & 0 \end{bmatrix}$, $\epsilon = 1,0$;

□ $A_0 = \begin{bmatrix} 0 & 1 \\ -0,5 & -0,5 \end{bmatrix}$, $A_1 = \begin{bmatrix} 0 & 0 \\ -\cos t & 0 \end{bmatrix}$, $\epsilon = 1,0$;

□ $A_0 = \begin{bmatrix} 1 & 1 \\ -0,5 & -0,5 \end{bmatrix}$, $A_1 = \begin{bmatrix} 0 & 0 \\ -0,5 \cos t & -0,5 \cos t \end{bmatrix}$, $\epsilon = 1,0$. ◆

Problem 10.4. Stability of a satellite

For small pitch angles the equation of motion of a pulsating gravitationally stabilized satellite on an elliptic trajectory is

$$\ddot{\theta}(t) - \frac{3k}{1 + e \cos t}\theta(t) = 0, \quad \theta(t) \ll 1. \tag{10.73}$$

1. What are the coefficients of the equivalent Mathieu's differential equation for small eccentricities $e \ll 1$?

   □ $\beta^2 = 3k, \quad \gamma = 3ke$;

   □ $\beta^2 = -3k, \quad \gamma = 3ke$;

   □ $\beta^2 = -3k, \quad \gamma = -3ke$.

2. Which stability boundaries are obtained for $e \ll 1$ in the region $-1 \leqslant k \leqslant 1$?

   □ $k_1 = 0, \quad k_{2,3} = \frac{1}{12}, \quad k_{4,5} = \frac{1}{3}, \quad k_{6,7} = \frac{3}{4}$;
   □ $k_1 = 0, \quad k_{2,3} = -\frac{1}{12}, \quad k_{4,3} = -\frac{1}{3}, \quad k_{6,7} = -\frac{3}{4}$;
   □ $k_1 = 0, \quad k_{2,3} = -\frac{1}{12} \mp \frac{1}{24}e, \quad k_{4,5} = -\frac{1}{3}, \quad k_{6,7} = -\frac{3}{4}$;
   □ $k_1 = 0, \quad k_{2,3} = -\frac{1}{12} \pm \frac{1}{24}e, \quad k_{4,5} = -\frac{1}{3}, \quad k_{6,7} = -\frac{3}{4}$.

   Hint: (10.51) should be used to derive the stability boundaries. ◆

11. Parametrically excited and forced vibrations

Parametrically excited vibrations occur in systems whose coefficients vary with time, mostly periodically. The presence of periodic coefficients influences the energy balance of the system. Parametrically excited vibrations can be damped, undamped, or dedamped. Further, one distinguishes between parametrically excited and forced parametrically excited vibrations. The corresponding important excitation functions are the impulse function, the step function, periodic excitation, and stochastic excitation.

11.1. Parametrically excited vibrations

The parametrically excited vibrations of a system with periodic coefficients

$$\dot{\boldsymbol{x}}(t) = \boldsymbol{A}(t)\boldsymbol{x}(t), \quad \boldsymbol{A}(t) = \boldsymbol{A}(t+T), \quad \boldsymbol{x}(t_0) = \boldsymbol{x}_0, \tag{11.1}$$

are determined by the general solution

$$\boldsymbol{x}(t) = \boldsymbol{x}(t^* + \nu T) = \boldsymbol{\Phi}(t_0 + t^*, t_0)\boldsymbol{\Phi}^\nu(t_0 + T, t_0)\boldsymbol{x}_0 \tag{11.2}$$

which was discussed in detail in Section 10.2. The steady-state behavior as $t \to \infty$ is described by the stability assertions given in Section 10.3. The parametrically excited undamped vibrations of a marginally stable periodic system are as a rule, *nonperiodic* processes. Periodic vibrations can occur only in systems whose eigenvalues satisfy certain specific conditions. By (10.33), for a single or multiple eigenvalue $\mu = 1$ there exists at least a one-periodic vibration $\boldsymbol{x}(t) = \boldsymbol{X}(t+T)$. For a simple or multiple eigenvalue $\mu = -1$, there exists at least a half-periodic vibration $\boldsymbol{x}(t) = \boldsymbol{x}(t+2T)$. These periodic parametrically excited vibrations arise precisely when the initial conditions coincide with the corresponding eigenvectors $\tilde{\boldsymbol{x}}_0$, according to (10.34). In general, to the pair of complex conjugate eigenvalues

$$\mu = \cos 2\pi k \pm \mathrm{i} \sin 2\pi k, \quad k = \tfrac{1}{1}, \tfrac{1}{2}, \tfrac{1}{3}, \ldots, \tag{11.3}$$

there belongs a k-periodic vibration. The initial conditions for the k-periodic vibrations can then be obtained from the eigenvalue problem

$$\left[\boldsymbol{E}-\boldsymbol{\Phi}^{1/k}(t_0+T,\, t_0)\right]\tilde{\boldsymbol{x}}_0=\boldsymbol{0}. \tag{11.4}$$

While for marginally stable time-invariant systems, free harmonic vibrations can be generated by a suitable excitation of the natural modes, for marginally stable periodic time-variant systems periodic vibrations can only be generated by suitable excitation if the eigenvalues satisfy condition (11.3).

11.2. Impulsive excitation

The state equation of a periodic time-variant vibrating system with impulsive excitation is

$$\dot{\boldsymbol{x}}(t)=\boldsymbol{A}(t)\boldsymbol{x}(t)+\boldsymbol{b}_{\mathrm{I}}\delta(t-t_{\mathrm{I}}),\quad \boldsymbol{A}(t)=\boldsymbol{A}(t+T),\quad \boldsymbol{x}(t_0)=\boldsymbol{x}_0, \tag{11.5}$$

where $\boldsymbol{b}_{\mathrm{I}}\delta(t-t_{\mathrm{I}})$ is the impulse function (7.1) and t_{I} is the instant of the impulse. Thus the general solution implies the *impulsive response*

$$\boldsymbol{x}_{\mathrm{I}}(t)=\boldsymbol{\Phi}(t,\, t_0)\boldsymbol{x}_0+\int_{t_0}^{t}\boldsymbol{\Phi}(t,\, \tau)\boldsymbol{b}_{\mathrm{I}}\delta(\tau-t_{\mathrm{I}})\mathrm{d}\tau. \tag{11.6}$$

To further evaluate the integral in (11.6), one has to take into account the properties (7.2) of the Dirac function. Thus one finally gets

$$\left.\begin{aligned} x_{\mathrm{I}}(t)&=\boldsymbol{\Phi}(t,\, t_0)\boldsymbol{x}_0 && \text{for } t<t_{\mathrm{I}},\\ x_{\mathrm{I}}(t)&=\boldsymbol{\Phi}(t,\, t_0)\left[\boldsymbol{x}_0+\boldsymbol{\Phi}(t_0,\, t_{\mathrm{I}})\boldsymbol{b}_{\mathrm{I}}\right] && \text{for } t>t_{\mathrm{I}}.\end{aligned}\right\} \tag{11.7}$$

This result also means that for time-variant systems the impulsive excitation function produces a stepwise change of the state variables at the instant of impulse. In mechanical systems, just as in the time-invariant case in (7.6), impulse forces cause a stepwise change of the velocities $\dot{y}(t)$ only.

The steady-state behavior of the impulsive response, i.e., the behavior of (11.7) as $t\to\infty$, again depends only on the stability of the system. The steady-state impulsive response $\boldsymbol{x}_{\mathrm{I}}(t\to\infty)$ vanishes for asymptotically stable systems, and increases to infinity for unstable systems.

11.3. Step excitation

The state equation of a periodic time-variant vibrating system with step excitation is

$$\dot{\boldsymbol{x}}(t)=\boldsymbol{A}(t)\boldsymbol{x}(t)+\boldsymbol{b}_{\mathrm{S}}s(t-t_{\mathrm{S}}),\quad \boldsymbol{A}(t)=\boldsymbol{A}(t+T),\quad \boldsymbol{x}(t_0)=\boldsymbol{x}_0, \tag{11.8}$$

where $\boldsymbol{b}_S s(t-t_S)$ is the step function (7.14) and t_S is the instant of the step. This implies, by (10.9), the *step response*

$$\boldsymbol{x}_S(t)=\boldsymbol{\Phi}(t,t_0)\boldsymbol{x}_0+\int_{t_0}^{t}\boldsymbol{\Phi}(t,\tau)\boldsymbol{b}_S s(\tau-\tau_S)\mathrm{d}\tau. \tag{11.9}$$

Using (7.15), one obtains for $t_S=t_0$ the response

$$\boldsymbol{x}_S(t)=\boldsymbol{\Phi}(t,t_0)\boldsymbol{x}_0+\int_{t_0}^{t}\boldsymbol{\Phi}(t,\tau)\boldsymbol{b}_S\mathrm{d}\tau. \tag{11.10}$$

The integral in (11.10) can be further evaluated with the aid of the transition property (10.6) and a time partition according to (10.31). Depending on the number of periods ν, one gets for vanishing initial conditions

$$\nu=0:\quad \boldsymbol{x}_S(t)=\boldsymbol{\Phi}(t_0+t^*,t_0)\int_{t_0}^{t_0+t^*}\boldsymbol{\Phi}(t_0,\tau)\boldsymbol{b}_S\mathrm{d}\tau,$$

$$\nu=1:\quad \boldsymbol{x}_S(t)=\boldsymbol{\Phi}(t_0+t^*,t_0)\Bigg[\boldsymbol{\Phi}(t_0+T,t_0)\int_{t_0}^{t_0+T}\boldsymbol{\Phi}(t_0,\tau)\boldsymbol{b}_S\mathrm{d}\tau+$$
$$+\int_{t_0}^{t_0+t^*}\boldsymbol{\Phi}(t_0,\tau)\boldsymbol{b}_S\mathrm{d}\tau\Bigg],$$

$$\nu=2:\quad \boldsymbol{x}_S(t)=\boldsymbol{\Phi}(t_0+t^*,t_0)\Bigg[\boldsymbol{\Phi}^2(t_0+T,t_0)\int_{t_0}^{t_0+T}\boldsymbol{\Phi}(t_0,\tau)\boldsymbol{b}_S\mathrm{d}\tau+$$
$$+\boldsymbol{\Phi}(t_0+T,t_0)\int_{t_0}^{t_0+T}\boldsymbol{\Phi}(t_0,\tau)\boldsymbol{b}_S\mathrm{d}\tau+\int_{t_0}^{t_0+t^*}\boldsymbol{\Phi}(t_0,\tau)\boldsymbol{b}_S\mathrm{d}\tau\Bigg], \tag{11.11}$$

or, for an arbitrary number of periods,

$$\boldsymbol{x}_S(t)=\boldsymbol{\Phi}(t_0+t^*,t_0)\Bigg[\sum_{\alpha=1}^{\nu}\boldsymbol{\Phi}^{\alpha}(t_0+T,t_0)\boldsymbol{c}_S(t_0+T,t_0)$$
$$+\boldsymbol{c}_S(t_0+t^*,t_0)\Bigg], \tag{11.12}$$

where the integral

$$\boldsymbol{c}_S(t_0+t^*,t_0)=\int_{t_0}^{t_0+t^*}\boldsymbol{\Phi}(t_0,\tau)\boldsymbol{b}_S\mathrm{d}\tau,\ t_0\leqslant t_0+t^*\leqslant t_0+T, \tag{11.13}$$

was introduced for abbreviation. It is also worth noting that the integral (11.13) needs only to be calculated for one period duration. By (11.12), the steady-state step response, $\nu \to \infty$, leads to the infinite matrix series

$$S_S = \Phi(t_0 + T, t_0) + \Phi^2(t_0 + T, t_0) + \Phi^3(t_0 + T, t_0) + \dots. \tag{11.14}$$

If one now assumes that the system is asymptotically stable, then the series (11.14) converges and one can subtract from (11.14) the series $S_S\Phi(t_0 + T, t_0)$. Thus one gets

$$S_S - S_S\Phi(t_0 + T, t_0) = \Phi(t_0 + T, t_0) + \lim_{\nu \to \infty} \Phi^\nu(t_0 + T, t_0). \tag{11.15}$$

But due to the asymptotic stability the limit in (11.15) vanishes and hence the sum of the infinite matrix series is the constant $n \times n$ matrix

$$S_S = \Phi(t_0 + T, t_0)[E - \Phi(t_0 + T, t_0)]^{-1}. \tag{11.16}$$

Thus for asymptotically stable systems the steady-state response is

$$x_{S\infty}(t) = \Phi(t_0 + t^*, t_0)[S_S c_S(t_0 + T, t_0) + c_S(t_0 + t^*, t_0)]. \tag{11.17}$$

The steady-state response to step excitation is a function of t^*, i.e., one gets a periodic response, $x_S(t) = x_S(t + T)$, unlike in the case of systems with time-invariant coefficients. Hence the step excitation can serve as a test function for determining whether the system has periodic parameters and how large the period is.

11.4. Periodic excitation

The forced vibrations of the parametrically excited system

$$\dot{x}(t) = A(t)x(t) + b_P(t), \quad A(t) = A(t + T), \quad x(t_0) = x_0, \tag{11.18}$$

can be determined for the periodic excitation functions whose periods are rational multiples of the period of the parameters

$$b_P(t) = b_P\left(t + \frac{r}{s}T\right), \qquad r = 1, 2, 3, \dots, \qquad s = 1, 2, 3, \dots. \tag{11.19}$$

Hence one gets from the general solution (10.9), taking into account (11.19), and using the time,

$$t = t_0 + r(t^* + \nu T), \quad \nu = 0, 1, 2, \dots, \tag{11.20}$$

and following a procedure similar to (11.11), the *response to periodic excitation* for vanishing initial conditions

$$x_P(t)=\Phi(t_0+rt^*,\,t_0)\left[\sum_{\alpha=1}^{\nu}\Phi^{\alpha r}(t_0+T,\,t_0)c_P(t_0+rT,\,t_0)\right.$$

$$\left.+c_P(t_0+rt^*,\,t_0)\right] \tag{11.21}$$

with the abbreviation

$$c_P(t_0+rt^*,\,t_0)=\int_{t_0}^{t_0+rt^*}\Phi(t_0,\,\tau)b_P(\tau)(\tau),$$

$$t_0\leqslant t_0+rt^*\leqslant t_0+rT. \tag{11.22}$$

The steady-state response for asymptotically stable systems is

$$x_{P\infty}(t)=\Phi(t_0+rt^*,\,t_0)\left[S_Pc_P(t_0+rT,\,t_0)+c_p(t_0+rt^*,\,t_0)\right] \tag{11.23}$$

with the constant $n\times n$ matrix

$$S_P=\Phi^r(t_0+T,\,t_0)\left[E-\Phi^r(t_0+T,\,t_0)\right]^{-1}. \tag{11.24}$$

This result shows that the steady-state response to the periodic excitation (11.19) is an rT-periodic function, $x_{P\infty}(t)=x_{P\infty}(t+rT)$. This means that only for integers r/s the period of the excitation coincides with the period of the response, as is the case for time-invariant systems. For noninteger values r/s the two periods are distinct.

Example 11.1. Forced parametrically excited vibrations of a periodically accelerated pendulum

The normalized state equation of the periodically accelerated pendulum, Example 10.2, is

$$\underbrace{\begin{bmatrix}\dot y\\ \ddot y\end{bmatrix}}_{\dot x(t)}=\underbrace{\begin{bmatrix}0 & 1\\ -(\beta^2+\gamma\cos t) & -2D\end{bmatrix}}_{A(t)}\underbrace{\begin{bmatrix}y\\ \dot y\end{bmatrix}}_{x(t)}+\underbrace{\begin{bmatrix}0\\ b_2\end{bmatrix}}_{b(t)}. \tag{11.25}$$

Suppose that this pendulum is subject to impulsive, step, and periodic excitations

$$\left.\begin{aligned}&b_{2I}(t)=\delta(t), && t_I=0\\ &b_{2S}(t)=s(t), && t_S=0\\ &b_{2P}(t)=\cos t, &&\end{aligned}\right\} \tag{11.26}$$

under the assumption that the vibrations begin from the position of equilibrium

$$x(t_0)=0, \qquad t_0=0. \tag{11.27}$$

The corresponding responses of the periodically accelerated pendulum can be seen in Fig. 11.1. Besides the transient responses, the steady-state responses are also shown. The steady-state impulsive response of the asymptotically stable periodically accelerated pendulum vanishes. The steady-state step response is a periodic function, as was shown in Section 11.2. By (11.19) and (11.26), the periodic excitation is characterized by the values $r=1$ and $s=1$. Thus, according to Section 11.4, the steady-state response to this excitation is characterized by the period of the excitation and of the system parameters. ◆

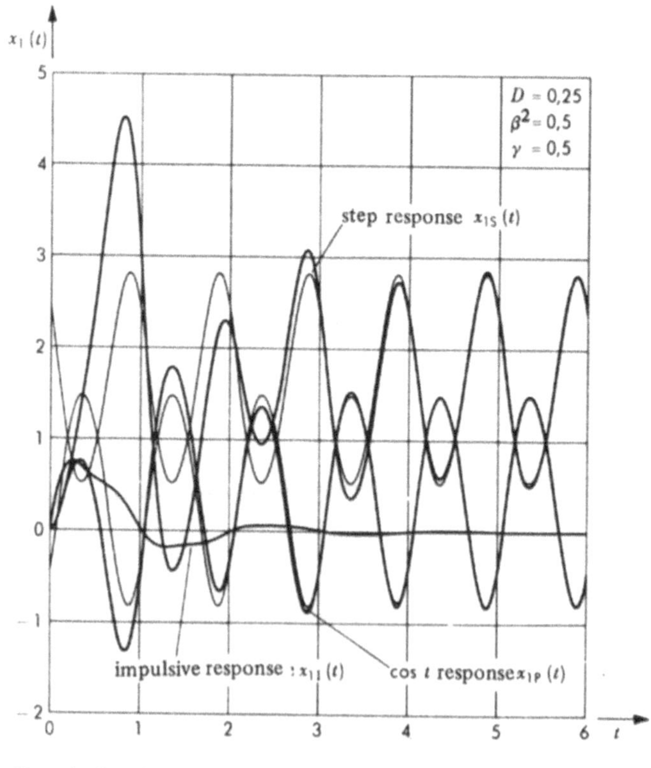

Fig. 11.1. Forced vibrations of a periodically accelerated pendulum: ——— transient response, ——— steady-state response

11.5. Stochastic excitation

The state equation of a *time-variant* system with stochastic excitation is

$$\dot{x}(t)=A(t)x(t)+b(t) \tag{11.28}$$

where $b(t)$ is a *nonstationary* process of white noise with vanishing mean value $m_b(t)=\mathbf{0}$ and time-variant intensity matrix $Q(t)$. Further, according to (9.23), the initial condition is introduced as a Gaussian random vector

$$x(t_0)=x_0\sim(m_0,\ P_0) \tag{11.29}$$

with the mean value m_0 and the covariance matrix P_0.

The *stochastic response* $x(t)$ is a nonstationary Gauss-Markov process which can be described by its mean vector $m_x(t)$ and the covariance matrix $P_x(t)$. By applying the stochastic integrals of Itô, one gets, by (9.25) and (9.26), the solutions in terms of the $n\times n$ fundamental matrix $\Phi(t,\ t_0)$:

$$m_x(t)=\Phi(t,\ t_0)m_0, \tag{11.30}$$

$$P_x(t)=\Phi(t,\ t_0)P_0\Phi^{\mathrm{T}}(t,\ t_0)+\int_{t_0}^{t}\Phi(t,\ \tau)Q(\tau)\Phi^{\mathrm{T}}(t,\ \tau)\mathrm{d}\tau. \tag{11.31}$$

These solutions also satisfy the deterministic differential equations

$$\dot{m}_x(t)+A(t)m_x(t), \qquad m_x(t_0)=m_0, \tag{11.32}$$

$$\dot{P}_x(t)=A(t)P_x(t)+P_x(t)A^{T}(t)+Q(t), \quad P_x(t_0)=P_0. \tag{11.33}$$

The Ljapunov matrix differential equation (11.33) is an extension of the algebraic Ljapunov matrix equation (9.32) which was applied for calculating the random vibrations of time-invariant systems. In the general time-variant case the solutions (11.30) and (11.31) for the mean vector and the covariance matrix can be practically determined only by integrating the differential equations (11.32) and (11.33). On the other hand for parametrically excited systems with *periodic* coefficients

$$A(t)=A(t+T), \quad Q(t)=Q(t+T), \tag{11.34}$$

the Floquet theory in Section 10.2 can be applied to reduce the integration to one period. First, for the mean vector (11.30) one can immediately use the fundamental matrix in the form (10.32). For the calculation of the covariance matrix (11.31) one may again use the *time partitioning* (10.31) and get, with $t_0 = 0$ and $\boldsymbol{P}_0 = \boldsymbol{0}$, the result

$$\boldsymbol{P}_x(t) = \boldsymbol{\Phi}(t^*, 0)\left[\sum_{\alpha=1}^{\nu} \boldsymbol{\Phi}^{\alpha}(T, 0)\boldsymbol{C}(T, 0)\boldsymbol{\Phi}^{\mathrm{T}\alpha}(T, 0) + \boldsymbol{C}(t^*, 0)\right]\boldsymbol{\Phi}^{\mathrm{T}}(t^*, 0), \tag{11.35}$$

where the $n \times n$ matrix integral

$$\boldsymbol{C}(t^*, 0) = \int_0^{t^*} \boldsymbol{\Phi}(0, \tau)\boldsymbol{Q}(\tau)\boldsymbol{\Phi}^{\mathrm{T}}(0, \tau)\mathrm{d}\tau \tag{11.36}$$

was used for abbreviation. The covariance matrix of the *stationary random vibrations*, $t \to \infty$, resp. $\nu \to \infty$, leads by (11.35) to the infinite matrix series

$$\boldsymbol{S}_{\mathrm{Z}} = \boldsymbol{\Phi}(T, 0)\boldsymbol{C}(T, 0)\boldsymbol{\Phi}^{\mathrm{T}}(T, 0) + \boldsymbol{\Phi}^2(T, 0)\boldsymbol{C}(T, 0)\boldsymbol{\Phi}^{2\mathrm{T}}(T, 0) + \dots. \tag{11.37}$$

Assuming asymptotic stability and subtracting from (11.37) the infinite matrix series $\boldsymbol{\Phi}(T, 0)\boldsymbol{S}_{\mathrm{Z}}\boldsymbol{\Phi}^{T}(T, 0)$, one gets

$$\boldsymbol{S}_{\mathrm{Z}} - \boldsymbol{\Phi}(T, 0)\boldsymbol{S}_{\mathrm{Z}}\boldsymbol{\Phi}(T, 0) = \boldsymbol{\Phi}(T, 0)\boldsymbol{C}(T, 0)\boldsymbol{\Phi}^{T}(T, 0) + \lim_{\nu \to \infty} \boldsymbol{\Phi}^{\nu}(T, 0)\boldsymbol{C}(T, 0)\boldsymbol{\Phi}^{\nu\mathrm{T}}(T, 0). \tag{11.38}$$

Due to the asymptotic stability the limit in (11.38) vanishes and one arrives at the algebraic Stein matrix equation

$$\boldsymbol{S}_{\mathrm{Z}} - \boldsymbol{\Phi}(T, 0)\boldsymbol{S}_{\mathrm{Z}}\boldsymbol{\Phi}^{\mathrm{T}}(T, 0) = \boldsymbol{\Phi}(T, 0)\boldsymbol{C}(T, 0)\boldsymbol{\Phi}^{\mathrm{T}}(T, 0), \tag{11.39}$$

whose solution yields the constant $n \times n$ matrix $\boldsymbol{S}_{\mathrm{Z}}$. Stein's matrix equation as well as Ljapunov's matrix equation belong to the class of linear matrix equations; some remarks concerning their solution can be found in Section 13.3. The covariance matrix of the stationary random vibration can be written in terms of the matrix $\boldsymbol{S}_{\mathrm{Z}}$ as

$$\boldsymbol{P}_{x\infty}(t) = \boldsymbol{\Phi}(t^*, 0)\left[\boldsymbol{S}_{\mathrm{Z}} + \boldsymbol{C}(t^*, 0)\right]\boldsymbol{\Phi}^{\mathrm{T}}(t^*, 0). \tag{11.40}$$

The covariance matrix (11.40) is a function of t^*, i.e., contrary to the case for systems with time-invariant coefficients, one obtains a periodic steady-state covariane matrix $\boldsymbol{P}_{x\infty}(t) = \boldsymbol{P}_{x\infty}(t + T)$. In general, one can show that the covariance matrix $\boldsymbol{P}_{x\infty}(t)$ of a periodic time-variant system $\boldsymbol{A}(t)=\boldsymbol{A}(t+T)$ is periodic, provided the period of the intensity matrix $\boldsymbol{Q}(t)$ is a rational multiple of the period of the system.

Example 11.2. Rotor blade vibrations of a helicopter

The rotor blade vibrations of a helicopter, Fig. 11.2, are described by the equation of motion

$$I\ddot{\varphi}(t)+d(\Omega t)\dot{\varphi}(t)+k(\Omega t)\varphi(t)=h(t) \tag{11.41}$$

Here the rotor blade angle $\varphi(t)$ is the position variable of the rigid rotor blade with moment of inertia I relative to the x-axis. Because of the rotation, the damping parameter $d(\Omega t)$ and the spring parameter $k(\Omega t)$ are periodic functions of time. In addition, the rotor blade is excited by stochastic forces $h(t)$ caused by air stream turbulences. It is assumed that these stochastic forces form a process of white noise. The parametrically excited vibrations of the angle $\varphi(t)$ can now be analysed by means of the covariance matrix.

After some simplification, (11.41) can be transformed into an equation approximating Mathieu's differential equation

$$\ddot{y}(t)+2D\dot{y}(t)+(\beta^2+\gamma\cos t)y(t)=n(t), \tag{11.42}$$

which is excited by a constant white noise $n(t)$ with intensity q.

Therefore the differential equation (11.33) for the covariance matrix $\boldsymbol{P}(t)$ is as follows:

$$\begin{bmatrix}\dot{P}_{11} & \dot{P}_{12}\\ \dot{P}_{12} & \dot{P}_{22}\end{bmatrix}=\begin{bmatrix}0 & 1\\ -(\beta^2+\gamma\cos t) & -2D\end{bmatrix}\begin{bmatrix}P_{11} & P_{12}\\ P_{12} & P_{22}\end{bmatrix}+$$
$$+\begin{bmatrix}P_{11} & P_{12}\\ P_{12} & P_{22}\end{bmatrix}\begin{bmatrix}0 & -(\beta^2+\gamma\cos t)\\ 1 & -2D\end{bmatrix}+\begin{bmatrix}0 & 0\\ 0 & q\end{bmatrix}. \tag{11.43}$$

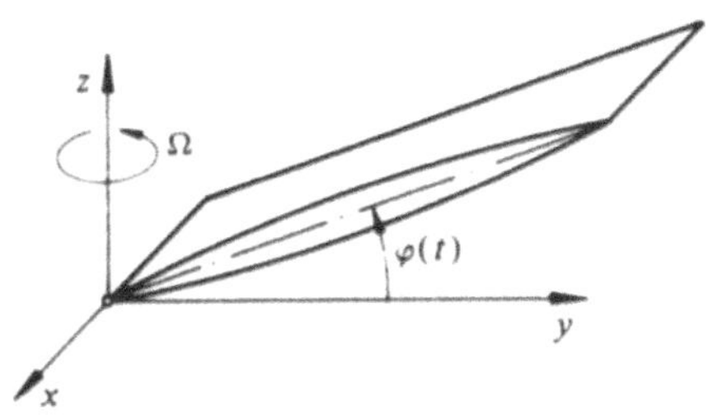

Fig. 11.2. Rotor blade of a helicopter

In other words, one has a third order system of differential equations

$$\left.\begin{aligned}\dot{P}_{11}(t) &= 2P_{12}(t),\\ \dot{P}_{12}(t) &= -(\beta^2+\gamma\cos t)P_{11}(t) - 2DP_{12}(t) + P_{22}(t),\\ \dot{P}_{22}(t) &= -2(\beta^2+\gamma\cos t)P_{12}(t) - 4DP_{22}(t) + q.\end{aligned}\right\} \tag{11.44}$$

Fig. 11.3 contains the variance P_{11} of the variable y, equal to the square of the standard deviation σ_y. One clearly recognizes that with growing time the variance becomes periodic, $P_{11}(t) = P_{11}(t+T)$. ◆

Problem 11.1. Periodic parametrically excited vibrations
For Mathieu's differential equation

$$\ddot{y}(t) + 2D\dot{y}(t) + (\beta^2 + \gamma\cos t)y(t) = 0, \tag{11.45}$$

the stability boundary is characterized by the following eigenvalues

$$\left.\begin{aligned}&\mu_{1,2} = \pm 1 && \text{for} \quad D = 0\\ &\mu_1 = \pm 1,\ \mu_2 < 1 && \text{for} \quad D > 0.\end{aligned}\right\} \tag{11.46}$$

In the undamped case, $D = 0$, one can find out by a limiting procedure $\gamma \to 0$ which stability boundaries, according to (10.51) or Fig. 10.3, represent half- or one-periodic natural vibrations. For $D = 0$ and $\gamma = 0$ Mathieu's differential equation approaches the equation of a single-dof-vibrating system

$$\ddot{y}(t) + \beta^2 y(t) = 0 \tag{11.47}$$

which immediately yields the period of the natural vibrations.

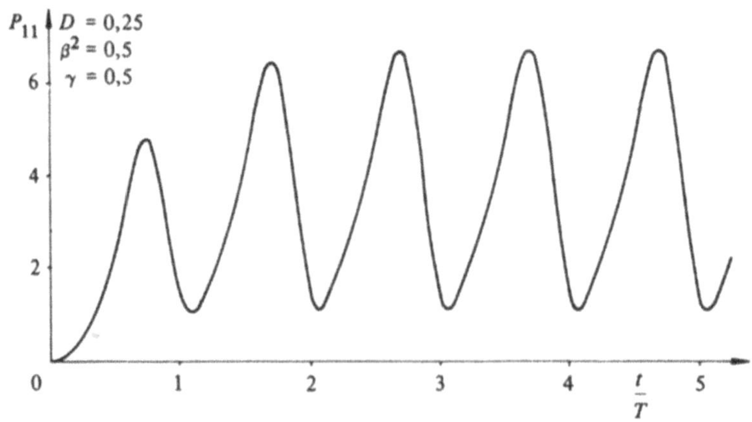

Fig. 11.3. Variance $P_{11}(t)$ of the parameter-excited random vibrations of a rotor blade

On which stability boundaries occur the following parametrically excited vibrations?

	half-period	one-period
$\beta_{2,3} = \frac{1}{2}$	□	□
$\beta_{4,5} = 1$	□	□
$\beta_{6,7} = \frac{3}{2}$	□	□
$\beta_{8,9} = 2$	□	□
$\beta_{10,11} = \frac{5}{2}$	□	□ ◆

Problem 11.2. Periodic parametrically excited forced vibrations

Consider a parametrically excited vibrating system satisfying Mathieu's differential equation

$$\ddot{y}(t) + 2D\dot{y}(t) + (\beta^2 + \gamma \cos t)\, y(t) = \cos\frac{1}{k}t, \tag{11.48}$$

with the excitation function $\cos(t/k)$. Some steady-state responses of the asymptotically stable system (11.48) are shown in Fig. 11.4.

Which of the responses correspond to rational and which to irrational values of the parameter k in the excitation function?

	rational	irrational
response 1	□	□
response 2	□	□
response 3	□	□ ◆

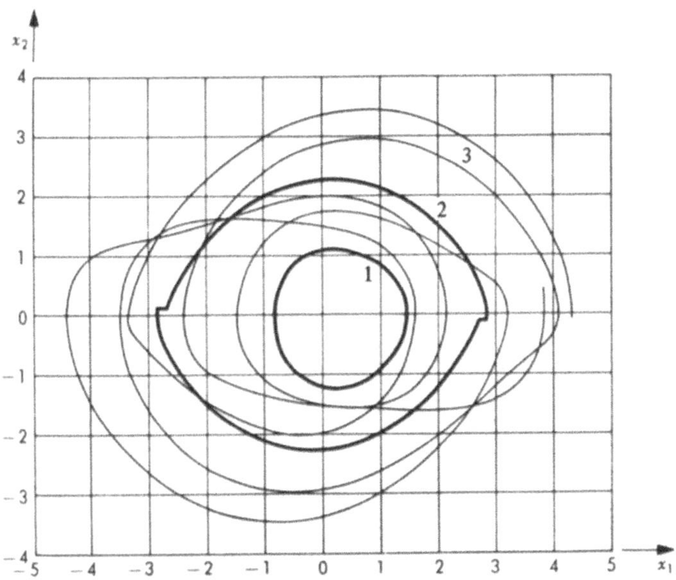

Fig. 11.4. Steady-state responses to the periodic excitation $\cos t/k$

PART IV

Mathematical background

The essential mathematical tool used in the analysis of finite dimensional linear vibrating systems is the matrix calculus. Hence an elementary knowledge of the matrix theory has been assumed throughout the book. In the subsequent sections there follows a short summary of those results in matrix theory which appear to be of special importance for the analysis of vibrations. Thus in Chapter 12 there is given an algebraic treatment of the properties of controllability and observability developed in control theory. These characteristics of the structure of a dynamic system are relevant for stability investigations (Chapter 5). Chapter 13 is in part devoted to the treatment of linear algebraic vector equations, as needed for the eigenvalue problem (Section 4.3) or for the frequency response calculations (Chapter 7 and 8), and in part to the Ljapunov equation which turns up in problems of stability, of optimal behavior of natural vibrations or of vibrations under disturbances, and in the covariance analysis (Section 5.2.3, 6.2.3, 8.7 and 9.4). In the concluding Chapter 14 attention is given to numerical aspects that have to be considered when solving linear systems of equations, determining eigenvalues and eigenvectors, and also when calculating frequency responses and fundamental matrices.

12. Controllability and observability

Controllability and observability are two of the most important notions in the newly developed theory of dynamic systems. They play an essential role in the design of control systems and allow insights into the physical problems. To put it simply, the property of *controllability* characterizes the possibility to influence the behavior of the dynamic system by control and disturbance input functions while the property of *observability* leads to conditions about the information on the state of the system that can be obtained by measurements. If a dynamic system is completely controllable, then all natural vibration of the system are excited by the input excitation function; correspondingly, all natural vibrations of a completely observable system can be determined by measurements.

For a more detailed explanation of these structural properties of dynamic systems, consider a linear time-invariant system in the state space representation

$$\dot{\boldsymbol{x}}(t)=\boldsymbol{A}\boldsymbol{x}(t)+\boldsymbol{B}\boldsymbol{u}(t),\quad \boldsymbol{x}(0)=\boldsymbol{x}_0, \tag{12.1}$$

with measurements

$$\boldsymbol{y}(t)=\boldsymbol{C}\boldsymbol{x}(t). \tag{12.2}$$

The vectors $\boldsymbol{x}$, $\boldsymbol{u}$, and $\boldsymbol{y}$ have the dimensions $n\times 1$, $r\times 1$, and $s\times 1$ respectively and describe the state, the input (control), and the output (measurement) of the system. The matrices $\boldsymbol{A}$, $\boldsymbol{B}$, and $\boldsymbol{C}$ have the dimensions $n\times n$, $n\times r$, and $s\times n$, and are called the system, the input (control), and output (measurement) matrix, respectively.

The system (12.1) is called *completely controllable* if for each arbitrary initial condition $\boldsymbol{x}(0)=\boldsymbol{x}_0$ and each arbitrary state $\boldsymbol{x}_1$ there exists a finite time $t_1>0$ and a control function $\boldsymbol{u}(t)$ defined on the time interval $[0,\ t_1]$ such that the trajectory of (12.1) starting at $\boldsymbol{x}_0$ attains for $t=t_1$ the value $\boldsymbol{x}_1$:

$$\boldsymbol{x}(t_1;\ \boldsymbol{x}_0, 0;\ \boldsymbol{u}(t))=\boldsymbol{x}_1. \tag{12.3}$$

This definition merely requires that by a suitable choice of control functions the system can be transfered in a finite time from an arbitrary initial state to an arbitrary final state; it is irrelevant along which trajectory the transition takes place. There are no restrictions on the control functions; arbitrary large amplitudes are admitted.

The system (12.1, 12.2) is called completely observable if for each arbitrary initial condition $\boldsymbol{x}(0)=\boldsymbol{x}_0$ there exists a finite time $t_1>0$ such that the initial condition $\boldsymbol{x}_0$ can be determined by the knowledge of the control function $\boldsymbol{u}(t)$ and the measurement function $\boldsymbol{y}(t)$ on the interval $[0, t_1]$.

The properties of controllability and observability were introduced in 1959/60 by Kalman. He also gave at this time criteria which permit to check on complete controllability and complete observability.

Kalman criteria

The dynamic system (12.1) is completely controllable if and only if

$$\operatorname{rank}\left[\boldsymbol{B} \,\vdots\, \boldsymbol{A}\boldsymbol{B} \,\vdots\, \boldsymbol{A}^2\boldsymbol{B} \,\vdots \ldots \vdots\, \boldsymbol{A}^{n-1}\boldsymbol{B}\right]=n \tag{12.4}$$

holds. The dynamic system (12.1, 12.2) is completely observable if and only if

$$\operatorname{rank}\left[\boldsymbol{C}^T \,\vdots\, \boldsymbol{A}^{\mathrm{T}}\boldsymbol{C}^{\mathrm{T}} \,\vdots\, \boldsymbol{A}^{\mathrm{T}^2}\boldsymbol{C}^{\mathrm{T}} \,\vdots \ldots \vdots\, \boldsymbol{A}^{T^{n-1}}\boldsymbol{C}^{\mathrm{T}}\right]=n \tag{12.5}$$

is satisfied.

Noncontrollable or nonobservable states $\boldsymbol{x}$ are characterized by the conditions

$$\boldsymbol{x}^{\mathrm{T}}\left[\boldsymbol{B} \,\vdots\, \boldsymbol{A}\boldsymbol{B} \,\vdots\, \boldsymbol{A}^2\boldsymbol{B} \,\vdots \ldots \vdots\, \boldsymbol{A}^{n-1}\boldsymbol{B}\right]=\mathbf{0} \tag{12.6}$$

or

$$\boldsymbol{x}^{\mathrm{T}}\left[\boldsymbol{C}^{\mathrm{T}} \,\vdots\, \boldsymbol{A}^{\mathrm{T}}\boldsymbol{C}^{\mathrm{T}} \,\vdots\, \boldsymbol{A}^{T^2}\boldsymbol{C}^{\mathrm{T}} \,\vdots \ldots \vdots\, \boldsymbol{A}^{T^{n-1}}\boldsymbol{C}^T\right]=\mathbf{0}. \tag{12.7}$$

The criteria of Kalman offer an algebraic characterization of the properties of linear time-invariant dynamic systems. However they don't give insight into the physical behavior of the vibrating system (12.1, 12.2). This is just attained by the criteria of Hautus which present connections between the natural vibrations of the homogeneous system $\dot{\boldsymbol{x}}(t)=\boldsymbol{A}\boldsymbol{x}(t)$ and the controllability by $\boldsymbol{B}\boldsymbol{u}(t)$, or the observability by $\boldsymbol{C}\boldsymbol{x}(t)$.

Hautus criteria

The dynamic system (12.1) is completely controllable if and only if each eigenvector of the transposed system matrix is not orthogonal to any column vector of the control matrix:

$$\left(\lambda_i\boldsymbol{E}-\boldsymbol{A}^{\mathrm{T}}\right)\tilde{\boldsymbol{x}}_i=\mathbf{0} \Rightarrow \boldsymbol{B}^{\mathrm{T}}\tilde{\boldsymbol{x}}_i \neq \mathbf{0}. \tag{12.8}$$

If, on the other hand, an eigenvector $\tilde{\boldsymbol{x}}_i$ with $\boldsymbol{A}^{\mathrm{T}}\tilde{\boldsymbol{x}}_i = \lambda_i \tilde{\boldsymbol{x}}_i$ satisfies $\boldsymbol{B}^{\mathrm{T}}\tilde{\boldsymbol{x}}_i = \mathbf{0}$, then the i-th natural vibration of the system $\dot{\boldsymbol{x}}(t) = \boldsymbol{A}\boldsymbol{x}(t)$ is not influenced by the excitation $\boldsymbol{B}\boldsymbol{u}(t)$: this natural vibration (and hence also the dynamic system) is not controllable.

The dynamic system (12.1, 12.2) is completely observable if and only if every eigenvector of the system matrix is not orthogonal to any row vector of the measurement matrix:

$$(\lambda_i \boldsymbol{E} - \boldsymbol{A})\tilde{\boldsymbol{x}}_i = \mathbf{0} \Rightarrow \boldsymbol{C}\tilde{\boldsymbol{x}}_i \neq \mathbf{0}. \tag{12.9}$$

On the other hand, if (12.9) fails by an eigenvector, i.e., $\boldsymbol{C}\tilde{\boldsymbol{x}}_i = \mathbf{0}$ holds, then the i-th natural vibration of the system $\dot{\boldsymbol{x}}(t) = \boldsymbol{A}\boldsymbol{x}(t)$ cannot be observed by the measurement function $\boldsymbol{y}(t) = \boldsymbol{C}\boldsymbol{x}(t)$; the dynamic system is not completely observable.

13. Matrix equations

Three types of linear algebraic equations are of importance in the treatment of linear vibrating systems: the linear vector equation

$$\boldsymbol{A}\boldsymbol{x} = \boldsymbol{b}, \tag{13.1}$$

the Ljapunov matrix equation

$$\boldsymbol{A}^{\mathrm{T}}\boldsymbol{X} + \boldsymbol{X}\boldsymbol{A} = -\boldsymbol{B} \tag{13.2}$$

and Stein's matrix equation

$$\boldsymbol{A}^{\mathrm{T}}\boldsymbol{X}\boldsymbol{A} - \boldsymbol{X} = -\boldsymbol{B}. \tag{13.3}$$

One section will subsequently be devoted to each of these.

13.1. The linear vector equation

If there is given a system of equations (13.1) with an $m \times n$ matrix $\boldsymbol{A}$ and an $m \times 1$ vector $\boldsymbol{b}$, then the questions arise if there exists an $n \times 1$ solution vector $\boldsymbol{x}$ for the system (13.1), and if yes, whether the solution is unique or if more solutions exist, and finally, how such solutions can be determined. This circle of problems can be tackled in various ways. Rank analysis, adjoint matrix, inverse matrix, Cramer's rule, and the generalized inverse matrix are the corresponding notions which will subsequently be related to each other.

Existence and uniqueness (method I)

The vector equation (13.1) has at least one solution if and only if

$$\operatorname{rank} \boldsymbol{A} = \operatorname{rank}[\boldsymbol{A} \vdots \boldsymbol{b}] \tag{13.4}$$

holds. The general solution of (13.1) is then obtained by a superposition of the general solution $\boldsymbol{x}_{\text{hom}}$ of the homogeneous system $\boldsymbol{Ax} = \boldsymbol{0}$ and a particular solution $\boldsymbol{x}_p$ of (13.1):

$$\boldsymbol{x} = \boldsymbol{x}_{\text{hom}} + \boldsymbol{x}_p. \tag{13.5}$$

The number of linearly independent solutions is determined by the column rank nullity

$$d = n - \text{rank } \boldsymbol{A}; \tag{13.6}$$

in particular, the solution is unique if and only if

$$\text{rank } \boldsymbol{A} = n \leqslant m \tag{13.7}$$

and for $n < m$ it is given by

$$\boldsymbol{x} = (\boldsymbol{A}^{\mathrm{T}}\boldsymbol{A})^{-1}\boldsymbol{A}^{\mathrm{T}}\boldsymbol{b} \tag{13.8}$$

while for $n = m$ it yields

$$\boldsymbol{x} = \boldsymbol{A}^{-1}\boldsymbol{b}. \tag{13.9}$$

In (13.4 to 7) the notions of the *rank of a matrix*, rank $\boldsymbol{A}$, and of the *inverse matrix*, $\boldsymbol{A}^{-1}$, were used. Here rank $\boldsymbol{A}$ is defined as the number of linearly independent row vectors of $\boldsymbol{A}$, or equivalent as the number of linearly independent column vectors of $\boldsymbol{A}$, or equivalent as the dimension of the largest nonvanishing minor of $\boldsymbol{A}$. While the rank is defined for arbitrary rectangular matrices ($m \neq n$), the inverse matrix can be introduced only for regular square matrices i.e., for matrices $\boldsymbol{A}$ with $m = n$ and det $\boldsymbol{A} \neq 0$. Here the (unique) inverse $\boldsymbol{A}^{-1}$ is formally defined by

$$\det \boldsymbol{A} \neq 0: \qquad \boldsymbol{A}^{-1}\boldsymbol{A} = \boldsymbol{A}\boldsymbol{A}^{-1} = \boldsymbol{E}. \tag{13.10}$$

One of the possibilities of calculating the inverse matrix involves the *adjoint matrix* adj $\boldsymbol{A}$ of the matrix $\boldsymbol{A}$. The elements of adj $\boldsymbol{A}$ are determined by the cofactors A_{ij} of the elements a_{ij} of $\boldsymbol{A}$. The cofactor A_{ij} is the related minor of order $n-1$ prefixed with the sign $(-1)^{i+j}$ where the minor is the subdeterminant of $\boldsymbol{A}$ obtained by striking out the i-th row and the j-th column from A. A transposed arrangement of these cofactors in a matrix yields the adjoint matrix

$$\text{adj } \boldsymbol{A} = [(\text{adj } A)_{ij}] = [A_{ji}]. \tag{13.11}$$

The adjoint matrix (13.11) is defined for regular (det $\boldsymbol{A} \neq 0$) as well as for singular (det $A = 0$) square matrices A. Due to the expansion theorem for determinants

$$\left.\begin{aligned}
&\sum_{j=1}^{n} a_{ij}A_{ij} = \det A, \quad && i = 1(1)n,\\
&\sum_{i=1}^{n} a_{ij}A_{ij} = \det A, \quad && j = 1(1)n,\\
&\sum_{j=1}^{n} a_{ij}A_{kj} = 0, && i \neq k,\\
&\sum_{i=1}^{n} a_{ij}A_{ik} = 0, && j \neq k,
\end{aligned}\right\} \tag{13.12}$$

one has the relation

$$\operatorname{adj} A \cdot A = A \cdot \operatorname{adj} A = \det A \cdot E. \tag{13.13}$$

Combining (13.13) with (13.10), one obtains

$$A^{-1} = \frac{\operatorname{adj} A}{\det A}. \tag{13.14}$$

provided det $A \neq 0$. The definition (13.11) and the property (13.13) imply the following properties for the adjoint matrix

$$\operatorname{rank} A = n \Leftrightarrow \operatorname{rank}(\operatorname{adj} A) = n, \tag{13.15a}$$

$$\operatorname{rank} A = n - 1 \Leftrightarrow \operatorname{rank}(\operatorname{adj} A) = 1, \tag{13.15b}$$

$$\operatorname{rank} A \leqslant n - 2 \Leftrightarrow \operatorname{adj} A = \mathbf{0}. \tag{13.15c}$$

In particular, in the case (13.15b), one can see that the homogeneous system of equations $Ax = \mathbf{0}$ possesses precisely one linearly independent solution vector which is characterized by any nonvanishing column of adj A; conversely, the homogeneous system of equations adj $Ax = \mathbf{0}$ possesses precisely $n - 1$ linearly independent solutions which are given by $n - 1$ linearly independent column vectors of A:

rank $A = n - 1$:

$$Ax = \mathbf{0} \Rightarrow x = \operatorname{adj} A \cdot c \neq \mathbf{0}, \tag{13.16a}$$

$$\operatorname{adj} A \cdot x = \mathbf{0} \Rightarrow x = Ac \neq \mathbf{0}, \tag{13.16b}$$

where c is an arbitrary $n \times 1$ vector.

For a system of equations (13.1) with a regular square matrix $\boldsymbol{A}$ there exists, by (13.9), a unique solution, and by (13.14) this solution is given by

$$\boldsymbol{x} = \frac{\operatorname{adj} \boldsymbol{A} \cdot \boldsymbol{b}}{\det \boldsymbol{A}}. \tag{13.17}$$

Applying (13.11), one can write (13.17) in the form of *Cramer's rule*:

$$\boldsymbol{x} = [x_i], \quad x_i = \frac{1}{\det \boldsymbol{A}} \sum_{j=1}^{n} b_j A_{ji}, \qquad i = 1(1)n. \tag{13.18}$$

It states that, to obtain the i-th coordinate of the solution vector, one has to form the quotient of two determinants, where the denominator equals the determinant of $\boldsymbol{A}$ and the numerator is the determinant of a matrix obtained from $\boldsymbol{A}$ by replacing the i-th column by $\boldsymbol{b}$.

With the exception of (13.16a), the above considerations yield no systematic solution procedure for (13.1) for singular matrices – in the nonrectangular $m \times n$ case for column singular matrices. For this purpose the generalized inverse of a matrix can be employed.

The *generalized inverse* or *pseudo-inverse* $\boldsymbol{A}^+$ of an arbitrary complex rectangular $m \times n$ matrix $\boldsymbol{A}$ is defined, according to Moore and Penrose, as the unique solution of the four simultaneous matrix equations

$$\boldsymbol{XA} = (\boldsymbol{XA})^*, \tag{13.19a}$$

$$\boldsymbol{AX} = (\boldsymbol{AX})^*, \tag{13.19b}$$

$$\boldsymbol{AXA} = \boldsymbol{A}, \tag{13.19c}$$

$$\boldsymbol{XAX} = \boldsymbol{X}. \tag{13.19d}$$

Here $\boldsymbol{C}^* = \overline{\boldsymbol{C}}^{\mathrm{T}}$ is the conjugate transpose of a matrix $\boldsymbol{C}$; for real matrices it yields $\boldsymbol{C}^* = \boldsymbol{C}^{\mathrm{T}}$. The unique solution of the system (13.19) is

$$\boldsymbol{X} = \boldsymbol{A}^+. \tag{13.20}$$

The pseudo-inverse matrix has the following properties:

$$(\boldsymbol{A}^+)^+ = \boldsymbol{A}, \tag{13.21a}$$

$$(A^*)^+ = (A^+)^*, \tag{13.21b}$$

$$(\lambda A)^+ = \lambda^+ A^+, \qquad \lambda^+ = \begin{cases} \dfrac{1}{\lambda} & \text{for} \quad \lambda \neq 0, \\ 0 & \text{for} \quad \lambda = 0, \end{cases} \tag{13.21c}$$

$$A^+ = (A^*A)^+ A^*, \tag{13.21d}$$

$$(A^*A)^+ = A^+(A^*)^+, \tag{13.21e}$$

$$(AA^*)^+ = (A^*)^+ A^+, \tag{13.21f}$$

$$\begin{aligned} \text{rank } A &= \text{rank } A^+ = \text{rank } A^+A = \text{rank } AA^+ = \text{tr } A^+A \\ &= \text{tr } AA^+, \end{aligned} \tag{13.21g}$$

$$AA^+ = A^+A \quad \text{if} \quad AA^* = A^*A. \tag{13.21h}$$

The generalized inverse can be calculated by making use of the fact that every $m \times n$ matrix A with rank $A = r$ ($r \leqslant \min(m, n)$) can be written as a product of a column-regular $m \times r$ matrix L and a row-regular $r \times n$ matrix R^*:

$$A = LR^*. \tag{13.22}$$

For instance, the matrix L can contain r linearly independent column vectors of A, and R indicates by which linear combination of these columns one can get A using L. By (13.22),

$$A^+ = R(R^*R)^{-1}(L^*L)^{-1}L^*. \tag{13.23}$$

In particular, this implies

$$A = lr^* \text{(dyadic)}: \qquad A^+ = \frac{rl^*}{(r^*r)(l^*l)}, \tag{13.24}$$

$$A \text{ column regular}: \quad A^+ = (A^*A)^{-1}A^*, \tag{13.25}$$

$$A \text{ row regular}: \quad A^+ = A^*(AA^*)^{-1}, \tag{13.26}$$

$$A \text{ regular}: \quad A^+ = A^{-1}. \tag{13.27}$$

For a unitary decomposition of A

$$A = U\Sigma V^* \tag{13.28}$$

with a unitary $m \times m$ matrix U and a unitary $n \times n$ matrix V,

$$U^{-1} = U^*, \quad V^{-1} = V^*, \tag{13.29}$$

and an generalized $m \times n$ diagonal matrix $\boldsymbol{\Sigma}$, one has

$$\boldsymbol{A}^{+} = \boldsymbol{V}\boldsymbol{\Sigma}^{+}\boldsymbol{U}^{*}. \tag{13.30}$$

For a singular quadratic matrix one may still apply the following assertions. If there exists a similarity transformation such that $\boldsymbol{A}$ can be written as

$$\boldsymbol{A} = \boldsymbol{T}\left[\begin{array}{c|c} \mathbf{0} & \mathbf{0} \\ \hline \mathbf{0} & \tilde{\boldsymbol{A}} \end{array}\right]\boldsymbol{T}^{-1} \tag{13.31}$$

where the submatrix $\tilde{\boldsymbol{A}}$ is regular, and

$$\boldsymbol{T} = [\boldsymbol{T}_1 \vdots \boldsymbol{T}_2], \qquad \boldsymbol{T}^{-1} = \left[\begin{array}{c} \boldsymbol{T}_3^{*} \\ \hline \boldsymbol{T}_4^{*} \end{array}\right] \tag{13.32}$$

then one has

$$\boldsymbol{A}^{+} = \left[\boldsymbol{T}_1(\boldsymbol{T}_1^{*}\boldsymbol{T}_1)^{-1}\boldsymbol{T}_1^{*} - \boldsymbol{E}\right]\boldsymbol{T}_2\tilde{\boldsymbol{A}}^{-1}\boldsymbol{T}_4^{*}\left[\boldsymbol{T}_3(\boldsymbol{T}_3^{*}\boldsymbol{T}_3)^{-1}\boldsymbol{T}_3^{*} - \boldsymbol{E}\right]. \tag{13.33}$$

Here $\boldsymbol{T}_1$ contains the linearly independent eigenvectors with eigenvalue $\lambda = 0$, $\boldsymbol{A}\boldsymbol{T}_1 = \mathbf{0}$, while $\boldsymbol{T}_2$ can be assigned to the eigenvectors and principal vectors with the nonvanishing eigenvalues, $\boldsymbol{A}\boldsymbol{T}_2 = \boldsymbol{T}_2\tilde{\boldsymbol{A}}$.

Since for Hermitean matrices, $\boldsymbol{A} = \boldsymbol{A}^{*}$, there always exists a representation (13.28) with $\boldsymbol{T} = \boldsymbol{U}$ and $\boldsymbol{T}^{-1} = \boldsymbol{U}^{*}$, one has

$$\boldsymbol{A} = \boldsymbol{A}^{*} = \boldsymbol{U}\left[\begin{array}{c|c} \mathbf{0} & \mathbf{0} \\ \hline \mathbf{0} & \tilde{\boldsymbol{A}} \end{array}\right]\boldsymbol{U}^{*} : \boldsymbol{A}^{+} = \boldsymbol{U}\left[\begin{array}{c|c} \mathbf{0} & \mathbf{0} \\ \hline \mathbf{0} & \tilde{\boldsymbol{A}}^{-1} \end{array}\right]\boldsymbol{U}^{*} = \boldsymbol{U}_2\tilde{\boldsymbol{A}}^{-1}\boldsymbol{U}_2^{*}, \tag{13.34}$$

which can also be read from (13.30).

The problem of the existence and uniqueness of the solution of the linear vector equation (13.1) can be now dealt with quite adequately using the pseudo-inverse introduced above.

Existence and uniqueness (method II)

The vector equation (13.1) has at least one solution if and only if

$$(\boldsymbol{E} - \boldsymbol{A}\boldsymbol{A}^{+})\boldsymbol{b} = \mathbf{0} \tag{13.35}$$

holds. The general solution of (13.1) is then given by

$$\boldsymbol{x} = \boldsymbol{A}^{+}\boldsymbol{b} + (\boldsymbol{E} - \boldsymbol{A}^{+}\boldsymbol{A})\hat{\boldsymbol{x}}, \tag{13.36}$$

where $\hat{\boldsymbol{x}}$ is an arbitrary $n \times 1$ vector. The solution is unique if and only if

$(\boldsymbol{E}-\boldsymbol{A}^{+}\boldsymbol{A})=\boldsymbol{0}$ holds, and then the solution is (13.8), by (13.21g) and (13.25).

Besides the assertion (13.35, 13.36), the generalized inverse finds an essential application in solving linear regression problems according to the *least squared error* minimization. The problem of finding a vector $\boldsymbol{x}_p$ satisfying

$$\|\boldsymbol{A}\boldsymbol{x}-\boldsymbol{b}\|_E^2=(\boldsymbol{A}\boldsymbol{x}-\boldsymbol{b})^{\mathrm{T}}(\boldsymbol{A}\boldsymbol{x}-\boldsymbol{b})\rightarrow \text{Minimum} \tag{13.37}$$

is solved by

$$\boldsymbol{x}_p=\boldsymbol{A}^{+}\boldsymbol{b}. \tag{13.38}$$

From

$$\boldsymbol{A}\boldsymbol{x}-\boldsymbol{b}=\boldsymbol{A}(\boldsymbol{x}-\boldsymbol{A}^{+}\boldsymbol{b})-(\boldsymbol{E}-\boldsymbol{A}\boldsymbol{A}^{+})\boldsymbol{b} \tag{13.39}$$

one obtains

$$\begin{aligned}\|\boldsymbol{A}\boldsymbol{x}-\boldsymbol{b}\|_E&=\|\boldsymbol{A}(\boldsymbol{x}-\boldsymbol{x}_p)\|_E+\|\boldsymbol{A}\boldsymbol{x}_p-\boldsymbol{b}\|_E\\&\geqslant\|\boldsymbol{A}\boldsymbol{x}_p-\boldsymbol{b}\|_E,\end{aligned} \tag{13.40}$$

where equality holds only for $\boldsymbol{A}\boldsymbol{x}=\boldsymbol{A}\boldsymbol{x}_p$. This can happen only if the solvability criterion (13.25) holds and $\boldsymbol{x}$ is chosen according to (13.36). Due to

$$\|\boldsymbol{x}\|_E=\|\boldsymbol{x}_p\|_E+\|(\boldsymbol{E}-\boldsymbol{A}^{+}\boldsymbol{A})\hat{\boldsymbol{x}}\|_E \tag{13.41}$$

the solution (13.38) still has the additional property

$$\|\boldsymbol{x}_p\|_E<\|\boldsymbol{x}\|_E \quad \text{for} \quad \boldsymbol{x}\neq\boldsymbol{x}_p. \tag{13.42}$$

13.2. The Ljapunov matrix equation

The bilinear matrix equation

$$\boldsymbol{A}^{\mathrm{T}}\boldsymbol{X}+\boldsymbol{X}\boldsymbol{A}=-\boldsymbol{B} \tag{13.43}$$

is called the algebraic Ljapunov equation since it appears in the stability theory founded by Ljapunov. But it also finds applications in optimization procedures and in the covariance analysis of random vibrations. In contrast to the equation discussed in the previous section, in (13.43) there appears the mathematical difficulty that the unknown matrix is multiplied by coefficient matrices from both the left and the right. Below conditions will be given concerning the existence and uniqueness of the solution $\boldsymbol{X}$, and also the methods of its calculation.

Existence

For given (real) $n \times n$ matrices $\boldsymbol{A}$ and $\boldsymbol{B}$ there exists at least one solution $\boldsymbol{X}$ of (13.43) if and only if there exists a (real) transformation matrix $\boldsymbol{T}$ such that

$$\boldsymbol{T}^{-1}\left[\begin{array}{c|c} \boldsymbol{A} & \boldsymbol{0} \\ \hline -\boldsymbol{B} & -\boldsymbol{A}^{\mathrm{T}} \end{array}\right]\boldsymbol{T}=\left[\begin{array}{c|c} \boldsymbol{A} & \boldsymbol{0} \\ \hline \boldsymbol{0} & -\boldsymbol{A}^{\mathrm{T}} \end{array}\right]. \tag{13.44}$$

This result becomes plausible if one writes (13.44) with the matrix

$$\boldsymbol{T}=\left[\begin{array}{c|c} \boldsymbol{E} & \boldsymbol{0} \\ \hline \boldsymbol{X} & \boldsymbol{E} \end{array}\right], \qquad \boldsymbol{T}^{-1}=\left[\begin{array}{c|c} \boldsymbol{E} & \boldsymbol{0} \\ \hline -\boldsymbol{X} & \boldsymbol{E} \end{array}\right] \tag{13.45}$$

and (13.43) is considered.

Uniqueness

The Ljapunov equation (13.43) has, for given (real) $n \times n$ matrices $\boldsymbol{A}$ and $\boldsymbol{B}$, a unique (real) solution $\boldsymbol{X}$ if and only if the matrices $\boldsymbol{A}$ and $(-\boldsymbol{A}^{\mathrm{T}})$ have no eigenvalues in common, i.e.,

$$\lambda_i + \lambda_j \neq 0, \quad i, j = 1(1)n, \tag{13.46}$$

where λ_k are the eigenvalues of $\boldsymbol{A}$. By (5.43), condition (13.46) is equivalent to

$$H_n = \det \boldsymbol{H} \neq 0, \tag{13.47}$$

where $\boldsymbol{H}$ is the Hurwitz matrix.

Consequently, if all eigenvalues of the matrix $\boldsymbol{A}$ have negative real parts, then the solution $\boldsymbol{X}$ of (13.43) is unique. Moreover, if the right side of (13.43) is symmetric, $\boldsymbol{B} = \boldsymbol{B}^T$, then the unique solution $\boldsymbol{X}$ is also symmetric.

Due to the numerous applications of the Ljapunov equation it is very important to devise methods of solving this equation analytically or numerically. Here, analytic procedures of finding the unique solution will be given; the numerical procedures will be dealt with in Section 14.2. Three analytical methods will be listed: the direct method of solution, the integral solution, and the matrix polynomial solution.

Direct solution procedure

This procedure is given under the assumption that $\boldsymbol{B}$ and thus $\boldsymbol{X}$ is a symmetric matrix. Then altogether there are $n(n+1)/2$ elements of $\boldsymbol{X}$ to be determined. If one arranges these elements in a vector

$$\boldsymbol{x} = [X_{11}\; X_{12} \ldots X_{1n}\; X_{22}\; X_{23} \ldots X_{2n}\; X_{33} \ldots X_{nn}]^{\mathrm{T}}, \tag{13.48}$$

containing only the elements of the upper triangular matrix of $\boldsymbol{X}$, and one forms for $\boldsymbol{B}$ a corresponding vector

$$\boldsymbol{b} = [B_{11}\; B_{12} \ldots B_{1n}\; B_{22}\; B_{23} \ldots B_{2n}\; B_{33} \ldots B_{nn}]^{\mathrm{T}}, \tag{13.49}$$

then (13.43) becomes equivalent to a linear vector equation

$$\hat{\boldsymbol{A}}\boldsymbol{x} = -\boldsymbol{b}. \tag{13.50}$$

The matrix $\hat{\boldsymbol{A}}$ is determined by the elements of $\boldsymbol{A}$ according to a unique logical prescription given in a general calculation program by Chen and Shieh. To solve equation (13.50) one can now apply the considerations in Sections 13.1 and 14.1.

Integral solution

If all eigenvalues of $\boldsymbol{A}$ have negative real parts, then the solution $\boldsymbol{X}$ of (13.43) can be written as a convergent improper integral

$$\boldsymbol{X} = \int_0^{\infty} \mathrm{e}^{\boldsymbol{A}^{\mathrm{T}}\tau} \boldsymbol{B}\, \mathrm{e}^{\boldsymbol{A}\tau} \mathrm{d}\tau. \tag{13.51}$$

This result can be easily verified. Since Re $\lambda_i < 0$, the integral (13.51) converges and one has

$$\begin{aligned}
\boldsymbol{A}^{\mathrm{T}}\boldsymbol{X} + \boldsymbol{X}\boldsymbol{A} &= \int_0^{\infty} \left[\boldsymbol{A}^{\mathrm{T}} \mathrm{e}^{\boldsymbol{A}^{\mathrm{T}}\tau} \boldsymbol{B}\, \mathrm{e}^{\boldsymbol{A}\tau} + \mathrm{e}^{\boldsymbol{A}^{\mathrm{T}}\tau} \boldsymbol{B}\, \mathrm{e}^{\boldsymbol{A}\tau}\boldsymbol{A}\right] \mathrm{d}\tau \\
&= \int_0^{\infty} \frac{\mathrm{d}}{\mathrm{d}\tau}\left[\mathrm{e}^{\boldsymbol{A}^{\mathrm{T}}\tau} \boldsymbol{B}\, \mathrm{e}^{\boldsymbol{A}\tau}\right] \mathrm{d}\tau \\
&= \mathrm{e}^{\boldsymbol{A}^{\mathrm{T}}\tau} \boldsymbol{B}\, \mathrm{e}^{\boldsymbol{A}t}\Big|_0^{\infty} = -\boldsymbol{B}.
\end{aligned} \tag{13.52}$$

Matrix polynomial solution

Using the Leverrier-Faddejev algorithm for the determination of the characteristic coefficients a_k, $k = 1(1)n$, (Section 4.3), one can represent the unique solution X of (13.43) by a finite matrix series. If one calculates, following the Leverrier-Faddejev algorithm, the quantities

$$\left.\begin{aligned} a_k &= -\frac{1}{k}\operatorname{tr}(AA_{k-1}), \quad a_0 = 1, \qquad A_0 = E, \\ A_k &= AA_{k-1} + a_k E, \quad k = 1(1)n, \end{aligned}\right\} \tag{13.53}$$

where for $k = n$ the relation $A_n = 0$ serves as a calculation check, and one uses the characteristic coefficients a_k to form the Hurwitz matrix

$$H = \begin{bmatrix} a_1 & 1 & 0 & 0 & \dots & 0 \\ a_3 & a_2 & a_1 & 1 & & \vdots \\ a_5 & a_4 & a_3 & a_2 & & \vdots \\ \vdots & & & \ddots & & \vdots \\ 0 & \cdots & \cdots & \cdots & \cdots & a_n \end{bmatrix}, \tag{13.54}$$

then for det $H \neq 0$ the unique solution of the Ljapunov equation (13.43) is given by

$$X = \sum_{k=0}^{n-1} c_{k+1} \sum_{l=0}^{2k} (-1)^l A_l^T B A_{2k-l}, \tag{13.55}$$

where c_i, $i = 1(1)n$, are the coordinates of the solution vector c of

$$Hc = \frac{1}{2}\begin{bmatrix} 1 \\ 0 \\ \vdots \\ 0 \end{bmatrix}, \tag{13.56}$$

thus corresponding to the cofactors of the first row of the Hurwitz matrix.

While the integral solution (13.51) is only of theoretical interest, the solutions (13.50) and (13.55) are quite useful for low order systems, as was seen in the investigations of Chapter 9. However for higher order systems one has to apply numerically stable solution algorithms, as shown in Section 14.2.

13.3. The Stein matrix equation

The bilinear matrix equation

$$A^T XA - X = -B \tag{13.57}$$

is called Stein's equation. It finds application in the stability theory of time-discrete systems as well as in the covariance analysis of random vibrations of periodic systems (Chapter 11). If the matrix $\boldsymbol{A}$ in (13.57) don't have an eigenvalue $\mu = 1$, then the matrix transformation

$$\left.\begin{aligned} \boldsymbol{A} &= (\boldsymbol{C} + a\boldsymbol{E})(\boldsymbol{C} - a\boldsymbol{E})^{-1}, \\ \boldsymbol{C} &= a(\boldsymbol{A} + \boldsymbol{E})(\boldsymbol{A} - \boldsymbol{E})^{-1} \end{aligned}\right\} \tag{13.58}$$

with $a \neq 0$ can be applied to reduce Stein's equation to the Ljapunov equation

$$\left.\begin{aligned} &\boldsymbol{C}^{\mathrm{T}}\boldsymbol{X} + \boldsymbol{X}\boldsymbol{C} = -\boldsymbol{D}, \\ &\boldsymbol{D} = 2a(\boldsymbol{A}^{\mathrm{T}} - \boldsymbol{E})^{-1}\boldsymbol{B}(\boldsymbol{A} - \boldsymbol{E})^{-1}. \end{aligned}\right\} \tag{13.59}$$

Thus the considerations of Section 13.2 can be carried over to Stein's equation.

If all eigenvalues of $\boldsymbol{A}$ have absolute values smaller than one, $|\mu_i| < 1$, then the solution $\boldsymbol{X}$ of (13.57) can be written as the sum of a convergent infinite series

$$\boldsymbol{X} = \sum_{k=0}^{\infty} \boldsymbol{A}^{\mathrm{T}^k}\boldsymbol{B}\boldsymbol{A}^k. \tag{13.60}$$

In Section 14.2 this solution of (13.57) will also be employed to solve Ljapunov's equation using the transformation (13.58).

14. Numerical aspects

For the analysis of vibrating systems with two or more degrees of freedom in general computational methods have to be applied. Only seldom are pencil and paper sufficient to determine the eigenvalues, natural modes, trajectories or frequency responses. Thus for higher system orders (usually $n \geqslant 4$) the aid of electronic computers is required. Here one distinguishes mainly between two types of machines, the analog computer and the digital computer which may also be combined to form a hybrid computer.

The *analog computer* works on the principle of a direct physical, and mostly electrical, analogy with the dynamic system. The physical quantities consisting of the information that has to be processed appear as continuous variables in certain value ranges. These signal values are connected to each other in a block-diagram, according to a simulation prescription, by elements such as potentiometers, summing amplifiers and multipliers, function generators and comparators. The result may be simulated time trajectories of the state coordinates or an amplitude frequency response. The analog computer is mainly applied for solving simultaneous differential equations. Details can be found in the literature mentioned in the references.

The *digital computer* processes the computed quantities as discrete values in a suitable number system. Its wide applicability is due to its ability to store and carry out programmed sequences of instructions. Thus a solution procedure must be formulated as a sequence of basic operations which the given digital computer can perform. For this the solution procedure has to be stated in a language comprehended by the computer (machine language) or in a programming language such as Fortran, Algol or Pascal which is translated subsequently by a special translating program (compiler) within the computer into machine language. After the program has been carried out by the computer, the results are printed or represented graphically by a suitable output unit.

Solution procedures of *numerical mathematics* are applied to the variety of problems in the mathematical-technical field and are often available in the computing centers as library programs. The following procedures are important to the investigation of linear vibrating systems:

a) numerical procedures to solve linear algebraic systems of equations;
b) numerical procedures for the determination of eigenvalues and eigenvectors;
c) numerical procedures for the calculation of the frequency response matrix and the fundamental matrix;
d) numerical integration procedures.

With the exception of c), the above are usually available among the standard procedures. The numerical procedures of matrix algebra are described especially well in various books and programmed in Fortran, Algol or Pascal. Excellent Algol procedures for the problems a) and b) can be found in the books of Dekker, Dekker and Hoffman, also Wilkinson and Reinsch; corresponding Fortran procedures can be read in Smith et al., Garbow et al. and in Dongarra et al. A catalogue of procedures for the matrix calculus has been compiled by Zielke.

In the subsequent sections short descriptions of the basic solution procedures are given, the details will be omitted. The reader will find the latter in monographs on matrices, numerical algebra, and integration procedures.

14.1. The linear vector equation

The solution of a linear vector equation

$$\boldsymbol{A}\boldsymbol{x} = \boldsymbol{b} \tag{14.1}$$

with a quadratic $n \times n$ matrix $\boldsymbol{A}$ proceeds usually by way of continued elimination of the unknowns until the system (14.1) is transformed into a so-called triangular system

$$\tilde{\boldsymbol{A}}\boldsymbol{x} = \tilde{\boldsymbol{b}} \tag{14.2}$$

with an upper triangular system matrix

$$\tilde{\boldsymbol{A}} = \begin{bmatrix} \tilde{a}_{11} & \tilde{a}_{12} & \cdots & \tilde{a}_{11} \\ & \tilde{a}_{22} & \ddots & \vdots \\ & \boldsymbol{0} & \ddots & \vdots \\ & & & \tilde{a}_{nn} \end{bmatrix}. \tag{14.3}$$

This elimination process consists of a sequence of elementary matrix transformations

$$\hat{\boldsymbol{A}} = \boldsymbol{T}\boldsymbol{A}, \qquad \tilde{\boldsymbol{b}} = \boldsymbol{T}\boldsymbol{b} \tag{14.4}$$

where the matrix $\boldsymbol{T}$ is regular. The process can be carried out by various methods, e.g., the Gauss elimination method or the Crout elimination method.

The solution of (14.3) proceeds by finding x_n from the last, then x_{n-1} from the second to last equation and so on, until x_1 is found from the first equation in (14.3).

Effective simplifications of these solution procedures can be obtained for symmetric matrices or for the so-called band matrices.

For overdetermined systems (14.1) with an $m \times n$ matrix $\boldsymbol{A}$, $m > r$ it is most convenient to decompose the matrix A according to (13.28),

$$\boldsymbol{A} = \boldsymbol{U\Sigma V^*}, \tag{14.5}$$

with

$$\boldsymbol{\Sigma} = \left[\begin{array}{cccc} \sigma_{11} & & \mathbf{0} & \\ & \sigma_{22} & & \\ \mathbf{0} & & \ddots & \\ & & & \sigma_{nn} \\ \hline & & \mathbf{0} & \end{array}\right]_{m \times n} \tag{14.6}$$

The solution $\boldsymbol{x}$ is then given by

$$\boldsymbol{x} = \boldsymbol{V\Sigma^+ U^* b} \tag{14.7}$$

The procedures for solving linear systems of equations are described in detail in Decker and in Wilkinson and Reinsch. Special attention is given to the numerical stability of the procedures which are written in Algol. Corresponding Fortran procedures are presented in Dongarra et al.

14.2. The Ljapunov matrix equation

The analytic solutions (13.50), (13.51), and (13.55) of Ljapunov's matrix equation

$$\boldsymbol{A}^{\mathrm{T}}\boldsymbol{X} + \boldsymbol{X}\boldsymbol{A} = -\boldsymbol{B} \tag{14.8}$$

are only partially suitable for numerical applications since they require too much computation time or else they do not allow for sufficient computational accuracy as a result of rounding errors. Thus, parallel to analytic solutions a series of numerical procedures for solving (14.8) is known. Among these will be a description of a direct, an iterative, and a transformation procedure.

The *direct solution procedure* is given by (13.48 to 50). For a symmetric matrix $B = B^T$ the bilinear system of equations (14.8) is transformed into a system (14.1) of order $n(n+1)/2$. Chen and Shieh have written a Fortran program carrying out this transformation. The advantage of this procedure is the simple programming; the disadvantage is the necessity of a large memory storage and the need for a large number of multiplicative operations which for large n is of the order of $n^6/24$. Thus the procedure only works fast and precisely for system orders of $n \leqslant 6, 7, 8$.

The *iterative solution procedure* makes use of the connection (13.58), (13.59) between the Ljapunov and Stein equations and of the solution (13.60). However it can only be applied for asymptotically stable system matrices A. According to Smith, the algorithm

$$\left.\begin{aligned} X_{\nu+1} &= X_\nu + \left[(A + aE)(A - aE)^{-1}\right]^{T^{2^\nu}} \\ &\qquad X_\nu \left[(A + aE)(A - aE)^{-1}\right]^{2^\nu}, \quad \nu = 1(1)\infty, \\ X_0 &= 2a(A^T - aE)^{-1} B (A - aE)^{-1}, \end{aligned}\right\} \tag{14.9}$$

converges to the solution

$$X = \lim_{\nu \to \infty} X_\nu. \tag{14.10}$$

Here a is a positive constant which is chosen so that on the one hand the matrix $A - aE$ can be conveniently inverted, and on the other hand the convergence in (14.9, 14.10) is fast. One may use for a one of the values

$$a \approx \sqrt[n]{\det(-A)}, \quad a \approx \sqrt{\frac{1}{n} \operatorname{Tr} A^T A}, \quad a \approx -\frac{1}{n} \operatorname{Tr} A. \tag{14.11}$$

An advantage of this procedure is its applicability to large system orders $n \geqslant 6, 7, 8$. A certain disadvantage lies in the restriction to asymtotically stable systems.

The *transformation procedure* consists of simplifying the form of (14.8) by an application of a similarity transformation to the system matrix A. Such a similarity transformation

$$\hat{A} = T^{-1} A T \tag{14.12}$$

takes (14.8) into

$$\hat{A}^T \hat{X} + \hat{X} \hat{A} = -\hat{B} \tag{14.13}$$

with

$$\left.\begin{aligned}\hat{\boldsymbol{X}} &= \boldsymbol{T}^{\mathrm{T}}\boldsymbol{X}\boldsymbol{T} = [\hat{\boldsymbol{x}}_1 \vdots \hat{\boldsymbol{x}}_2 \vdots \ldots \vdots \hat{\boldsymbol{x}}_n],\\ \hat{\boldsymbol{B}} &= \boldsymbol{T}^{\mathrm{T}}\boldsymbol{B}\boldsymbol{T} = [\hat{\boldsymbol{b}}_1 \vdots \hat{\boldsymbol{b}}_2 \vdots \ldots \vdots \hat{\boldsymbol{b}}_n].\end{aligned}\right\} \tag{14.14}$$

According to Kreisselmeier or Mayer-Spasche, a stable procedure consists of a transformation to the (lower) Hessenberg form:

$$\hat{\boldsymbol{A}} = \begin{bmatrix} h_{11} & h_{12} & & & \\ & & & \boldsymbol{0} & \\ h_{21} & h_{22} & h_{23} & \ddots & \\ \vdots & \vdots & \vdots & & \ddots \\ h_{n-1,1} & h_{n-1,2} & h_{n-1,3} & \cdots & h_{n-1,n} \\ h_{n1} & h_{n2} & h_{n3} & \cdots & h_{nn} \end{bmatrix}, \tag{14.15}$$

see Section 14.3. The equation (14.13) can then be solved recursively columnwise:

$$\hat{\boldsymbol{x}}_n = \boldsymbol{G}^{-1}\boldsymbol{d}, \tag{14.16a}$$

$$\hat{\boldsymbol{x}}_{i-1} = -\frac{1}{h_{i-1,i}}\left(\hat{\boldsymbol{A}}^{\mathrm{T}}\hat{\boldsymbol{x}}_i + \sum_{k=i}^{n} h_{ki}\hat{\boldsymbol{x}}_k + \hat{\boldsymbol{b}}_i\right), \quad i = n(1)2. \tag{14.16b}$$

Here $\boldsymbol{G}$ and $\boldsymbol{d}$ are determined as follows from formula (14.16b): If one takes $\hat{\boldsymbol{x}}_n = \boldsymbol{0}$, then one obtains from (14.16b) after the intermediary steps for $i = n,\ n-1, \ldots, 2$ finally for $i = 1$ the vector $\boldsymbol{d} = -\hat{\boldsymbol{x}}_0$; if one however puts $\hat{\boldsymbol{x}}_n = \boldsymbol{e}_k$ (k-th unit vector) and $\hat{\boldsymbol{B}} = 0$, then one obtains from (14.16b) the k-th column of $\boldsymbol{G}$ as $(\boldsymbol{G})_k = \hat{\boldsymbol{x}}_0$.

The procedure must be slightly modified if $h_{i-1,i}$ vanishes for some i. The algorithm (14.16) then has to be carried out sectionwise. The procedure can be used in all cases when there exists a unique solution of (14.8). It consists of well-known and numerically very stable procedures and is especially recommended when Hessenberg's form has to be determined for other reasons, e.g., for solving the eigenvalue problem.

Comparisons between various numerical procedures for solving Ljapunov's equation have been carried out by Pace and Barnett, by Hagander, and also by Koupan and Müller.

14.3. Eigenvalues, eigenvectors

The numerical calculation of all the eigenvalues of a matrix $\boldsymbol{A}$ is based on the invariance (4.94) of the eigenvalues under similarity transformations. The aim is then to find a similarity transformation which reduces A to a triangular form; the eigenvalues of A are then the elements on the main diagonal. To reduce the number of steps in the calculation and to make the procedure numerically stable, the transformation matrices are chosen so that the necessary inverse matrices can be given explicitly without further numerical calculation.

The following method is usually used in digital computations: Following the *procedure of Hessenberg* or the *procedure of Householder* the matrix is transformed to a *quasi-triangular form*

$$\hat{A} = T^{-1} A T = \begin{bmatrix} * & * & \cdots & * \\ * & * & \cdots & * \\ & \mathbf{0} \quad * & \ddots & \vdots \\ & & * & * \end{bmatrix} \tag{14.17}$$

which is also called the (upper) Hessenberg form. Next there follows a sequence of *LR-transformations of Rutishauser* which takes the quasi-triangular matrix (14.17) to an (upper) triangular hypermatrix

$$\hat{\hat{A}} = V^{-1} \hat{A} V = \begin{bmatrix} * & * & \cdots & \cdots & \cdots & \cdots & \cdots & * \\ & * & & & & & & \vdots \\ & & & * & * & & & \vdots \\ & & & * & * & \ddots & & \vdots \\ & \mathbf{0} & & & & & * & * \\ & & & & & & * & * \end{bmatrix}. \tag{14.18}$$

Along the main diagonal one finds the real eigenvalues and two-dimensional submatrices of the type

$$D = \begin{bmatrix} a & b \\ c & d \end{bmatrix} \tag{14.19}$$

for the complex conjugate eigenvalues which are calculated from

$$\lambda^2 - (a + d)\lambda + (ad - bc) = 0. \tag{14.20}$$

Thus the *eigenvalues* are determined.

The eigenvectors corresponding to the known eigenvalues which are all assumed distinct, are first calculated for the transformed matrix (14.17):

$$(\lambda_i E - \hat{A})\tilde{x}_i = \mathbf{0}, \qquad i = 1(1)n. \tag{14.21}$$

Since the solution of a homogeneous system of equations is determined only up to a constant multiplier, in determining $\tilde{\hat{x}}_i$ one first adopts $\tilde{\hat{x}}_{ni} = 1$. Then (14.21) is a triangular system of the form (14.2) for the coordinates $\tilde{\hat{x}}_{1i}, \ldots, \tilde{\hat{x}}_{n-1,i}$ and thus it can be solved recursively. If this procedure leads to a contradiction, then one has to set $\tilde{\hat{x}}_{ni} = 0$, and then repeat the procedure for a system (14.21) whose order is now reduced by one. As a result, one obtains the eigenvectors $\tilde{\hat{x}}_i$, $i = 1(1)n$, which still have to be transformed back to the original coordinate system by the transformation $\boldsymbol{x} = \boldsymbol{T}\hat{\boldsymbol{x}}$.

This basic method of determining the eigenvalues and eigenvectors has been repeatedly improved. Details can be found in Wilkinson and in Zurmühl. The corresponding Algol procedures are presented in Dekker and Hoffman, and in Wilkinson and Reinsch. Fortran procedures have been written by Smith et al., and by Garbow et al.

14.4. Frequency response

The complex frequency response vector (7.42)

$$\boldsymbol{g} = (\mathrm{i}\Omega \boldsymbol{E} - \boldsymbol{A})^{-1}\boldsymbol{b} \tag{14.22}$$

can be calculated by solving the linear system of equations

$$(\mathrm{i}\Omega \boldsymbol{E} - \boldsymbol{A})\boldsymbol{g} = \boldsymbol{b} \tag{14.23}$$

as shown in Section 14.1. The system (14.23) has to be solved for each desired value of $\Omega = \Omega_k$. For instance, if amplitude frequency responses $a_i(\Omega)$ (7.45) are required for a frequency interval $\bar{\Omega} = [\Omega_1, \Omega_2]$, then one has to solve (14.23) for a sufficiently dense set of discrete frequency values, e.g., for $\Omega_{k+1} = \Omega_1 + (\Omega_2 - \Omega_1)/Nk$, $k = 0(1)n$, and for each of these to calculate the amplitude according to (7.45). This procedure demands much computation time but cannot be avoided in the case when the system matrix depends on the frequency, $\boldsymbol{A} = \boldsymbol{A}(\Omega)$.

However, if the matrix $\boldsymbol{A}$ does not depend on the frequency the following procedure to determine $\boldsymbol{g}(\Omega)$ in a given frequency range $\bar{\Omega}$ is to be recommended. Assuming that the eigenvalues λ_l of $\boldsymbol{A}$ are distinct, one determines the eigenvectors $\tilde{\boldsymbol{x}}_l$, $l = 1(1)n$, according to Section 14.3. One can then write (14.22) as

$$\begin{aligned} \boldsymbol{g} &= \left[\boldsymbol{X}(\mathrm{i}\Omega \boldsymbol{E} - \boldsymbol{X}^{-1}\boldsymbol{A}\boldsymbol{X})\boldsymbol{X}^{-1}\right]^{-1}\boldsymbol{b} \\ &= \boldsymbol{X}(\mathrm{i}\Omega \boldsymbol{E} - \boldsymbol{\Lambda})^{-1}\boldsymbol{X}^{-1}\boldsymbol{b} \\ &= \boldsymbol{X}\,\mathrm{diag}\left\{\frac{1}{\mathrm{i}\Omega - \lambda_l}\right\}\boldsymbol{X}^{-1}\boldsymbol{b}. \end{aligned} \tag{14.24}$$

where $\boldsymbol{X}$ is the modal matrix as in (4.71). Now, if the excitation vector depends on the frequency,

$$\boldsymbol{b}(\Omega)=\boldsymbol{b}_0+\Omega\boldsymbol{b}_1+\Omega^2\boldsymbol{b}_2, \tag{14.25}$$

then the vectors

$$\hat{\boldsymbol{b}}_j=\boldsymbol{X}^{-1}\boldsymbol{b}_j, \qquad j=0,1,2, \tag{14.26}$$

can be calculated independently of Ω, according to Section 14.1. Then for each frequency $\Omega=\Omega_k$ each coordinate of $\tilde{b}_{jl}$ has to be weighted by the factor $\Omega_k^j/(\mathrm{i}\Omega_k-\lambda_l)$; the so obtained weighted vectors have to be summed and the vector sum has to multiplied by $\boldsymbol{X}$ to finally obtain $\boldsymbol{g}(\Omega_k)$:

$$\boldsymbol{g}(\Omega_k)=\boldsymbol{X}\left[\sum_{j=0}^{2}\frac{\Omega_k^j}{\mathrm{i}\Omega_k-\lambda_l}\hat{b}_{jl}\right]. \tag{14.27}$$

Although the problem of finding all eigenvalues and eigenvectors is much more cumbersome than the solution of a linear equation, the solution (14.24) for many different frequencies is more than justified by the great simplicity of the operations (14.26, 14.27). For the calculation of amplitudes and phase frequency responses the above procedure requires essentially less computing time than solving (14.23) for various frequencies.

14.5. Fundamental matrix

In calculating the fundamental matrix

$$\boldsymbol{\Phi}(t)=\mathrm{e}^{\boldsymbol{A}t} \tag{14.28}$$

one has to consider whether $e^{\boldsymbol{A}t}$ is required only for one single fixed time $t=T$, or if $e^{\boldsymbol{A}t}$ has to be known as a function of t.

In the first case it is advisable to write the time T as

$$T=2^k h. \tag{14.29}$$

Then $e^{\boldsymbol{A}\mathrm{T}}$ is obtained by squaring k times $\boldsymbol{\Phi}(h)=e^{\boldsymbol{A}h}$:

$$\mathrm{e}^{\boldsymbol{A}\mathrm{T}}=\boldsymbol{\Phi}(T)=\boldsymbol{\Phi}(2^k h)=\boldsymbol{\Phi}^{2^k}(h)=\Big(\underbrace{\ldots(\boldsymbol{\Phi}(h))^2\ldots}_{k}\Big)^2. \tag{14.30}$$

The size of h is chosen small enough so that $\boldsymbol{\Phi}(h)$ is well approximated by a finite segment of the exponential series, e.g.,

$$\boldsymbol{\Phi}(h) \approx \boldsymbol{E} + \frac{1}{1!}\boldsymbol{A}h + \frac{1}{2!}\boldsymbol{A}^2h^2 + \frac{1}{3!}\boldsymbol{A}^3h^3 + \frac{1}{4!}\boldsymbol{A}^4h^4 \tag{14.31}$$

or, in general,

$$\left.\begin{aligned} \boldsymbol{\Phi}_{k+1}(h) &= \boldsymbol{\Phi}_k(h) + [\boldsymbol{\Phi}_k(h) - \boldsymbol{\Phi}_{k-1}(h)]\boldsymbol{A}\frac{h}{k+1}, \quad k = 1(1)k_0, \\ \boldsymbol{\Phi}_{-1}(h) &= \boldsymbol{0}, \quad \boldsymbol{\Phi}_0(h) = \boldsymbol{E}, \\ \boldsymbol{\Phi}(h) &\approx \boldsymbol{\Phi}_{k_0+1}(h), \end{aligned}\right\} \tag{14.32}$$

where k_0 is chosen so that

$$\|\boldsymbol{\Phi}_{k_0}(h) - \boldsymbol{\Phi}_{k_0-1}(h)\| < \epsilon \|\boldsymbol{\Phi}_{k_0}(h)\| \tag{14.33}$$

holds for a prescribed accuracy number ϵ. If the approximation (14.31) is chosen for $\boldsymbol{\Phi}(h)$, then the error is of the order h^5. This also corresponds to the accuracy of the Padé approximation

$$\boldsymbol{\Phi}(h) \approx \left(\boldsymbol{E} - \frac{h}{2}\boldsymbol{A} + \frac{h^2}{12}\boldsymbol{A}^2\right)^{-1}\left(\boldsymbol{E} + \frac{h}{2}\boldsymbol{A} + \frac{h^2}{12}\boldsymbol{A}^2\right), \tag{14.34}$$

which involves only $\boldsymbol{E}$, $\boldsymbol{A}$, and $\boldsymbol{A}^2$, but also requires solving the linear system of equations

$$\left(\boldsymbol{E} - \frac{h}{2}\boldsymbol{A} + \frac{h^2}{12}\boldsymbol{A}^2\right)\boldsymbol{\Phi}(h) = \left(\boldsymbol{E} + \frac{h}{2}\boldsymbol{A} + \frac{h^2}{12}\boldsymbol{A}^2\right). \tag{14.35}$$

Now suppose that the fundamental matrix has to be found for arbitrary times. Then there are two possibilities of finding a solution. By (4.90), one has

$$\boldsymbol{\Phi}(t) = \boldsymbol{X}\,\mathrm{e}^{\boldsymbol{J}t}\boldsymbol{X}^{-1}, \tag{14.36}$$

where $e^{\boldsymbol{J}t}$ is known explicitly from (4.88, 4.89) and $\boldsymbol{X}$ is the generalized modal matrix (4.84). Since in the calculation of eigenvectors one can make use of numerically stable and well tested procedures only if the eigenvalues are distinct, the solution (14.36) is to be recommended only for the most frequent Case I when the system matrix is similar to a diagonal one. Then, by (4.78),

$$\begin{aligned}\boldsymbol{\Phi}(t) &= \boldsymbol{X}\,\mathrm{diag}\{\mathrm{e}^{\lambda_i t}\}\,\boldsymbol{X}^{-1}\\ &= \sum_{i=1}^{n} \mathrm{e}^{\lambda_i t}\tilde{\boldsymbol{x}}_i\,\tilde{\boldsymbol{y}}_i^{\mathrm{T}},\end{aligned} \tag{14.37}$$

where $\tilde{\boldsymbol{x}}_i$ are the column vectors of $\boldsymbol{X}$ and $\tilde{\boldsymbol{y}}_i^{\mathrm{T}}$ are the row vectors of $\boldsymbol{X}^{-1}$. Since $e^{\lambda_i t}$ are standard functions which the digital computer immediately provides on request, only the determination of λ_i and $\tilde{\boldsymbol{x}}_i\,\boldsymbol{y}_i^{\mathrm{T}}$, $i = 1(1)n$, is necessary to calculate the fundamental matrix according to (14.37) at every instant t.

The second possibility to calculate the fundamental matrix $\boldsymbol{\Phi}(t)$ explicitly for arbitrary instants t is based on the theorem (4.130) of Cayley and Hamilton. According to (4.131), one has

$$\boldsymbol{\Phi}(t) = \alpha_0(t)\boldsymbol{E} + \alpha_1(t)\boldsymbol{A} + \ldots + \alpha_{n-1}(t)\boldsymbol{A}^{n-1}, \tag{14.38}$$

where the coefficient functions $\alpha_i(t)$ can be obtained from the system of equations (4.132). By inverting the generalized Vandermonde matrix appearing there, one can express the $\alpha_i(t)$ as linear combinations of the elementary functions $t^s\,\mathrm{e}^{\lambda_i t}$. Thus the matrix polynomial (14.38) can also be evaluated at every time instant. Since only the eigenvalues, and not the eigenvectors were used here, the procedure can also be applied for multiple eigenvalues.

14.6. Analog integration of the fundamental matrix

The fundamental matrix $\boldsymbol{\Phi}(t, t_0)$ of a time-variant, and of course also a time-invariant dynamic system, can always be calculated by an analog computer from the matrix differential equation

$$\dot{\boldsymbol{\Phi}}(t, t_0) = \boldsymbol{A}(t)\boldsymbol{\Phi}(t, t_0), \quad \boldsymbol{\Phi}(t_0, t_0) = \boldsymbol{E}. \tag{14.39}$$

For this purpose one reduces (14.39) to the vector differential equation

$$\dot{\boldsymbol{\varphi}}(t, t_0) = \boldsymbol{A}(t)\boldsymbol{\varphi}(t, t_0), \tag{14.40}$$

which is solved for the n different initial conditions

$$\boldsymbol{\varphi}(t_0, t_0) = \boldsymbol{e}_i, \qquad i = 1(1)n, \tag{14.41}$$

where $\boldsymbol{e}_i$ are the unit vectors; the solution corresponding to $\boldsymbol{e}_i$ is the i-th column of the fundamental matrix.

For the integration of time-variant systems one uses the linear and nonlinear elements of the analog computer shown in Fig. 14.1. Combining these elements, one can design a block-diagram providing a complete simulation of the system (14.40). As a result, one obtains the trajectory of the state coordinates which can be read from oscillographs, XY plotters, or digital printers.

For example, in Fig. 14.2 there can be seen the block-diagram for the damped Mathieu differential equation

$$\ddot{y}(t) + 2D\dot{y}(t) + (\beta^2 + \gamma \cos t)\, y(t) = 0. \tag{14.42}$$

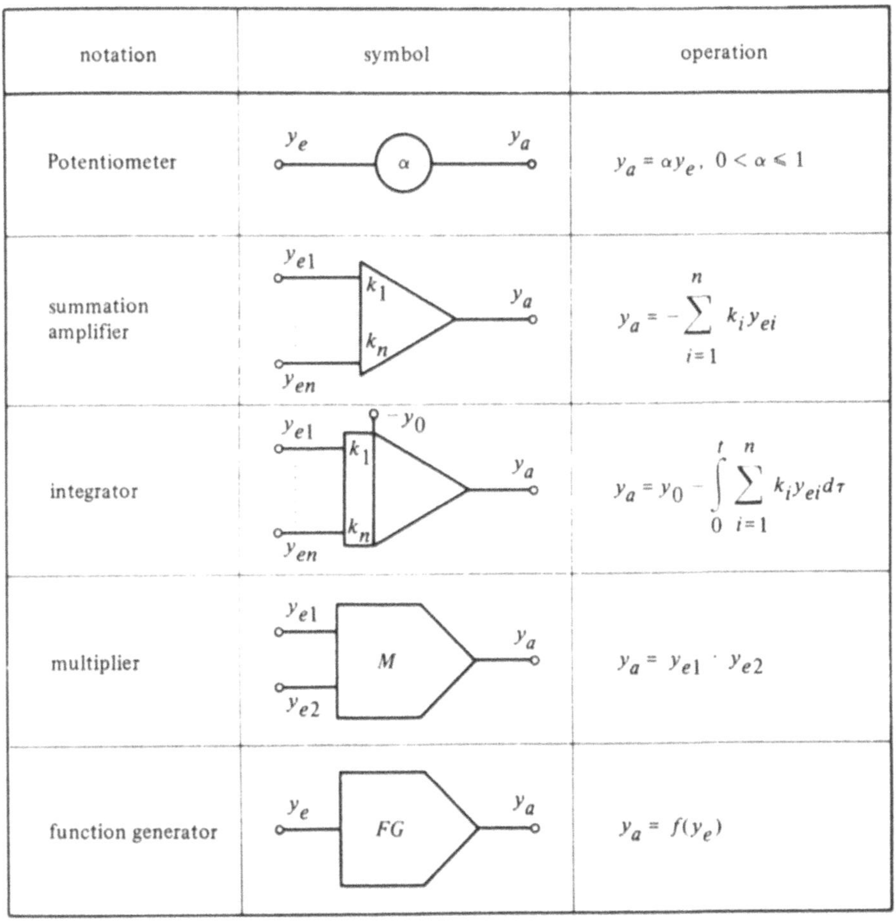

Fig. 14.1. Linear and nonlinear elements of an analoge computer

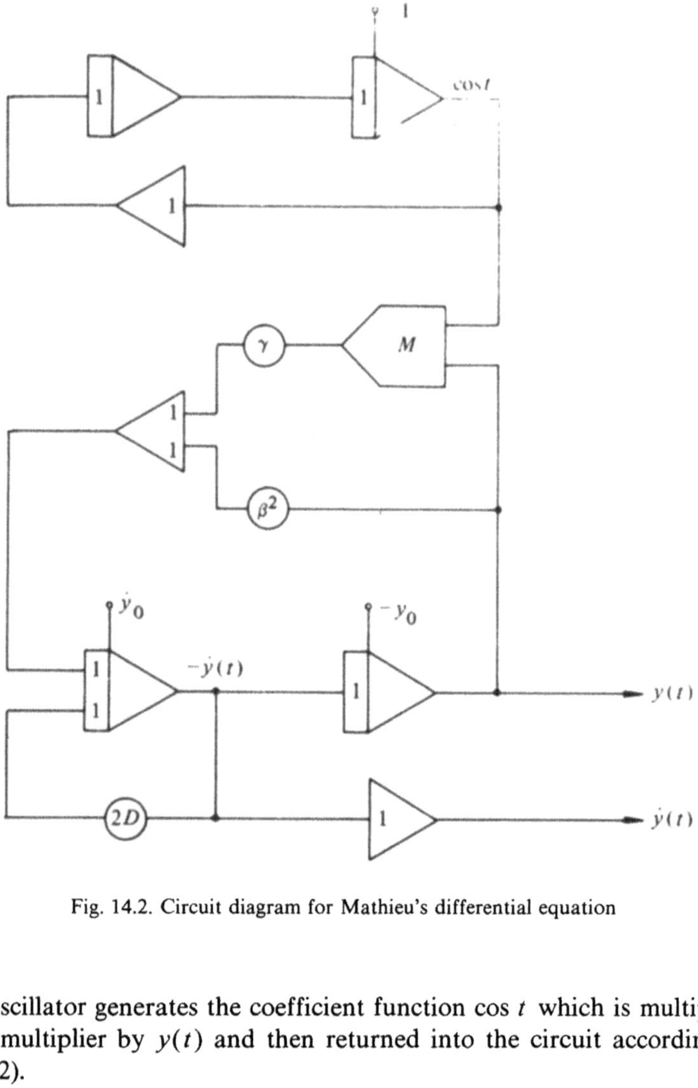

Fig. 14.2. Circuit diagram for Mathieu's differential equation

An oscillator generates the coefficient function cos t which is multiplied in a multiplier by $y(t)$ and then returned into the circuit according to (14.42).

14.7. Digital integration of the fundamental matrix

In general, the digital calculation of the fundamental matrix $\boldsymbol{\Phi}(t, t_0)$ of a time-variant system cannot be based just on algebraic quantities such as eigenvalues and eigenvectors. One has to apply numerical procedures such as presented in Collatz, Knapp and Wanner, Wanner, or Stoer and Burlisch. For instance, consider the classical Runge-Kutta procedure. It solves a nonlinear system of differential equations

$$\dot{\boldsymbol{x}}(t) = \boldsymbol{a}(\boldsymbol{x}(t), t), \quad \boldsymbol{x}(t_0) = \boldsymbol{x}_0 \tag{14.43}$$

by performing a step-by-step calculation of the states $x(t_{k+1}) = x(t_k + \Delta t)$ for $t_k = t_0 + k(\Delta t)$, $k = 1, 2, 3, \ldots$, starting from the state x_0. The procedure is applied until the desired terminating time is reached. For the state $x(t_{k+1})$ one has, according to Runge and Kutta, approximately

$$x(t_k + \Delta t) \approx x(t_k) + \tfrac{1}{6}\left[\Delta x_k^{(1)} + 2\Delta x_k^{(2)} + 2\Delta x_k^{(3)} + \Delta x_k^{(4)}\right], \qquad (14.44)$$

where the quantities $\Delta x_k^{(i)}$, $i = 1(1)4$, are determined as follows:

$$\left.\begin{aligned}
\Delta x_k^{(1)} &= a(x(t_k), t_k)\Delta t,\\
\Delta x_k^{(2)} &= a\left(x(t_k) + \tfrac{1}{2}\Delta x_k^{(1)}, t_k + \tfrac{1}{2}\Delta t\right)\Delta t,\\
\Delta x_k^{(3)} &= a\left(x(t_k) + \tfrac{1}{2}\Delta x_k^{(2)}, t_k + \tfrac{1}{2}\Delta t\right)\Delta t,\\
\Delta x_k^{(4)} &= a\left(x(t_k) + \Delta x_k^{(3)}, t_k + \Delta t\right)\Delta t.
\end{aligned}\right\} \qquad (14.45)$$

The order of magnitude of the error in (14.44) is $(\Delta t)^5$.

If one applies the Runge-Kutta procedure (14.44, 14.45) to linear homogeneous time-variant systems

$$\dot{x}(t) = A(t)x(t), \qquad x(t_0) = x_0 \qquad (14.46)$$

for the determination of the corresponding fundamental matrix $\Phi(t, t_0)$ one finds that

$$\begin{aligned}
\Phi(t_k + \Delta t, t_0) \approx \Big\{ E &+ \frac{1}{1!}\frac{1}{6}\left[A(t_k) + 4A\left(t_k + \tfrac{1}{2}\Delta t\right) + A(t_k + \Delta t)\right]\Delta t\\
&+ \frac{1}{2!}\frac{1}{3}\Big[A\left(t_k + \tfrac{1}{2}\Delta t\right)A(t_k) + A^2\left(t_k + \tfrac{1}{2}\Delta t\right)\\
&+ A(t_k + \Delta t)A\left(t_k + \tfrac{1}{2}\Delta t\right)\Big](\Delta t)^2\\
&+ \frac{1}{3!}\frac{1}{2}\Big[A^2\left(t_k + \tfrac{1}{2}\Delta t\right)A(t_k)\\
&+ A(t_k + \Delta t)A^2\left(t_k + \tfrac{1}{2}\Delta t\right)\Big](\Delta t)^3\\
&+ \frac{1}{4!}A(t_k + \Delta t)A^2\left(t_k + \tfrac{1}{2}\Delta t\right)A(t_k)(\Delta t)^4\Big\}\Phi(t_k, t_0).
\end{aligned}$$

$$(14.47)$$

For time-invariant systems (14.47) reduces to (14.31). This means, for time-invariant systems the Runge-Kutta procedure is equivalent to terminating the expansion into the exponential series at the fourth power term. For time-variant systems there is formed in (14.47) an average of the matrices $\boldsymbol{A}(t_k)$, $\boldsymbol{A}(t_k + \frac{1}{2}\Delta t)$, and $\boldsymbol{A}(t_k + \Delta t)$, where the influence of the middle value $\boldsymbol{A}(t_k + \frac{1}{2}\Delta t)$ is especially strong.

References

Andronow, A.A., and Witt, A.A.: Theorie der Schwingungen, Part 1 and 2, 2nd edition. Akademie-Verlag, Berlin 1965 and 1969.

Brommundt, E.: Vibrations of Continuous Systems. Courses and Lectures No. 1. CISM, Udine 1969.

Bronstein, I.N., and Semendjajew, K.A.: Taschenbuch der Mathematik, 6th edition. Verlag H. Deutsch, Zürich-Frankfurt a. M. 1966.

Bucy, R.S., and Joseph, P.D.: Filtering for Stochastic Processes with Application to Guidance. Wiley, New York 1968.

Bunke, H.: Gewöhnliche Differentialgleichungen mit zufälligen Parametern. Akademie-Verlag, Berlin 1972.

Caughey, T.K., and O'Kelly, M.E.J.: Classical Normal Modes in Damped Linear Dynamic Systems. ASME Transact., Ser. E, J. Appl. Mech. 32 (1965) 583–588.

Cesari, L.: Asymptotic Behavior and Stability Problems in Ordinary Differential Equations. 2nd edition. Springer-Verlag, Berlin-Heidelberg-New York 1963.

Chen, C.F., and Shieh, L.S.: A Note on Expanding $PA = A^T P = -Q$. IEEE Trans. AC-13 (1968) 122–123.

Coddington, E.A., and Levinson, N.: Theory of Ordinary Differential Equations. McGraw-Hill, New York-Toronto-London 1955.

Collatz, L.: Differentialgleichungen, 5th edition. B.G. Teubner, Stuttgart 1973.

Crandall, S.H.: Random Vibrations. MIT Press, Cambridge 1963.

Den Hartog, J.P.: Mechanical Vibrations, 4th edition. McGraw-Hill, New York-Toronto-London 1956.

Dekker, T.J.: Algol 60 Procedures in Numerical Algebra, Part 1. Mathematisch Centrum Amsterdam, 1968.

Dekker, T.J., and Hoffmann, W.: ALGOL 60 Procedures in Numerical Algebra, Part 2. Mathematisch Centrum Amsterdam 1968.

Deutsche Normen, DIN 1311: Schwingungslehre, Blatt 1–4, Beuth-Vertrieb, Berling-Köln 1961.

Dongarra, J.J., Moler, C.B., Bunch, J.R., and Stewart, G.W.: LINPACK Users' Guide. SIAM, Philadelphia 1979.

Drenick, R.F.: Die Optimierung linearer Regelsysteme. Oldenbourg-Verlag, München-Wien 1967.

Eykhoff, P.: System Identification. Wiley, London-New York-Sydney 1974.

Fabian, L.: Zufallsschwingungen und ihre Behandlung. Springer-Verlag, Berlin-Heidelberg-New York 1973.

Fertis, D.G.: Dynamics and Vibration of Structures. Wiley-Interscience, New York-London-Sidney-Toronto 1973.

Fischer, U., and Stephan, W.: Prinzipien und Methoden der Dynamik. VEB Fachbuchverlag, Leipzig 1972.

Fischer, U., and Stephan, W.: Schwingungen. VEB Fachbuchverlag, Leipzig, and Birkhäuser-Verlag, Basel-Boston-Stuttgart 1981.

Floquet, G.: Sur les équations différentielles linéaires à coefficients périodiques. Ann. Ecole Norm. (2) **12** (1883) 47–89.

Forbat, N.: Analytische Mechanik der Schwingungen. VEB Deutscher Verlag der Wissenschaften, Berlin 1966.

Gantmacher, F.R.: Matrizenrechnung I and II, 2nd edition Deutscher Verlag der Wissenschaften, Berlin 1965 and 1966.

Garbow, B.S., Boyle, J.M., Dongarra, J.J., and Moler, C.B.: Matrix Eigensystem Routines – EISPACK Guide Extension. Lecture Notes in Computer Science, Vol. 51, Springer-Verlag, Berlin-Heidelberg-New York 1977.

Giloi, W., and Lauber, R.: Analogrechnen. Springer-Verlag, Berlin-Göttingen-Heidelberg 1963.

Gröbner, W.: Matrizenrechnung. BI-Hochschultaschenbuch 103/103a, Mannheim 1966.

Hagander, P.: Numerical Solution of $A^TS+SA+Q=0$. Inform. Sciences **4** (1972) 35–50.

Hahn, W.: Stability of Motion. Die Grundlehren der mathematischen Wissenschaften, Vol. 138. Springer-Verlag, Berlin-Heidelberg-New York 1967.

Hautus, M.L.J.: Controllability and Observability Conditions of Linear Autonomous Systems. Indagationes Mathematicae **31** (1969) 443–448.

Heinhold, J., and Kulisch, U.: Analogrechnen. BI-Hochschultaschenbuch 168/168a. Mannheim-Zürich 1969.

Hermite, Ch.: Sur le nombre des racines d'une équation algébrique comprises entre deux limite donnée. J. Reine Angew. Math. **52** (1856) 39–51.

Hurwitz, A.: Über die Bedingungen unter welchen eine Gleichung nur Wurzeln mit negativen reellen Teilen Besitzt. Math. Ann. **46** (1895) 273–284.

Huseyin, K.: Vibrations and Stability of Multiple Parameter Systems. Sijthoff & Noordhoff, Alphen aan den Rijn 1978.

Isermann, R.: Prozessidentifikation. Springer-Verlag, Berlin-Heidelberg-New York 1974.

James, H.M., Nichols, N.B., and Phillips, R.S.: Theory of Servomechanisms. McGraw-Hill, New York 1947.

Jury, E.I.: Inners and Stability of Dynamic Systems. Wiley-Interscience, New York-London-Sydney-Toronto 1974.

Kalman, R.E., Ho, Y.C., and Narendra, K.S.: Controllability of Linear Dynamical Systems. Contrib. Diff. Equations **1** (1962) 189–213.

Klotter, K.: Technische Schwingungslehre, Vol. 1, Part A and B, 3rd edition, 1978 and 1980; Vol. 2, 2nd edition, 1960 (reprint 1981). Springer-Verlag, Berlin-Göttingen-Heidelberg.

Knapp, H., and Wanner, G.: Numerische Integration gewöhnlicher Differentialgleichungen; Einschrittverfahren. In: Laugwitz, D. (ed.): Überblicke I. BI-Hochschultaschenbuch 161/161a, Mannheim-Zürich 1968, p. 87–114.

Knobloch, H.W., and Kappel, F.: Gewöhnliche Differentialgleichungen. B.G. Teubner, Stuttgart 1974.

Kotowski, G.: Lösungen der inhomogenen Mathieuschen Differentialgleichung mit periodischer Störfunktion beliebiger Frequenz. Z. Angew. Math. Mech. **23** (1943) 213–229.

Koupan, A., and Müller, P.C.: Zur numerischen Lösung der Ljapunovschen Matrizengleichung $A^TP+PA=-Q$. Regelungstechnik **24** (1976) 167–169.

Krebs, V.: Das Gleichgewichtstheorem – eine grundsätzliche Aussage über das Verhalten von Regelkreisen, Part 1 und 2. Regelungstechnik und Prozess-Datenverarbeitung **21** (1973) 25–27 and 56–59.

Kreisselmeier, G.: A Solution of the Bilinear Matrix Equation $AY+YB=-Q$.SIAM J. Appl. Math. **23** (1972) 334–338.

Kwakernaak, H., and Sivan, R.: Linear Optimal Control Systems. Wiley-Interscience, New York-London-Sydney-Toronto 1972.

Lakshimikantham, C., and Wan, F.Y.M.: An Algebraic Method for Linear Dynamical Systems with Stationary Excitations. J. Sound Vibration **28** (1973) 105–120.

Lancaster, P.: Theory of Matrices. Academic Press, New York-London 1969.
Liénard, A., and Chipart, M.H.: Sur le signe de la partie réelle des racines d'une équation algébrique. J. Math. Pures Appl. **10** (1914) 291–346.
Lippmann, H.: Schwingungslehre. BI-Hochschultaschenbuch 189/189a. Mannheim 1968.
Ljapunov, A.M.: Problème générale de la stabilité de mouvement. Ann. Fac. Sci. Toulouse **9** (1907) 203–474 (French translation of the original Russian paper in Comm. Soc. Math. Charkow, 1893).
Magnus, K.: Schwingungen, 2nd edition. B.G. Teubner, Stuttgart 1969.
Magnus, K., Lückel, J., Müller, P.C., and Schiehlen, W.: Moderne Schwingungstechnik. Technische Universität München and Münchner Volkshochschule, Kontaktstudium Q 12, München 1971.
Magnus, K., and Müller, H.H.: Grundlagen der Technischen Mechanik. B.G. Teubner, Stuttgart 1974.
Mahrenholtz, O.: Analogrechnen in Maschinenbau und Mechanik. BI-Hochschultaschenbuch 154/154a, Mannheim-Zürich 1968.
McCallion, H.: Vibration of Linear Mechanical Systems. Longman, London 1973.
Meirovitch, L.: Analytical Methods in Vibration. Macmillan, New York 1967.
Meyer-Spasche, R.: A Constructive Method of Solving the Liapunov Equation for Complex Matrices. Numer. Math. **19** (1972) 433–438.
Meyer, E., and Guicking, D.: Schwingungslehre. Vieweg, Braunschweig 1974.
Müller, P.: Die Berechnung von Ljapunov-Funktionen und von quadratischen Regelflächen für lineare, stetige, zeitinvariante Mehrgrössensysteme. Regelungstechnik **17** (1969) 341–345.
Müller, P.C.: Solution of the Matrix Equation $AX+XB=-Q$ and $S^TX+XS=-Q$. SIAM J. Appl. Math. **18** (1970) 682–687.
Müller, P.C.: Asymptotische Stabilität von linearen mechanischen Systemen mit positiv semidefiniter Dämpfungsmatrix. Z. Angew. Math. Mech. **51** (1971) T197–T198.
Müller, P.C.: Stabilität und Instabilität bei linearen, zeitinvarianten, dynamischen Systemen. Z. Angew. Math. Mech. **54** (1974) T55–T56.
Müller, P.C.: Stabilität und Matrizen. Springer-Verlag, Berlin-Heidelberg-New York 1977.
Müller, P.C.: Lineare stationäre Zufallsschwingungen bei farbigem Rauschen. Z. Angew. Math. Mech. **58** (1978) T164–T165.
Müller, P.C., Popp, K., and Schiehlen, W.O.: Berechnungsverfahren für stochastische Fahrzeugschwingungen. Ing.-Arch. **49** (1980) 235–254.
Pace, I.S., and Barnett, S.: Comparison of Numerical Methods for Solving Liapunov Matrix Equations. Internat. J. Control **15** (1972) 907–915.
Papoulis, A.: Probability, Random Variables and Stochastic Processes. McGraw-Hill, New York 1965.
Pontrjagin, L.S.: Gewöhnliche Differentialgleichungen. VEB Deutscher Verlag der Wissenschaften, Berlin 1965.
Roberson, R.E., Debra, D., and Wittenburg, J.: Gyrodynamical Applications to Spacecraft. Courses and Lectures No. 154. CISM, Udine 1975.
Routh, E.J.: A Treatise on the Stability of a Given State of Motion (Adams Price Essay). Macmillan, London 1877.
Sagirow, P.: Dynamik technischer Systeme. Universität Stuttgart, Stuttgart 1974.
Schiehlen, W.: Zustandsgleichungen elastischer Satelliten. Z. Angew. Math. Phys. **23** (1972) 575–586.
Schiehlen, W.: Technische Dynamik. B.G. Teubner, Stuttgart 1985
Schiehlen, W.: Zur Untersuchung von Zufallsschwingungen. Z. Angew. Math. Mech. **54** (1974) T64–T65.
Schiehlen, W.: Parametererregte Zufallsschwingungen. Z. Angew. Math. Mech. **55** (1975) T67–T68.

Schiehlen, W., and Kolbe, O.: Ein Verfahren zur Untersuchung von Regelsystemen mit periodischen Parametern. Regelungstechnik **15** (1967) 451–455.
Schiehlen, W.O., and Kreuzer, E.J.: Symbolic Computerized Derivation of Equations of Motion. In: Magnus, K. (ed.): Dynamics of Multibody Systems. Springer-Verlag, Berlin-Heidelberg-New York 1978, p. 290–305.
Schwarz, H.R., Rutishauser, H., and Stiefel, E.: Numerik symmetrischer Matrizen. B.G. Teubner, Stuttgart 1968.
Schwarz, R.G.: Identifikation mechanischer Mehrkörpersysteme, Fortschr.-Ber. VDI-Z. Reihe 8, Nr. 30, VDI-Verlag, Düsseldorf 1980.
Schweitzer, G., Schiehlen, W., Müller, P.C., Hübner, W., Lückel, J., Sandweg, G., and Lautenschlager, R.: Kreiselverhalten eines elastischen gelagerten Rotors. Ing.-Arch. **41** (1972) 110–140.
Smith, B.T., Boyle, J.M., Garbow, B.S., Ikebe, Y., Klema, V.C., Moler, C.B.: Matrix Eigensystem Routines – EISPACK Guide. Lecture Notes in Computer Science, Vol. 6, Springer-Verlag, Berlin-Heidelberg-New York 1974.
Smith, R.A.: Matrix Equation XA + BX = C. SIAM J. Appl. Math. **16** (1968) 198–201.
Stoer, J.: Einführung in die Numerische Mathematik I. Heidelberger Taschenbuch 105, Springer-Verlag, Berlin-Heidelberg-New York 1972.
Stoer, J., and Bulirsch, R.: Einführung in die Numerische Mathematik II. Heidelberger Taschenbuch 114, Springer-Verlag, Berlin-Heidelberg-New York 1973.
Strobel, H.: Systemanalyse mit deterministischen Testsignalen. VEB Verlag Technik, Berlin 1968.
Strutt, M.J.O.: Lamésche, Mathieusche und verwandte Funktionen in Physik und Technik. Verlag von J. Springer, Berlin 1932.
Thomson, W., and Tait, P.G.: Treatise on Natural Philosophy. Vol. I, Part I. University Press, Cambridge, 1879, p. 391.
Thomson, W.T.: Theory of Vibration with Applications. Prentice Hall, Englewood Cliffs 1972.
Vernon, J.B.: Linear Vibration Theory. Wiley, New York-London-Sydney 1967.
Wanner, G.: Integration gewöhnlicher Differentialgleichungen, BI-Hochschultaschenbuch 831/831a, Mannheim-Zürich 1969.
Wedig, W.: Zufallsschwingungen. In: VDI-Bericht Nr. 221. VDI-Verlag, Düsseldorf 1975.
Wilkinson, J.H.: The Algebraic Eigenvalue Problem. Clarendon Press, Oxford 1965.
Wilkinson, J.H., and Reinsch, C.: Linear Algebra; Handbook for Automatic Computation, Vol. II. Springer-Verlag, Berlin-Heidelberg-New York 1971.
Willems, J.I .: Stability Theory of Dynamical Systems. Nelson, London 1970.
Zeller, K.: Automatische Bestimmung der Parameter von Rotorsystemen. Dissertation, TU München, 1975.
Zielke, G.,: ALGOL-Katalog – Matrizenrechnung. Oldenbourg-Verlag, München-Wien 1972.
Zurmühl, R.: Matrizen und ihre technischen Anwendungen, 4th edition. Springer-Verlag, Berlin-Göttingen-Heidelberg 1964.

Index

GPSR Compliance

The European Union's (EU) General Product Safety Regulation (GPSR) is a set of rules that requires consumer products to be safe and our obligations to ensure this.

If you have any concerns about our products, you can contact us on ProductSafety@springernature.com

In case Publisher is established outside the EU, the EU authorized representative is:

Springer Nature Customer Service Center GmbH
Europaplatz 3
69115 Heidelberg, Germany

Batch number: 10583998

Printed by Printforce, the Netherlands